MEDIA OWNERSHIP

Research and Regulation

New Media: Policy and Social Research Issues
Ronald E. Rice, series editor

Double Click: Romance and Commitment Among Couples Online
Andrea Baker

The Role of Culture and Political Institutions in Media Policy: The Case of TV Privatization in Greece
Theomary Karamanis

Media Policy for the 21st Century in the United States and Western Europe
Yaron Katz

Telecommunications and Development in China
Paul S.N. Lee (ed.)

Impact and Issues in New Media: Toward Intelligent Societies
Paul S.N. Lee, Louis Leung and Clement Y.K. So (eds.)

Global Trends in Communication Research and Education
Kenneth W.Y. Leung, James Kenny, and Paul S.N. Lee (eds.)

Communication Technology and Society: Audience Adoption and Uses
Carolyn A. Lin and David J. Atkin (eds.)

Foundations of Communication Policy: Principles and Process in the Regulation of Electronic Media
Philip M. Napoli

Media Ownership: Research and Regulation
Ronald E. Rice (ed.)

Virtual Politicking: Playing Politics in Electronically Linked Organizations
Celia T. Romm

Automating Interaction: Economic Reason and Social Capital in Addressable Networks
Myles A. Ruggles

The Playful Audience: From Talk Show Viewers to Internet Users
Tony Wilson

MEDIA OWNERSHIP

Research and Regulation

edited by

Ronald E. Rice

University of California-Santa Barbara

HAMPTON PRESS, INC.
CRESSKILL, NJ 07626

Printed in the United States of America

Library of Congress Cataloging-in-Publication Data

Media Ownership : research and regulation / edited by Ronald E. Rice.
p. cm. -- (The Hampton Press communication series. New Media)
Includes bibliographical references and index.
ISBN 978-1-57273-684-9 (hbk.) -- ISBN 978-1-57273-685-6 (pbk.)
1. Mass media--Ownership. 2. Telecommunication--Ownership. I. Rice, Ronald E.
P96.E25M428 2007
338.4`730223--dc22
2007041000

Hampton Press, Inc.
23 Broadway
Cresskill, NJ 07626

Dr. Constance Penley
Co-Director of the Carsey-Wolf Center for Film,
Television, and New Media
Professor of Film Studies, University of California, Santa Barbara
It is my great pleasure and honor working with you on our many
Center activities

Arthur N. Rupe
Entrepreneur, philanthropist, concerned citizen
It is my professional good fortune to occupy your endowed
Rupe Chair in the Social Effects of Mass Communication

CONTENTS

IV. REGULATORY AND LEGAL ISSUES

V. ETHICAL AND ACCESS ISSUES

List of Tables

List of Figures

FOREWORD
CHANGING MEDIA OWNERSHIP

Christopher H. Sterling

Growing up in Wisconsin in the 1950s, I was glued to the handful of TV channels located in Madison, a mid-size university town and state capital. We had one educational channel (although I watched only if really desperate), plus an affiliate for each of the then three national networks. Only one was a VHF channel—and in those pre-cable days it dominated the market's audience and advertising revenue. These four stations had different owners–and thus each offered a separate media "voice." So did most of the six or eight Madison radio stations. Federal Communications Commission (FCC) rules then prevented ownership of more than one station of a type in a single market. A "distant signal" meant waiting for *The New York Times* copies to arrive after lunch at the local drug store. (Back then, of course, we lacked such competing options as cable, satellites, videotape, or any of today's digital options. Today's students have trouble even imagining such a desert and are amazed we survived.) In the Madison of a half century ago, then, a media "outlet" and a "voice" were generally one and the same. How things have changed! Now, a relative handful of owners control the larger number of stations serving the city and a cornucopia of newer services are available as well—there are far more outlets than voices.

This new anthology makes a very useful contribution to the fast growing number of studies and legal decisions about the issues involved with media ownership. The issue is anything but new—critiques of media control date back to a simpler world of newspapers, movies, and radio—and even to 19th-century tracts railing against telegraph and telephone monopolies that urged government control. As early as 1922, one radio monthly was already asking "Who Shall Own the Stations?" although service was barely 2 years old. The Hearst chain was one of the first to combine newspapers and broadcast stations in the 1930s and such expanding economic links across media raised policymaker concerns. Sometimes they took action—in the 1940s, government antitrust actions broke up restrictive membership requirements of the Associated Press, NBC's ownership of two radio networks, and vertically integrated Hollywood studios and theater chains.

The rapid rise of television soon raised similar issues as the medium appeared to be the proverbial license to print money, channels were hard to come by, and three networks dominated the field from the start. The post-war consolidation of daily newspaper ownership was another input to Harvey Levin's (1960) *Broadcast Regulation and Joint Ownership of Media*, an early examination of newspaper and broadcasting cross-ownership policy. Levin noted that the press was a pioneering owner of both radio and television, in part to protect their investment in the older news medium. Bryce Rucker's (1968) *The First Freedom* argued that continuing media consolidation was raising serious policy issues. In 1974, the Rand Corporation undertook an intensive review of the literature on media ownership, and suggested still more work was needed. Five years later, the Federal Trade Commission held a long symposium on media ownership trends and published two volumes of proceedings. That same year, Ben Compaine (a contributor to the present volume) edited a descriptive book of text and tables that outlined *Who Owns the Media?* (revised versions appeared in 1982 and again in 2000). Media critic Ben Bagdikian has in the past two decades issued five revisions of *The Media Monopoly,* his hard-hitting expose, each edition of which expresses growing concern about the issue. In recent years, many more analyses have appeared, some arguing that ownership concerns are a mere tempest in the media teapot, but more of them stressing the threats of growing conglomeration. Robert McChesney, to name one of the most prolific authors (and another contributor here), has become a key spokesman among those concerned about and attempting to change media consolidation policy to encourage greater diversity.

Why all this ownership debate over the years? We seldom hear expressions of concern about ownership in other businesses. After all, consolidation has become a widespread business phenomenon in America. Why, then, the focused concern, even anguish, about the same trend in the media? In this book, for example, McChesney's chapter outlines a looming media ownership crisis on several levels. Not at all, responds Compaine's chapter, which argues (as he has elsewhere) that concentration is often lower in media than in other businesses, and that nothing is amiss. Finding a silver lining amidst the disagreement, Phillip Napoli and Nancy Gillis outline a host of relevant study opportunities for media researchers. Assessing the value of various diversity indexes and how to weigh the importance of different media may have important policy impacts. Indeed, the June 2003 FCC order to eliminate or loosen many remaining broadcast ownership regulations (a decision often referenced in the pages that follow) raised a firestorm of comment, complaint, and research studies—until legal action stopped the FCC in its tracks.

Summing up the substantial literature, critics of media consolidation generally focus on two primary arguments. The first and by far the most important centers on the news and public affairs content provided by most daily media (newspapers, broadcast stations, cable networks). Surveys show that Americans get most of their news and public affairs information from these media, and that such knowledge informs their voting and other means of participating in a democratic society. Varied media "voices" are a vital part of that process. Federal courts have held that a robust and competitive media market encouraging multiple points of view or voices is a central part of the political process. Consolidation's critics argue that as news media play such a unique social and political role, they should not be dismissed as mere "toasters with pictures," to cite former FCC chairman Mark Fowler's memorable 1980s phrase about television.

A secondary argument is economic, centering on advertising-supported print and broadcast media. Advertising is a central bridge linking people and markets, and competing media outlets (stations, newspapers, cable, Web sites) provide many of those links. Lack of competition—as exemplified with the consolidation of daily newspapers to monopoly status in most markets—has made clear the danger of advertising rates priced out of reach of local businesses, contributing to the growth of national chains. Yet, publicly held media firms are coming under severe pressure to increase profits—often by severe cuts in their news budgets. The Knight-Ridder newspaper chain discussed by Ann Louise Bardach in this book is one example–and the company was soon put up for sale. Knight-Ridder was not making enough to satisfy its investors. Critics ask how much profit constitutes "enough"? Any owner deals with economic concerns—in this case, the purchase, operation, and revenues of a media outlet. Once an owner controls multiple media outlets (especially in the same metropolitan area), however, diversity of viewpoints and competitive advertising rates usually suffer accordingly. Although we live in an age of growing media plenty–indeed, hundreds of channels—these are increasingly controlled by a dwindling number of owners.

For most Americans, of course, the media problem is not a dearth but a flood of information. Aside from all the traditional media modes, the wide-open world of the Internet provides billions of pages of information and hundreds of new blogs added daily, But how many are widely read? Research shows that the Web sites used most often (such as cnn.com) are largely owned by firms already entrenched in print and electronic media. Any consideration of media ownership has to take into account this seeming oxymoron of plenty (channels) versus scarcity (owners).

Researchers and policymakers (such as Diane Watson and Shawn Chang on the pages that follow) continue to focus principally upon ownership of media outlets or the transmission "pipes" (stations, channels, cables, satellite links, newspapers, Web sites). Amy Mitchell offers a wealth of current information on this topic, demonstrating how much is changing and how quickly. Transformation of radio ownership since 1996, for example, has allowed one company voice (typically Clear Channel or Cumulus) to control up to eight radio outlets per market. Other observers focus on the potential bottlenecks of various "portals" (the means through which we access electronic services–from remote controls to TiVos, to Web search engines and Internet portals). Still others argue that most media content (often more repetitive than really diverse) is also controlled by a relative handful of companies. A few critics have questioned whether media firms should be allowed to continue controlling both content and distribution (as most do today). One can even consider ownership of the equipment with which we access electronic media—such as the combination of various set-top "boxes" connected to a television receiver. Many devices (including cable TV boxes and TiVos) are provided on a subscription basis from media owners. Such equipment helps to illustrate growing convergence across the media (content), telecommunications (distribution), and information (the computer-driven box technology) industries, as Patricia Aufderheide emphasizes.

Of primary concern to most media observers, however, is media ownership within individual markets (as opposed to trends nationwide). As we all read or tune our media on a local basis (even network signals come through a local broadcast, cable, or satellite portal), what is available elsewhere is generally of little concern. Two exceptions come to mind—the Internet as a national or global service, and audio- or video-streaming, although that is not yet widespread. Some analysts point out that it is also becoming harder to determine when media are really competing in the same marketplace. Are newspapers and radio stations interchangeable to readers or advertisers as once presumed? Is a Web-based news service (such as cnn.com) or a blog comparable to such legacy media as newspapers or local television stations? Can satellite radio offering 150 channels on a national basis take the place of many local radio stations? The answers are not yet clear—and are, in any case, fast changing.

To "ownership" in the traditional media outlet sense, must now be added intellectual property controls, thanks to the recent strengthening of U.S. copyright law for content owners. Pushed by movie studios, networks, and publishers, the "limited time" called for in the Constitution now extends past the typical lifetime. Mark Rose points out some of the complexities inherent in developing copyright law, while Sarah Pritchard argues that these tighter copyright restrictions are limiting access to the information necessary for students and researchers. Globalization has added yet another layer to the picture, creating something of a three-dimensional chess game with the growth of cross-border media links. Denise Bielby and Molly Moloney explore many of these.

And, of course, technology keeps changing the definitions. As recently as a decade ago—before development of the commercial Internet—ownership seemed a more stable concept. Now we are witnessing the decline of traditional media as the country's long-time center—the once dominant role of network television newscasts, for example. Their control seems less important today because their audience reach is less than half of what it was two decades ago. The one-time monolithic audience (it wasn't, but compared to conditions today, it surely seems so) for such programs has fragmented—to cable, the Internet, and other places. As one result, we are seeing the near disappearance of radio news in many markets. Thousands of stations offer no news or public affairs at all. Daily newspaper circulation is down and keeps declining as younger would-be readers turn to video, videogames, or the Internet. One result of the push to retain readers is a "down market" approach catering more to interest in crime, fires, and pop culture (the once important weekly newsmagazines are a classic example of this trend).

These technology-driven changes also call into question the conglomeration trend of recent years. The disastrous 2000 merger of AOL and Time Warner is a prime example of what can—and did—go wrong, as Jennifer Holt analyzes. The division of Viacom into two separate media companies in 2006 may be one sign of greater focus and specialization to come.

Government can play any of several roles in this ongoing debate—some of them explored in the chapters that follow. History suggests regulatory plans often crash against economic realities (as with the media push to networking and other modes of content sharing, as opposed to the "ideal" of individual local media). The regulator's job is made more complex in this multi-channel era of splintered audiences. Where we once had a single medium or channel (such as a movie theater), now we often have several (multiplex screens), although usually under the same ownership. Yet history demonstrates the limits of laws or rules that are based too often on a compartmentalized view of markets and thus impose uneven competitive playing fields. Finally, any governmental action takes time—often too much in a fast-changing media marketplace.

So, what is the outlook? More than ever, there is a continuing need for the policy research community and media regulators to constructively interact. Pat Aufderheide makes this argument well in an early chapter here—indeed this entire anthology is a big step in the right direction as it reflects varied approaches and points of view. We need more neutral research into the linkages between ownership and content patterns. As recent court decisions have made clear, any policy change needs solid evidentiary underpinning, not mere regulatory opinion—or wishful thinking. It should be remembered that once media ownership patterns are established, they are very hard to undo. Ask those in the radio business.

PREFACE

I joined the University of California, Santa Barbara (UCSB) in Spring 2004. I was very fortunate to be hired as the Arthur N. Rupe Chair in the Social Effects of Mass Communication in the Department of Communication. The endowment generates a modest amount of additional funds each year, some of which is intended to support a small conference each other year (beginning with Spring 2004) on some important media-related topic of interest to both the academic and public community. Both before and during my initial term there, faculty in the Department mentioned that there was a new center being developed—the Carsey-Wolf Center for Film, Television, and New Media—and they encouraged me to find out more about it, and possibly get involved.

At that time, Dr. Constance Penley of the Film Studies Department was the director of the center, and Dr. Miriam Metzger of the Department of Communication was the associate director. Connie and Miriam tracked me down and asked me if I would be interested in revising and submitting a proposal with them to the Critical Issues in American Endowment of the College of Letters and Science. This is a $25,000 grant provided once a year to a university group with an innovative combination of presentations, activities, noted participants, additional funding, and student involvement. We felt that the controversial Federal Communications Commission rules of 2003 and the Third Circuit Court's rejection of those rules signaled a critical junction in media ownership trends, and thus a critical issue in America. As we were interested in the complex interaction of industry, policy, technological, cultural, production, creative, social, and research aspects of media ownership, this topic also clearly seemed to be an appropriate issue for us, the center, and the university, which emphasizes interdisciplinarity, to take on.

I immediately saw the connections with the goals of the Rupe conference, so proposed that we combine our plans for a year-long media ownership series with a culminating spring Rupe conference on media ownership. We developed the proposal, entitled "Media Ownership: Research and Regulation," and submitted it in late Spring 2004.

In late Summer 2004, Dr. David Marshall, the dean of the Division of Humanities and Fine Arts, and Dr. Melvin Oliver, the Dean of the Division of Social Sciences informed us that we had been awarded the 2004–2005 grant. With the $25,000 grant from the Critical Issues in America Endowment, and a $15,000 contribution from the Rupe Chair, the center provided a year-long series in 2004–2005 titled "Media Ownership: Research and Regulation." Many of the contributors to this book were participants in the series. The series also included freshmen seminars that Dr. Penley taught in the winter term and I taught in the spring term.

We specifically included in our Critical Issues in America grant funding to have all the presentations filmed, edited, and provided to UCTV for cable, satellite, and online distribution. So each of these presentations is available as online streaming video (available from the center Web site at http://www.cftnm.ucsb.edu/events), and available in rotation on cable and satellite channels from UCTV (see http://www.uctv.tv).

Included in the grant proposal was a commitment to publish the presentations in a book for my series for Hampton Press, *New Media: Policy and Social Research Issues*. I have worked with each presenter to develop and revise his or her research and arguments into chapters for this edited book, *Media Ownership: Research and Regulation*. While I was finishing the editing process, I discovered three other extremely rigorous projects related to media ownership that were just being completed—on ownership of and access to media, on media industry financial influence on politics, and on current changes in media ownership—by Dr. Benjamin Compaine, John Dunbar, and Amy Mitchell. They were very gracious to prepare, in a very short period of time, summary versions of their projects to contribute to this book.

This book now consists of a very comprehensive, innovative, empirical, and analytical collection of chapters. It has been an extremely educational, professional, humane, and pleasurable experience working with each of the contributors. One important aspect of the year-long series was that we always had a lovely dinner with great Santa Barbara wine after each event, inviting the current presenters and all prior presenters, to develop a sense of community and ongoing discussion.

A NOTE ON THE CARSEY-WOLF CENTER FOR FILM, TELEVISION, AND NEW MEDIA

Based on our very collegial and successful collaboration on the Critical Issues grant, Drs. Penley and Metzger invited me to become the co-director of the center, which I did as of September 2004. As a result of our collaboration, the center currently has three main ongoing programs:

The *Media Ownership Project* consists of courses, lecture series, and conferences on media ownership that serve to advance research and public knowledge about the important issue of media consolidation and its effects on diversity of media opinion and content, creativity, competition, and democratic access to the marketplace of ideas.

The *Environmental Media Initiative* (EMI) takes advantage of the diverse group of interdisciplinary UCSB faculty working in the areas of environmental science and media. The Initiative focuses on teaching, research, and production aimed to improve our understanding and management of the many relationships between the environment and the media. The EMI considers how the environment is conceptualized, portrayed, and communicated through media channels, and how media channels and components themselves are designed, implemented, evaluated, and in turn influence public opinion on environmental issues and environmental policy.

The *UCSB Media Internship Program* offers a highly selective competition to identify both undergraduate and graduate students who want to develop practical applications of the knowledge they have gained in their scholarly studies. The center matches these outstanding students with host organizations of the center's Advisory Board.

ACKNOWLEDGMENTS

I greatly appreciate the help, advice, and socializing from Constance Penley, my co-director (see the dedication); Arthur N. Rupe (see the dedication); The Critical Issues in America Endowment of the College of Letters and Science, and its anonymous donor who endowed the fund; The Interdisciplinary Humanities Center, which has provided office space, administrative, technical, publicity and scheduling help; Sarah Bequette, former center administrator, for her help in managing event scheduling, finances, internships, and publicity; Matt Harnack, senior graphic artist and Web designer, for developing and maintaining the web site, posters, publicity; Nicole Klanfer, associate dean of development, who has provided extraordinary support, energy, and ideas to the center, both in terms of planning meetings and events, as well as cultivating and fostering the amazingly generous donations from our Advisory Board to the center. Due to her tireless and gracious energies, the center has more than $9 million in capital funding, the required amount for our portions of the new buildings to be completed sometime in fall of 2009; David Marshall, dean of Arts and Humanities, and Melvin Oliver, dean of Social Sciences, who have both provided ongoing conceptual, practical, and financial support to the center, and who awarded us the Critical Issues in America Grant; The Center's Advisory Board members, who take time out from their demanding jobs as heads of major media organizations; Miriam Metzger, associate director, who helped get the center off the ground and co-wrote the proposal that funded the series; Ray Tracy, television services, who, with his superb staff, created perfectly lit and thoughtfully edited programs for UCTV from our events; Faye Nennig and Elizabeth Fisk, Department of Communication, for help in financial administration and event scheduling; Gretchen Falvo, University Events, for her detailed advice on restaurants, receptions, and university events; The Film Studies Department, which has supported, and even encouraged, Dr. Penley's extensive commitment to the center and the media ownership series; and The Department of Communication, which provides administrative support to the Rupe Chair. I especially appreciate the encouragement and suggestions of Professors Dave Seibold, Michael Stohl, and Howard Giles concerning the Rupe

Chair and the center; Daisy Lemus, Graduate Student, Department of Communication, for event planning and coordination, and help in formatting each of the chapters; Paul Myers, graduate student, Department of Communication, for incredibly fast and accurate transcripts of several presentations; Melodie Tao, undergraduate student, Department of Communication, for her help in organizing the media ownership web site listing; Barbara Bernstein, publisher of Hampton Press, for supporting the idea of this edited book, and her tolerance as the contents kept expanding; Mariann Hutlak for her production management; and Claire B. Johnson for her patience on evenings and weekends while I was glued to the computer.

I

OVERVIEW AND PUBLIC INTEREST ISSUES

1

CENTRAL CONCEPTS IN MEDIA OWNERSHIP RESEARCH AND REGULATION

Ronald E. Rice

Trends and developments in social values, political ideologies, media policies, economic conditions, globalization, media technologies, and telecommunications networks have all interacted to generate significant changes in the nature of media industries, production, content, distribution, exhibition, and use. This book considers a wide variety of perspectives on one of those developments: changes in and consequences of media ownership, concentration, and conglomeration. This chapter highlights some of the central concepts in this debate, and some of the major perspectives and arguments of the contributors to this book.

MOTIVATION AND CONTEXT

In 2003, the Federal Communications Commission (FCC) approved sweeping changes in rules on media ownership that would allow forms of vertical and horizontal integration of the media and entertainment industries that have not been seen since before the trust-busting "Paramount decrees" of 1948. The consolidation possible under those proposed new rules would be even greater than before because it would take place on an unprecedented global scale and would have significant implications for the future of broadband access, the digital spectrum on which all media are converging.

The FCC's move in June 2003 not only energized the public in an unprecedented way, but it motivated lawmakers to question FCC responsibilities and policy. Despite the fact that FCC Chairman Michael Powell resisted holding public forums on the controversial regulatory proposals that were being considered by the FCC, several groups were successful in providing commentary to some of the FCC commissioners in a handful of public settings, either at press conferences or on university campuses. The organizations and nonprofit groups opposing the deregulation of ownership made a strong effort to present evidence of the harmful effects of media concentration on creativity and democracy.

However, it was alarming to see how little they were able to turn to academia for research on media ownership—largely because academia has been strangely silent on this issue. And this silence is the case even in disciplines that would be expected to address this crucial topic, such as film and television studies, communication, English, political science, sociology, and history. Compounding the problem of the lack of authoritative research on the social and political effects of media concentration was (and is) the general lack of press coverage on the regulatory debates. Not surprisingly, the conglomerates that own the media outlets and that lobby the FCC to loosen ownership rules see little benefit in having readers and viewers who are informed about the consequences of vastly increased media concentration. A Pew Excellence in Journalism study conducted a few months before the 2003 FCC vote found that 72% of Americans knew "nothing at all," whereas only 4% indicated they knew "a lot" about the proposed sweeping changes (Journalism.org, 2003).

Under pressure, the FCC itself commissioned 12 studies on topics ranging from viewpoint diversity in cross-owned newspapers and television stations, to radio market structure and music diversity. Almost all of them concluded that media concentration does not eliminate diversity, giving the appearance that the FCC had stacked the regulatory deck. The studies, carried out for the most part by FCC staffers, were sharply criticized by both scholars and the industry press for their limited scale and lack of methodological rigor.

As soon as the FCC decided to the change ownership rules, a broad and diverse range of civic groups, under the name of Prometheus Radio Project, filed suit to overturn this set of regulations. In June 2004, the U.S. Court of Appeals in Philadelphia (Third Circuit) agreed with the critics' assessment of the FCC's research, staying the implementation of the rules after finding that the FCC's studies employed contestable assumptions and inconsistencies. The court also found the FCC's "diversity index" used to assign weight to media outlets in a given market to be arbitrary and capricious, and strongly criticized the FCC for failing to release the diversity index to the public. The court also rejected industry pro-consolidation arguments that media ownership limits were inappropriate or unconstitutional. The U.S. Supreme Court decided not to review the Third Circuit's ruling.

It is important to note that the Third Circuit Court's primary basis for rejecting the FCC ownership rules had to do with the empirical basis for their proposed changes. The two court rulings do not halt or even reject any such changes per se; rather, they have returned the process to the FCC and the legislature. The FCC could revise, drop, or resubmit the proposed rules with different supporting arguments, so this issue will continue. Indeed, the Telecommunications Act of 1996 requires the FCC to examine its rules every 2 years. However, part of the FCC's proposed rules change was one that Congress had already passed in December 2003 and January 2004—The National Television Ownership rule, raising the cap on the percentage of U.S. homes that any one company's stations reach from 35% to 39% (lower than the 45% proposed by the FCC but just high enough to allow two media conglomerates under the cap, Watson & Chang, Chapter 7, this volume).

As indicated by this controversy, many people are deeply concerned about the increasing concentration of media production and distribution in the hands of a few, highly interconnected corporations. They worry about the effect of such consolidation on diversity of opinion and content, creativity, commercialization, and democratic access to the marketplace of ideas.

Given this crystallizing event, there are two fundamental questions driving the various chapters in this book. The first is *whether media ownership is becoming more or less concentrated* (and how this varies across media sectors). This can and should be primarily an empirical question, given definitions of media sectors, ownership, concentration, and trends (Compaine, Chapter 9; Mitchell, Chapter 10; Noam, Chapter 8, this volume). The second is *what roles media have played, and do and should play, in the marketplace, society and culture, and what consequences follow from those.* This seems to be primarily a political and social question, given perspectives used to analyze and interpret media in a society such as the United States that claims democratic values (Aufderheide, Chapter 3; Bardach; Chapter 11; Lievrouw; Chapter 18; McChesney, Chapter 2; Warner; Chapter 5). And both of these questions can be understood and answered only through broader historical, political, legal, and social analysis (Aufderheide, Chapter 3; Bielby & Moloney, Chapter 13; Dunbar, Chapter 12; Epstein, Chapter 15; Harwood, Chapter 15; Holt, Chapter 6; Rose, Chapter 4; McChesney, Chapter 2; Napoli & Gillis, Chapter 14; Rose, Chapter 4; Warner, Chapter 5; Watson & Chang, Chapter 7).

MEDIA OWNERSHIP LIMITS

A guiding principle in the founding of the nation and the Constitution was to locate power in the people, thus requiring *local and decentralized power.* This is especially true concerning media, so that they might best serve individuals and communities, and both monitor and challenge centralized government. Distrust of central control of public media is to some extent grounded in the media practices of local towns and communities during the U.S. revolution (Warner, Chapter 5).

This conflicts with a *centralization or concentration of media ownership*. Yet, centralized control may be required to avoid local confusion and even destruction of the common, public good, or, centralized corporate power, not foreseen by the founding fathers. So, for example, in 1927 the federal government took over control of assigning broadcast frequencies, to avoid interference from multiple stations using the same frequencies within and across communities. Furthermore, motivated by the concern with *localism*, it allocated the limited number of both radio and television frequencies to local market areas, and imposed limitations on broadcast power, so that the frequencies could be re-used in different local communities. Providing these limited, yet essentially public, frequencies was couched in terms of the common good: Licenses were given to stations for supposedly limited but renewable periods, conditioned on the station operating in the *public interest, convenience, and necessity.*

In line with the goal of supporting local control and orientation, the FCC initially limited ownership according to the 7–7–7 rule: no company could own more than seven AM, seven FM, and seven TV stations, with no two in the same market. Holt (Chapter 6)] highlights the central role, ideologies, and significant cases of *anti-trust policy* underlying such media ownership regulation (and, as well, the reinterpretation of antitrust policy and goals during the Reagan Administration). These and related media ownership rules under antitrust and FCC policies have been significantly motivated by the goals of *diversity of voices* and *marketplace of ideas*, in particular as necessary to foster the access to information and opinions crucial to a democracy. But, as corporations had obtained the status of *corporate personhood* in the late 19th century, with rights and protections to those guaranteed to individuals, the question of First Amendment implications on affecting diversity of content in the media was relevant. Thus, media regulation has in general focused on structural instead of content limits on ownership (with some exceptions, such as libel, obscene and indecent content, children's programming, false advertising, etc.), presuming that diversity of owners and creators would generate their own diversity of content (Epstein, Chapter 15, this volume; Horwitz, 2005).

But, of course, a central problem is just what *diversity* means (ownership, content, programming, media portrayals, outlets, voices, audiences, exposure, composition of media workforce), how it's measured, and whether these regulatory attempts have helped prevent reductions in diversity (Compaine, Chapter 9; Napoli, 1999, 2001, 2003; Napoli & Gillis, Chapter 14). In their chapter, Napoli and Gillis develop and justify a rigorous and reliable diversity index that takes into content (such as time or space for local news or public affairs), motivations for using different media, potential or actual audience reach, effects of media use (e.g., agenda-setting or civic involvement).

With the Telecommunications Act of 1996, declining support for arguments based on minority or civil rights (Horwitz, 2005), the spread of digitization (with implications for convergence of content across what had been separate technologies and thus regulatory arenas), and the development of the Internet as a worldwide distribution channel, the debate over diversity has become much more complex. For example, the same content can appear in multiple formats and combinations across multiple media, across multiple audiences and locales.

Horwitz noted that *scarcity* of the spectrum (available, noninterfering broadcast frequencies), conceptualized as a public resource, had, until the rise of many new media forms, been another primary motivation for regulating broadcast media. As Epstein emphasizes, because broadcast frequencies were considered a public resource, media ownership (at least broadcast media) is not a right; it is a government (i.e., public) benefit. But the confusion and interference stemming from new radio stations independently choosing what frequencies to use, and how strong their signals were, before they were formally allocated and licensed by the Radio Act of 1927, was a second major motivation. That is, although the frequencies might be public and thus "free," unregulated domination of frequencies also constitutes repression of freedom of speech of others. Epstein distinguishes between *allocational* scarcity (frequency, spectrum) from *economic* scarcity (the immense economic resources required to own and operate media). Scarcity arguments provided legal precedent for government regulation of some aspects of media, especially broadcast media.

Note the paradox of (a) concern by many authors, citizens, politicians, and lobbyists about control of media by government, and the central role of the First Amendment in protecting free speech as one of the checks on Congressional power, while (b) government policy and FCC rules have been, until recently, the primary forces behind preventing corporate control of media. Early U.S. legislation was specifically designed to limit commercial control of media (such as through the 1790 limited copyright and patent law; Warner, Chapter 5). Part of the motivation for the Communication Act of 1934 was a concern about joint ownership of wireless and wired media resulting in reducing competition in, and gaining monopoly control over, radio communication. This role of government in preventing control of media by corporations was stated most famously as: "Freedom to publish is guaranteed by the Constitution, but freedom to combine to keep others from publishing is not. Freedom of the press from governmental interference under the First Amendment does not sanction repression by private interests" (*Associated Press v. United States*, 1945, p. 20).

Furthermore, the concept of *ownership* itself can be misleading. One argument against ownership limits is that media corporations are, in fact, "owned" by stockholders. However, Noam (2006) showed that most media stock is held by large investment institutions, not individuals; and most copyrights are owned by corporations, not the original creators. Furthermore, Potter (2005) emphasized the distinction between *ownership* (generally, stockholders, unless it is privately held) and *control* of a media company (generally the CEO and top-level executives; also affected by Wall Street expectations, changing technology, standards-setting bodies, government regulation, and competitors).

Copyright in some ways lies at the heart of media ownership issues. Ownership consists of control of distribution channels (media outlets), and of content. But content is controlled by those who own the copyright. If no one "owned" any particular content, then all the focus would be on controlling the distribution channels. Within copyright, there are two main debates: how digitization and networking affects control over copyright and content, and the length of the copyright (much extended under recent regulatory regimes, and much more expansive than initial U.S. copyright restrictions; Rose, Chapter 4).

As Rose argues, there is a tension in copyright definitions and legislation, between the need to protect the author's investment in their creative product (although most current copyrights are owned by media corporations, obtained from the original author as part of production and distribution contracts), and the need for society to be able to learn and develop through shared information—that is, *information as a private property* and as a *public good*. This tension is built right into the history, political and economic contexts, and wordings of initial copyright acts in England (the Statute of Anne in 1710) and the United States (both in the Constitution and in the Copyright Act of 1790). Rose argues that these cannot be resolved as mutually exclusive principles, but must be managed as interacting principles. Rose also shows how distinctions between private and public aspects of published material; the changing importance of printer, bookseller and author; and the conceptualization of the length and purposes of copyright, all developed over time (Warner, Chapter 5). Although the very definitions of what is patentable or copyrightable continues to expand, copyright now also confronts the ability and possibly right of media users to *reconfigure* media technologies and *remediate* content (Lievrouw, Chapter 18).

Horwitz (2005) provided a cogent analysis of how recent court decisions have strongly affected FCC attempts to implement rules designed to maintain or improve diversity. Both as an outcome of the institutionalization of Reagan-era deregulatory perspectives through court appointees, and an increasing emphasis on strict and explicit justifications for diversity rulings, the FCC found itself on the losing side of a variety of court decisions, and thus began moving toward less stringent ownership caps because they could not provide the empirical evidence supporting the supposed relationships between, and consequences of, ownership and diversity. Due to changes associated with social, economic and political ideologies, and industry pressures, other media policies were also "deregulated" (Epstein, Chapter 15; Holt, Chapter 6), such as the Fairness Doctrine and the Fin-Syn rules.

These altered media policies, and technological developments, helped unleash the wave of mergers and acquisitions that are the source of current concern about media ownership and concentration, and the impetus for this book (Croteau & Hoynes, 2006, Compaine, Chapter 9; Holt, Chapter 6; Mitchell, Chapter 10; Napoli & Gillis, Chapter 14; Noam, Chapter 8; Potter, 2005).

"Deregulated" is in quotes for a reason. As Horwitz (2005), Napoli (2001), McChesney (2004, Chapter 2, this volume), Starr (2004), Warner (Chapter 5), Watson and Chang (Chapter 7), and others argued, prior government regulatory polices created, subsidized, shaped and protected different media at different times (thus rejecting the argument that there ever was or can be a "free market"). Indeed, the FCC and government media policy has often served marketplace interests even while invoking the rhetoric and goals of "public interest" and "diversity"—such as by providing spectrum for free, rarely rejecting license renewals, reducing protections such as the Fairness Doctrine, and so on. Indeed, any initial limits or decreases in ownership limits should be placed in the context of government policy fundamentally choosing a commercial broadcasting model, and providing free public resources to for-profit organizations (Croteau & Hoynes, 2006).

Table 1.1 summarizes past and current national, local, and cross-ownership media ownership levels, and some relevant major legislation.

TABLE 1.1. *Summary of Relevant Legislation and Media Ownership Limits*

Major Relevant (related in some way to ownership) Acts, Decisions, and Policies

- 1710: Statute of Anne, England's copyright Act
- 1790: U.S. Copyright Act (justified explicitly in Constitution, Article I) – further public learning through making knowledge available to the public after 14 years of copyright ownership, renewable once
- 1792: U.S. Postal Act—heavily subsidizes delivery of newspapers and magazines
- 1886: Supreme Court (*Santa Clara County v Southern Pacific Railroad*) decision used as basis for claiming "corporate personhood," whereby corporations had essentially the same rights as a natural person, including First and Fourteenth Amendment protections. The 1978 Supreme Court *First National Bank of Boston v. Bellotti* ruling asserted that corporations can influence ballot initiatives and other political campaigns under their First Amendment rights (see http://www.corporatepolicy.org/issues/ constit.htm; Perrow, 2002; reclaimdemocracy.org; Thoennes, 2005). Although industry had been lobbying representatives, influencing newspapers, and suppressing workers' opposition, for such rulings for most of the 19th century, in both colonial time and in the new Republic, corporations had to obtain and request renewal of state charters, apply scaled voting to insure equal shareholder voting rights, avoid interlocking directorates, and so on. These and related rulings emphasize again the multiple layers of influences that media organizations have on society. Media corporations have the protections and rights as an individual (free speech, due process, equal protection), but also control the channels through which they both conduct their influence as well as which they might be opposed or criticized.
- 1890: The Sherman Antitrust Act makes illegal any restraint or monopolization of trade or commerce among the states or with foreign nations.

- 1914: The Clayton Act of 1914 (with extensions by the 1936 Robinson-Patman Price Discrimination Act; the 1976 Hart-Scott-Rodino Antitrust Improvements Act; and the Celler-Kefauver Amendment in 1950) prohibits price discrimination (including inducing or receiving that) between different purchasers, and engage in bribery, kickbacks, price-fixing, and similar activities, especially when it lessens competition or increases monopoly.
- 1927: Federal Radio Act.
- 1934: Federal Communications Act (with amendments in 1952, 1982, and 1993).
- 1948: The Paramount decision rejected the ability of the major studios to control movie studies, distribution networks, and theatres, stating that this was monopolistic and thus contrary to antitrust legislation, reduced competition, and threatened citizen's First Amendment rights to diverse information. As a result, studios divested themselves of theatres and stopped certain trade practices, and the principles of separating content from distribution, and limiting vertical integration, were central to FCC media regulation for decades to follow.
- 1949: Fairness Doctrine established, grounded in the Communications Act of 1937, where "equal opportunity" had to be offered to political candidates by a radio station that had provided broadcast access to any other candidate for the same office. The Fairness Doctrine extended this general principal to news programs, documentaries and interviews. Media companies, as well as journalists, were concerned about the First Amendment implications, so in many cases declined to provide any coverage of some controversial topics, which became known as the "chilling effect." By 1985, the FCC began disavowing the justification for the Doctrine, and a court ruling in 1987 indicated that as the Doctrine was not law, but FCC policy, the FCC was not obligated to enforce it, so the FCC removed it. Interestingly, in the Spring of that year, Congress had actually voted to make the Doctrine law, but that was vetoed by Reagan, and again by the first Bush Administration.
- Public Broadcasting Acts: 1952 Comparable Use Amendment to the 1934 Act; 1962: Educational Television; 1967: Public Broadcasting; 1971: Federal Election Campaign Act; 1974: Repealing Certification Requirements; 1992: Public Telecommunications Act
- 1970: The Financial Interest and Syndication Rules were established to foster content diversity, prevented television networks from owning parts of companies providing them programming, but these were relaxed in 1991, ruled against by the courts in 1993, and repealed by the FCC in 1996. (For a review of effects, see Chapter 13 by Bielby & Moloney, and Chapter 6 by Holt.)
- 1972: "Must Carry" rules requiring cable systems to carry all local television stations, which helped UHF stations survive and thrive, and to serve as a means for the development of Turner's CNN in 1980 and subsequent media empire, and Murdoch's Fox network.
- 1983: approval by Department of Justice of the founding of the Tri-Star studio, which combined cable, broadcast, and film interests—the first major signal of regulatory acceptance of vertical integration and the decline of antitrust concerns.
- 1984: The Cable Communications Policy Act
- 1985: Department of Justice no longer enforces the Paramount Decree, allowing vertical integration in the media industries once again, and a massive set of acquisitions and mergers.
- 1992: Financial Interest and Syndication rules rejected by U.S. Court of Appeals, and thus repealed and modified by FCC in next few years.

- 1992: The Cable Television Consumer Protection and Competition Act. The *Cable Act's* "must-carry" rules required cable systems to carry local broadcaster content, scaled according to the cable system's capacity: "low burden" —fewer than 12 channels; intermediate burden—12–36 channels; and greatest burden—than 36. Systems with fewer than 300 subscribers are exempt. The act also included public access set-asides.
- 1996: Data Investment and Intellectual Property Antipiracy Act.
- 1998: Digital Millennium Copyright Act (DMCA)—extends copyright protection to include production and dissemination of technology that could overcome copyright protection devices, while limiting liability of online providers of users' copyright infringement.
- 1999: Bono Copyright Term Extension Act—extended copyright protection an additional 20 years to 70 years after the death of the author.
- 1999: Collections of Information Antipiracy Act.

Limits on National Broadcast Ownership

- Early on, national multiple television station ownership was limited to *three stations*. This limit was expanded to seven AM radio stations, seven FM radio stations, and seven television stations (no more than five VHF stations) in 1953; to 12–12–12 in 1984; and 30–30–12 in 1992, with a limit on television stations of 25% of the U.S. TV households. The Telecommunications Act of 1996 lifted these caps further, by removing any national caps for radio, with local limits of between 5 and 8 stations depending on market size, and the television cap of 12 as long as a company does not reach more than 35% of the national audience. Now, a company may own *television broadcast stations that reach no more than 39%* of the national television audience (the potential audience of a UHF station counts half against the cap).
- Initially, broadcast stations could *not affiliate with an entity owning more than one network*. Radio was freed from this restriction in 1977. This limit was reduced for television in the Telecommunications Act of 1996, whereby the entity could not control more than one of the four largest networks, or one of those and either the UPN or WB network. This was further relaxed in 2001 to allow one of those four to have a relation with UPN and/or WB.

Limits on Cable Ownership

- Cable had limits placed on it by the Telecommunications Act of 1996. No single cable company could be involved in cable systems passing more than 30% of nationwide households (this was changed to 36.7% in 1999 due to a new formula). And no cable company could have ownership interests with no more than 40% of programming on its cable systems, or, if their systems had more than 75 channels, they had to set aside 45 channels for nonaffiliated content. Both of these were overturned (found not sufficiently justified) by the courts in 2001.

Limits on Local Broadcast Ownership

- Before the Telecommunications Act of 1996, an entity could not control or own (typically, at least 5%) more than *one television station* in the same signal area, and no more than two AM and two FM stations in a market, and those could jointly exceed

25% of the market's audience share. In 1999, this was raised to *two television stations* in the same market area, as long as one was not in the four highest ranked stations, and if at least eight other stations in the market are independent owned and full power. (There are some exceptions relating to new stations, nonoperating stations, bankrupt stations, and lack of interest by outside owners.)

- A company may own up to five *radio* stations in markets with fewer than 14 total stations (but not more than 50% of the total) as long as no more than 3 of them being the same service of AM or of FM, up to 6 in a market of 15 to 29 stations and no more than 4 of AM or of FM, up to 7 in a market of 30 to 44 stations and no more than 5 of AM or of FM, and as many as 8 in markets with 45 or more stations and no more than 5 of AM or of FM.

Limits on Cross-Ownership

- Initially, one company could not own *a radio and a television station* in the same substantial area. In 1989, this was relaxed in the top 25 markets if there would still be at least 30 separately owned independent media voices (radio, television, and some local cable stations or newspapers).
- Beginning in 1975, a company that owns a *television or radio station* may not own a *daily newspaper* in the same market (excepting most pre-existing arrangements). A company can own a *television station* (or two, if it meets the FCC's local television ownership "duopoly" rule described above) and *up to 7 radio stations* as long as 20 independent "voices" (separately owned radio or television stations) remain after the purchase. In smaller markets, a company may own a television station and up to four radio stations if at least 10 independent voices remain.
- Beginning in 1970, no company could own *a cable system and a broadcasting network*; but this was reduced in 1992 and eliminated in 1996. Furthermore, no company could own, in the same local market, *a cable system and a broadcast television station*. And, no company could own, in the same local market, *a cable system and a telephone system*; but the 1996 Act removed this limitation.

CONCERNS: CONCENTRATION, IMPACTS, AND INFLUENCE

The evidence for whether media concentration is increasing or decreasing, as well as whether the effects of theses trends are positive or negative, is, to many minds, mixed. Furthermore, Noam puts current views of concentration in context, through a brief historical review showing that media were much more concentrated in the U.S. during the 18th and 19th centuries.

The primary measure of concentration is the Herfindahl-Hirschman Index (see Compaine, Chapter 9; Napoli & Gillis, Chapter 14; Noam, Chapter 8). Noam's extensive analysis concludes that using this measure, overall, the entire information sector is less concentrated than in 1984, less than in 2001, but more than in 1996. However, the data provide different conclusions for each of four major sectors. Mass media were more concentrated than 1988, but was the least concentrated of the four sectors, with the highest being in the telecommunication and the information technology sectors. More familiar distinctions, such as among TV networks, radio networks, cable, and so forth, again provide different results, with only radio markedly increasing in national concentration. However, concentration among the five most vertically integrated firms has definitely risen since 1988. Noam intriguingly argues that the move toward digital convergence will raise the level of concentration in the traditional media up to the higher levels in the telecommunications and the information technology sectors.

Compaine forcefully argues that the diversity of sources has much expanded (such as total number of radio stations, the rise of public broadcasting which is equivalent to a major network, and new services such as satellite digital radio and Internet radio, especially using wireless protocols), and the control of audience markets by major media has much decreased (as measured by network television ratings). Indeed, some argue that the more media outlets there are, the more fragmented the audience becomes, the more likely it is that niche (such as minority) content and ownership can succeed. Looking abroad, Bielby and Moloney (Chapter 13) also show that even globalized, internationally conglomerate media are often challenged and reshaped by national and cultural business practices, story and character norms, technical standards, linguistic preferences, desire for co-productions, national policies about production locations and programming quotas, familiar genres, and format, and so on.

However, although the number and variety of channels and content seems to remove scarcity as a basis for regulating media, it also creates an overwhelming *media smog,* with complex levels of bias and accuracy, possibly much less variety in voices than outlets, and less control available to the individual (Aufderheide, Chapter 3). The range and complexity of our choices require, and concentration allows, media to both *funnel* and *filter* what is available to us. Increased concentration also increases control of the funnels and filters of the behind-the-scenes infrastructure, software programs, interfaces, display technologies, content, and intellectual property.

Mitchell (Chapter 10) provides very detailed descriptions of transactions, business strategies, executives' interpretations, and pressures within the nine sectors of the news industry during 2004. Although there continues to be buying and selling, the bulk of conglomeration occurred in the years immediately following the Telecommunications Act of 1996, and has recently been somewhat suspended as corporations await the long-term implications of the rejection of the 2003 FCC rules raising ownership caps, as there are fewer obvious acquisitions available, and as the allure of the promises of synergy has worn off (Holt, Chapter 6).

But there are other, broader, more philosophical and ideological arguments for or against media concentration. Bagdikian (1983) was an early voice of concern about the concentration of power, information and culture in media conglomerates, especially when news media are incorporated into *vertically integrated* organizations that are necessarily motivated (and legally obliged) by market and profit concerns. This then has severe implications for biases and pressures on news content, and on the entire political process. For example, Mitchell's data document the decline in news as significant component (in terms of both expenses and mention) of media corporations.

Croteau and Hoynes (2006) and Dunbar (Chapter 12) report the staggering monetary influence media industries have in terms of campaign contributions, lobbying, junkets, and revolving doors. Dunbar documents that more than $1 billion from 1998 to 2004 was spent by the media industry on lobbying, campaign contributions, junkets for lawmakers and staff; and nearly 400 former FCC or congressional employees responsible for regulating the media went to work for the media industry. His cases such as Sinclair Broadcasting in the television sector, revolving door industry professionals involved in affecting the Telecommunications Act of 1996, and the cable industry's fight against á *la carte* cable channel selection, provide compelling examples of the forms and extent of industry influence on media policy. Also see Aufderheide, Bardach, and McChesney, for implications on *media bias* and access to information; and Lievrouw, Rose, and Warner, for implications on copyright and access to information. Media conglomeration is also global in scale, with corporations attempting to control markets and audiences in different countries and regions; US media companies dominate much of this (Bielby & Moloney, Chapter 13). Fears about cultural homogenization and weakening of national industries join the concerns about diversity, access, and participation already noted.

Many are concerned that increasing concentration decreases content quality, news content, and diversity, and both increases and homogenizes the shaping of cultural images and values (Bettig & Hall, 2003). Some studies (see Potter, 2005, for a brief review; and Compaine, Chapter 9) do not find much evidence for such declines, and less for causality. Compaine summarizes a major empirical study concluding that media concentration did not much affect diversity of content or quality of news. Indeed, Potter summarized Einstein's (2004) argument that a much greater influence is the fundamentally commercial nature of the media—that reliance on advertising leads to limitations in content, program length, and controversy, all to reach, and avoid offending, valued audience segments (Aufderheide, Chapter 3; McChesney, Chapter 2).

However, Croteau and Hoynes (2006) documented a wide range of incidents, grouped into four categories: homogenization and imitation (imitation, declining localism), trivialization and sensationalism (superficial and confrontational content, scandal and light news), media (especially news) constrained by commercial interests, and media constrained by censorship and conflicts of interest (both self-censorship and corporate censorship in the service of commercial interests, and advertiser influence; see also Alterman, 2003, p. 25).

Consider the example of *The L.A. Times*. The newspaper had already gone through a crisis of commercialization after an economist and former marketing executive for General Mills took over as chairman and CEO in 1997. His implementation of severe cost-cutting, the removal of the "wall" between business and editorial departments, the development of new paper sections light on reporting but attractive to audiences (and thus advertisers), the integration of links between advertisers and story content, and an arrangement with the Staples Center to split advertising profits from heavy news coverage about the center, all lead to reporter resignations, embarrassing coverage by its own reporters, and a subsequent steep decline in stock value. The Tribune Company took advantage of this opportunity to buy *The L.A. Times*, adding it to its ownership of other papers (including the flagship *Chicago Tribune*), television and radio stations, and other media (Croteau & Hoynes, 2006). At the time, newspaper staff and media critics were optimistic about this turn of events.

Auletta (2005) picks up the story at this point. He provided a detailed examination of the pressures of the "free market," corporatism, and media conglomerates on the news media. Although at first the new management of *The L.A. Times* allowed and supported increased investments in reporting, overseas bureaus, and investigative reporters, in 2003 the Tribune management began placing pressure on the editor, managing editor, and reporters to cut costs, reduce news space, consolidate the Washington DC bureaus, and dampen coverage of controversial issues. The corporate mission for profit maximization, driven by the inherent logic and potential competitors of the marketplace, was to maintain the nearly 30% profit margins from the Tribune's other newspapers, and follow the model of its overall multimedia business that had $5.7 million revenues in 2004 (second only to the Gannett newspaper chain, with $6.6 million). This orientation and associated organizational strategies appear to be inconsistent with what Dean Baquet, the managing editor of *The L.A. Times*, considers to be the main job of good newspapers: to help readers understand the world (Auletta, 2005).

Bardach's summary and transcription of the experiences and insights of the managing editors from *The New York Times*, *Slate*, and U.S. edition of the *Financial Times* defies the easy binary contrast between advantages and disadvantages of media concentration. Rather, they noted the demoralization of journalists and journalism, the welcome if still disorganized challenge from online bloggers, and the increasing fragmentation of news sources, audiences and advertising revenue. On the other hand, these three text news media are probably the least likely to suffer from corporate pressures on news coverage and resources, as the two U.S. media are owned by families and not by either public or conglomerate organizations, and the newly acquired *Slate* has not been intruded upon by its British headquarters. However, they and Bardach note the numerous examples of cost-cutting, profit maximization, and declining quality of news coverage at other newspapers, both major and minor.

Another concern is reduced access to media, both in terms of meaningful ownership and production of content (Lievrouw, Chapter 18; Potter, 2005). But this is debated as well. The rise of computing components, digital content and transmission, Internet outlets, and alternative media has both increased control over ownership of and voice in the major mass media, as well as provided many more opportunities for individual content production in low-cost print and online media (Aufderheide, Chapter 3; Lievrouw, Chapter 18). McChesney, (Chapter 2)asserts that media ownership patterns as well as other social trends are generating several *crises of the media*, relating to debating and covering war, covering elections, and *hypercommercializing* of all aspects of society through the media.

The crucial point here is that no longer are media separate from business and the political process—they *are* major industries and *are* political actors. After all, free speech and a free press were central elements of the Bill of Rights specifically to monitor and challenge centralized power, then meant as the government. Concentrated and corporatized media are less and less able or willing to perform the central role of providing a check to centralized power, whether government or industry (i.e., the press as the fourth estate, serving a watchdog function). Furthermore, the media control access to political debates, and public exposure, so that voices (such as aspiring or opposing political candidates) become dependent on the favor and orientation of those same media outlets.

Clearly, this threatens crucial values and processes of any democracy, and challenges founding social values. These forces are also reshaping what it means to be a democracy and what those guiding social values are. And that is part of the wider debate as to what kinds of cultural, informational and symbolic representations, and social identities, a society wants and feels are useful, valuable, and meaningful (Harwood, Chapter 16; McChesney, Chapter 2). From this perspective, the number of owners or outlets is almost irrelevant; the logic of the marketplace is the prime regulator, generating consequences that are perfectly understandable and desirable. As Horwitz (2005) noted, even antitrust regulation is largely designed to foster economic efficiency, not to promote social equity, except for selected emphases on communities, small businesses, and the like.

THE MARKET MODEL AND THE PUBLIC SPHERE MODEL

Thus, there seem to be at least two very different perspectives on the question of the media in general, and the role and implications of media ownership in particular (Croteau & Hoynes, 2006): the *marketplace perspective* and the *public sphere perspective*. (There are of course many other ways of labeling this underlying distinction and tension, but these two terms seem appropriate and useful.) Both emphasize their own sets of factors and outcomes, and ignore others. As the FCC has intended to foster competition and innovation among media, as well as fulfilling the *public interest*, there is a built-in tension and contradiction between market interests and public sphere interests.

The Marketplace Model

The first perspective emphasizes the role and influence of media in the *marketplace.* It looks at media as essentially no different from other businesses operating in a competitive, commercial marketplace. Under this perspective, commercial media companies must follow market logic—maximize revenues and resources, protect against competitors, minimize costs and risks, and seek new markets and audiences—to both satisfy investors and avoid extinction. Furthermore, they are required by law to do this, or they may be seen as not fulfilling their fiduciary responsibility to their stockholders. As a sign of the changing underlying ideology of the role of the FCC in particular, and government in general, embedded in The Telecommunications Act of 1996 is the belief that the U.S. government's role is to support rather than hinder media industries in the marketplace, because a *free market* works best (most efficiently) without control or interference (regulation).

Under the market perspective, maximizing profit is a valid, desirable, motivating, and socially beneficial goal. To the extent that media industries have unique economic advantages (the repeat use of content across different media, copyright ownership), traditional advantages (*economies of scale* and *scope*), and competition is reduced (through *horizontal* and *vertical integration*), they should have higher profits. Indeed, return on revenues average around 14% (7% for audiorecordings and 10% for books, to 20% for radio and 22% for television broadcasting) while return on assets average around 16% (from 9% for film, and cable/pay television, to 24% for magazines and 26% for newspapers; Potter, 2005). These are extraordinary profits for mature industries, and, as seen earlier, put tremendous pressures on traditional news divisions to cut costs and reduce news collection and coverage.

Horizontal integration means owning companies operating at the same point in the "value" chain of media inputs, production, distribution, exhibition and sale (i.e., a publisher acquiring more newspapers) which increases concentration in the market. *Vertical integration* means owning companies either prior or subsequent in the "value" chain (i.e., a cable operator buying a film company to obtain content for distribution over cable channels). *Conglomerate merger* means acquiring both other media companies and/or non-media companies (i.e., a conglomerate that includes newspapers, radio stations, and manufacturing companies). The reverse may also happen, where non-media corporations acquire media companies (Potter, 2005) because of their much higher profit ratio.

Economies of scale means that marginal costs of a product or service become lower than average costs, because of the diminishing effect of large capital costs for initial production, large production runs, and leveraging suppliers (paying less for a given amount of resources) because of large orders. Media companies especially benefit from economies of scale because additional uses of content once produced, and additional audience viewers, are nearly costless (as the image, text, or sound is what is being charged for multiple times, not actually the material it comes in or through). The production and use of digital content increases this benefit even more, as storing, distributing and displaying the image, text or sound is nearly costless, and the content can be disassembled and used in a variety of ways, unlike material-based content, such as a printed magazine or book. Thus, as Potter (2005) and others noted, media companies are especially motivated to increase the number of their consumers (audience). However, larger and more complex organizations require greater coordination costs, so there may be limits to economies of scale. For example, experiments of large media conglomeration such as Disney/ABC, AOL/Time Warner or Vivendi have essentially failed (Holt, Chapter 6). *Economies of scope* means that with a wider range of media outlets, and the ability to recombine the product (media content), media companies can create more variations and recombinations of initially produced product, use information about the product and audience across media outlets and audiences, and take advantage of alliances and cross-market opportunities (Aufderheide, Chapter 3; Harwood, Chapter 16).

Synergy can be thought of as the popular term for implications of horizontal and vertical integration, and economies of scale and scope. It means bringing media components together, allowing conglomerates to cross-develop, -promote, -license, -distribute, and -combine media products (see examples from the movie *Titanic* and the Disney empire in Croteau & Hoynes, 2006). Ownership concentration allows both control and acquisition of supplies at below-market cost, multiple uses and combinations of media content, multiple distribution outlets, and multiple revenue streams. Particular content can be recombined into a wide variety of media products, thus gaining multiple returns from the initial production costs. Media products from separate industries can be combined into more integrated products (such as theme parks and video games). Different media outlets can promote and license the content from other outlets (such as placing actors in the ballpark from films or television programs owned by the same company that is televising the game or even owns the team or the sports stadium, or placing products within television, film and videogame programming; see Croteau & Hoynes, 2006).

There is of course considerable competition among media businesses, for creative and production talent, for audiences through content and media, and for advertisers through offering access to specific valued consumer audiences. Potter (2005) noted, based on 2004 Census data, that the total spent on media in 2003—$700 billion—was more than the total spent by the U.S. government on education, job training, social services, transportation (highways, airports, trains), natural resources and environment, agriculture, international affairs, and national defense; and 75% of the amount spent on media goes to indirect costs, primarily advertising (see also Croteau & Hoynes, 2006, for a summary of the financial growth of media). Constant and pervasive brand image is critical to media markets, because, to the extent that most content for similar audiences, and features for similar media products, are similar, companies attempt to distinguish themselves among potential audience groups on image, brand, logo, or style (Potter, 2005), and to increase attention to and consumption of content in another medium. Not often mentioned in discussions of increasing media concentration is the concomitant increased concentration among advertisers and advertising agencies, which then also are less interested in local media and local audiences (Potter, 2005). Furthermore, to the extent that the percent of gross national product spent on mass media is relatively constant (around 3%), companies must compete for a fairly constant percent of available resources within an economy. So the stakes are high, the profits extraordinary, and the competition severe.

Advantages and Disadvantages of the Market Model. The market model of course offers a variety of advantages (Croteau & Hoynes, 2006).

- When markets are competitive and not hindered by regulation promote organizational efficiency.
- Markets are responsive to both technological developments (what can be supplied) and consumer preferences (what is demanded).
- They are flexible, not delayed by centralized decision making or rigid long-term plans.
- Markets promote ongoing innovation in products and services.
- They enable the production, marketing, and distribution of media products like material goods.

But, markets also have disadvantages and limits:

- They do not correspond to the one-person, one-vote philosophy of democracy—rather, influence is proportional to financial power; inequalities in control are reinforced through increased financial leverage and influence over the content produced and distributed through media.
- Markets do not discriminate between positive and negative social consequences, especially long-term or collective ones.
- Markets are driven solely by products and services that can be profitable—thus needs of small consumer groups, of those with little income or marketing salience, or whose social benefits cannot be easily appropriated by the providers, will not be valued or produced.
- Markets do not fully represent the benefits and costs of all products and services—consider the widespread consequences of pollution or smoking, which generates significant costs to individuals and society that are not part of the production costs or consumer price.
- They are seldom "free": Inherent market tendencies (oligopoly and monopoly, risk avoidance, hidden prices due to vertical and horizontal integration, short-term criteria such as quarterly dividends stifling long-term develop-

ments) can distort the market itself. Indeed, a successful strategy of marketplace dominators is to reduce competition from other companies.
- Rarely do markets fulfill the classical economic theory assumptions of perfect information to buyers and sellers, costless access to information about the value of products and services, low barriers to entry, and no one (or few) producers or no one (or few) consumers can influence the market.

Some of these limits are exacerbated when the informational, or non-material, aspects of media are considered in the market process. When this happens the following can occur:

- Markets do not easily handle the unique aspects of information goods, as opposed to material goods—their value is uncertain over time and dependent on the user, they cannot be easily kept from being used multiple times by multiple users, and they may be licensed or sold multiple times after initial production costs are recovered.
- Content, especially with digitization, can be used (i.e., charged for) multiple times across multiple media. This is fundamentally different from material goods (such as an apple), which, once sold, belong to the owner. Books, for example, seem to be a material good, but consumers only purchase the right to keep that one copy. They do not buy the content—that's what copyright is all about—and in fact the media company owning the copyright can resell versions of that content in a wide variety of ways. Furthermore, to the extent that a company is vertically integrated, it can promote that content, and use different combinations of that content, across its multiple distribution networks and media outlets.
- Involving media may not have the direct relationships with consumers (buyers) required by the classical market model—consider that all media using advertising to support their costs (which varies by medium) are selling consumers to the advertisers, so the content becomes an intermediary whose primary function is to help the media business attract certain audience segments valuable to certain other businesses.

In Chapter 2, McChesney, in particular, points out that many arguments for decreased media regulation are myths, or simplistic and uninformed assumptions. He asserts, instead, that (a) the media did not arise from a "free market"; (b) the current commercial and conglomerate media system was not intended or even conceptualized by U.S. founding principles; (c) current professional journalism standards are not strong enough to counter media corporatism and the logic of the marketplace; (d) the current media system does not necessarily give people want they want—it is after all oriented first toward advertisers, and second to only some segments of the audience; furthermore, it creates and then reinforces preferences; and (e) new technology—both in terms of decreased scarcity and increased interactivity—is not sufficient to overcome media conglomerate trends and implications.

The Public Sphere Model

The second perspective, based on founding democratic principles of the United States, as well as initial FCC criteria, emphasizes the role and influence of media in the *public sphere.* Much of the media regulation just noted, especially by the FCC, is grounded, whether well-founded or not, in public sphere principles of diversity, localism, access to information for an informed citizenry necessary for democracy, monitoring and challenging of centralized power, free speech and freedom of the press, and scarcity of the public resource of broadcast frequencies. The concept of the public sphere not only includes the public interest and democracy, but also the more general Habermasian goal of fostering continuous public dialogue (Croteau & Hoynes, 2006; Rose, Chapter 4). Croteau and Hoynes argued the media should serve the public interest in four ways:

1. *Diversity* of cultural and political views and experiences.
2. *Innovation* in both the *technology and content* of the media.
3. *Substance of content,* whether in news or entertainment, that considers significant social issues.)
4. *Separation of the sources and channels* that provide coverage and information, so as to prevent centralization of power, whether in the government or the marketplace.

The public sphere model of course demands much more explicit norms, goals, and values being applied to decisions as to what kinds of structures, content, and social implications are desirable. Conversely, the market model implicitly embeds preferences and assumptions about norms, goals and values, but limits their scope to internal market efficiencies and narrowly defined social benefits. It would seem that in addition to the familiar public sphere themes, more general themes of equity, ethics, and fairness should be considered (Harwood, Chapter 16). For example, consumers can be seen as primarily a means to an end *for* the media organization, which a Kantian imperative would challenge, whereas citizens can be seen as primarily an end fostered *by* the media organization. So, for example, a lower rate of profit might be a valuable exchange for greater localism and diversity (Bardach, Chapter 11; Harwood, Chapter 16).

As many note (Horwitz, 2005), the marketplace model inherently biases content toward that which is attractive to large audiences with disposable income, and also fosters concentration of control of both media distribution and content. So the commercial media system is potentially (some say inherently) a threat to diversity, the marketplace of ideas, and free speech. Thus, the role of the FCC (until recently) in attempting to limit common and cross-media ownership of both distribution and content. Under the market model, the public consists of consumers. Under the public sphere model, the public consists of citizens.

Other Comparisons

There are other ways to compare underlying assumptions about media ownership. Warner contrasts *networking* vs. *broadcasting*, most obviously highlighted by the rise of the networked computer (most importantly, the Internet, but also all forms of information sharing across digital devices), but in fact inherent in social traditions and media forms of U.S. history. Networking is manifested by, and fosters, one-to-one and many-to-many, symmetrical, decentralized and participatory communication. Broadcasting is associated with one-to-many, asymmetrical, centralized, and institutionalized production of communication. Warner emphasizes that the form of networking or broadcasting is not inherent in any particular communication technology; how those technologies are used is shaped by political, social, legal, rhetorical, cultural and economic forces and decisions, with significant implications and consequences for those same factors. An excellent example is of the role of U.S. Postal regulation in subsidizing and thus fostering widespread distribution and economic survival of newspapers and other journalism (see Chapters 2 and 5). The two fundamentally different ways of structuring and perceiving media are also implicated in the current regulatory and industry debates (such as, e.g., copyright aspects of peer-to-peer music sharing and webcasting, or the socially fragmenting or integrating functions of blogs and portals).

Conceptualizing and regulating media from a broadcast perspective may be increasingly less relevant, however. Interactivity has changed "audiences" into "users" who create, publish, and exchange, as well as consume, content. In Chapter 18, Lievrouw describes how more personal, digital and networked media are being integrated into traditional media industries. More importantly, perhaps, is that more oppositional, alternative, critical, counter-commercial, participatory, artistic, and ironic media forms and uses—what is called *oppositional new media* (culture jamming, alternative computing, mediated sociality and mobilization, and Indymedia) —are also developing. These are firmly based in the networking perspective, but also explicitly oppose the commercializing, homogenizing, and institutional nature of traditional media. But the very attributes of networking, digitization, programmability, personalization and media vertical integration have also increased the frequency, pervasiveness, and utility of collecting personal data, and corresponding concerns about individual privacy (Chapter 18).

A related tension is between *Big Media* (mainstream, vertically integrated, little participation by audience, potentially better control over accuracy, high institutional costs, reinforcing ruling politics) and *Little Media* (alternative, distributed, personalized, audiences as both users and producers, little control over accuracy, low cost, challenging ruling politics) (Chapter 3). But the Big Media are making major inroads into acquisition, control over, and presence in Little Media.

CONCLUSION

Curran (2000), Entman and Wildman (1992), Epstein, and Horwitz (2005), among others, argue for a mixed media approach, combining the benefits of both the market and the public sphere models. Epstein argues that there is legal and Constitutional basis for developing a system where media corporations can trade off public access for increased ownership. Horwitz (2005) argues that the very concept of *the marketplace of ideas* rhetorically biases the debate toward the market model, so proposals should avoid this term and emphasize issues of diversity and integration between the two media models. Croteau and Hoynes (2006) propose a variety of possible public sphere-oriented media policies, some linked to license renewal assessment. These might include a minimum amount of public affairs and citizen-oriented content, mandated public access programming, free air time during election periods, application of the former Fairness Doctrine, support for low-power radio, converting one of the forthcoming multiplexed digital channels from each station to public access, insuring universal access to the Internet, municipalities offering wireless broadband, maintain postal subsidies for low-circulation publications, expand public broadcasting in ways that avoid the market model, greater coverage of media ownership and policy issues by journalists, greater involvement of citizens in media issues, greater development of and access to alternative media. Media debates need to be popularized, and media users need to become involved and informed (Chapter 2). And, as Aufderheide, Lievrouw, Rose, and Warner argue in their chapters, there needs to be room for both big and little media, mainstream and oppositional media, networking and broadcasting, and sharing and owning.

Table 1.2 provides a list of Web sites providing extensive additional information, arguments, and resources about media ownership research and regulation.

Table 1.2. *Web Sites on United States Media Ownership: Economic Issues, Education/News, Government Policy, Legal Issues, Marketing Aspects, Media Conglomerate Tracking, Public Involvement*

Economic Issues

- Alliance for Public Technology (a nonprofit membership organization "concerned with fostering access to affordable and useful information and communication services and technologies by all people"): http://www.apt.org/about
- The Angry Economist: http://angry-economist.russnelson.com/media-concentration.html
- The Gathering Storm Over Media Ownership: http://www.alternet.org/story/15654

Education/ News

- Action Coalition for Media Education (Teaching media education: knowledge, skills and activism): http://www.acmecoalition.org/
- *Broadcasting & Cable* Online (media industry news): http://www.broadcastingcable.com
- Copyright and the Boy Scouts (Merit badge for copyright!): http://news.com.com/2061-10796_3-5693563.html?part=rss&tag=5693563&subj=news
- Cross Media Ownership: http://www.digital-law.net/benderdiss/book.html
- I Want Media: http://www.iwantmedia.com/consolidation.html
- Journalism.org (Research, Resources and Ideas to Improve Journalism; see their "State of the Media" annual report on ownership, issues, etc. across various media, including online journalism): http://www.journalism.org
- Local Media: One of the most comprehensive directories for local, citizen-oriented, and progressive media groups and organizations (micropower, radio, news, internet radio and tv, legal resources, local music sites and band web pages, etc.): http://www.citizensmedia.org/linkset.html
- Media Concentration: http://www.ala.org/ala/oif/ifissues/mediaconcentration.htm
- Media Education Foundation (study guides, handouts, downloadable materials, videos on media ownership, informed citizens, media content, etc.): http://www.mediaed.org
- Media Ownership Overview: http://www.womenarts.org/advocacy/MediaOwnership.htm
- Media Ownership Timeline (Summary of major deals March 1989 – December 2002): http://www.hollywoodreporter.com/thr/icopyright_display.jsp?vnu_content_id=1820240
- Media Ownership: http://www.wan-press.org/rubrique112.html
- Mediawatch (PBS NewsHour coverage of information industries; see list of journals, media criticism, media studies centers, media organizations): http://www.pbs.org/newshour/media/resources.html
- Museum of Broadcast Communications Encyclopedia of Television (see entries for media policy issues): http://www.museum.tv/archives/etv/
- Project Censored (tracks major stories not covered by corporate media): http://www.projectcensored.org/
- Technology and the Limits of Media Ownership: http://news.com.com/Technology+and+the+limits+of+media+ownership/2010-1071_3-5072792.html
- The Western Media: http://www.krysstal.com/democracy_media.html
- World Newspapers (A site that lists magazines, research, articles and links on media criticism): http://www.world-newspapers.com/media.html

Government Policy

- BNA's Web Watch: http://www.bna.com/webwatch/mediaownership.htm
- Center for Responsive Politics (a nonpartisan, nonprofit research group based in Washington, that tracks money in politics, and its effect on elections and public policy): http://www.opensecrets.org/
- Center on Democratic Communications of the National Lawyers Guild (devoted to protecting the right of all people to a system of media and communications based on principles of democracy and cultural and informational self-determination, not dominated by commercial concerns): http://www.nlgcdc.org/

- The Communications Act (Pike & Fischer's listing of the major amendments, 1934-1996): http://www.pf.com/telecomAct1934TOC.asp?; part of its comprehensive summary of Communications Law and Regulation: http://www.pf.com/commLawISP.asp
- FCC—Media Ownership Policy Reexamination: http://www.fcc.gov/ownership/
- Media Law and Policy: http://www.obs.coe.int/medium/law.html.en
- Media Ownership Rules Under Review: http://www.journalism.org/resources/research/reports/ownership/rulechart.asp
- The Cross Media Ownership Campaign: http://www.xmedia.org.au/
- Tracking the Payback: http://www.opensecrets.org/payback/issue.asp?issueid=MO1&CongNo=108

Legal Issues

- Campaign for Press and Broadcasting Freedom: http://keywords.dsvr.co.uk/freepress/body.phtml?doctype=&id=156
- Center for American Progress: http://www.americanprogress.org/site/pp.asp?c=biJRJ8OVF&b=108399
- Domination Fantasies: http://www.reason.com/0401/fe.bc.domination.shtml
- FortCulture — on Digital Rights and Media ownership: http://www.fortculture.org/
- Media Access Project (A non-profit public interest telecommunications law firm; see their Issues on Media Consolidation page): http://www.mediaaccess.org/programs/diversity/fcc.html
- On Media Concentration and Diversity Question (very comprehensive review of FCC media ownership regulations and opposing philosophies): http://communication.ucsd.edu/people/ConcentrationpaperICA.htm
- Sfgate: http://www.sfgate.com/cgi-bin/article.cgi?file=/c/a/2003/05/12/MN53063.DTL

Marketing Aspects

- Adbusters Media Foundation (an alternative and oppositional forum for social activist movements through media): http://www.adbusters.org
- Advertising Age Online (access to wide variety of updated media tables and charts): http://adage.com
- National Alliance for Media Arts Culture (Dedicated to the support and advocacy of independent film, video, audio and online/multimedia arts): http://www.namac.org/

Media Conglomerate Tracking

- CPT's Selected Data—Concentration in Telecommunications and Media Markets: http://www.cptech.org/telecom/seldata.html
- A Survey of Who Owns Europe: http://www.ifj.org/pdfs/europeownershipupdate2003.PDF#search='media%20concentration,%20europe'
- American Press Institute's The Media Center Convergence Tracker (search engine for mergers, concentration, etc.): http://www.mediacenter.org/convergencetracker/search/index.cfm
- Columbia Journalism Review's Who Owns What (summarizing more than 50 of the biggest media companies): http://www.cjr.org/tools/owners/index.asp

- Corporate Influence in the Media: http://www.globalissues.org/HumanRights/Media/Corporations/Owners.asp
- Diversity, Democracy and Access—Is Media Concentration a Crisis?: http://www.mediachannel.org/ownership/front.shtml
- Media Concentration: http://www.moveon.org/moveonbulletin/bulletin7.html
- MediaChannel (A lovely/scary ownership interconnection chart of the six largest media conglomerates): http://www.mediachannel.org/ownership/chart.shtml and more detailed text summaries at: http://www.mediachannel.org/ownership/granville.shtml
- MediaGiants—Who Owns What?: http://www.thinkandask.com/news/mediagiants.html
- National Organization for Women on Media Control: http://www.nowfoundation.org/issues/communications/tv/mediacontrol.html
- PBS Frontline—Media Giants: http://www.pbs.org/wgbh/pages/frontline/shows/cool/giants/
- Stop Sleeping: http://home.earthlink.net/~platter/media-ownership.html
- The Big Ten: http://www.thenation.com/special/bigten.html
- The European Commission and Regulation of the Media Industry: http://www.medialaw.ru/laws/other_laws/european/e-eh.htm
- The FCC and Media Cross Ownership: http://www.polarity1.com/pcfcc.html
- The Global Network for Democratic Media: http://www.mediachannel.org/ownership/
- The Nation Magazine's Description of the Holdings of the 10 Biggest Media Conglomerates (AOL/Time Warner, AT&T, General Electric, The News Corporation, Viacom, Walt Disney, Bertelsmann, Vivendi/Universal, Walt Disney, Sony): http://www.thenation.com/special/bigten.html
- US Public Interest Research Group—Media Ownership and Concentration: http://www.pirg.org/consumer/media/
- Well-Connected (Tracking the Broadcast, Cable and Telecommunications Industries): http://www.openairwaves.org/telecom/default.aspx
- Who Owns the Media (Ownership databases; ownership charts; ownership resources; media, politics and money; interlocking boards of directors): http://www.freepress.net/ownership/resources.php

Public Involvement

- Alliance for Community Media (ACM; on behalf of millions of Americans nationwide, ACM supports competition, diversity, and localism in media. The ACM supports requirements that companies using public rights-of-way and public spectrum provide the means for local community use of media.): http://www.alliancecm.org
- Benton Foundation (on Media Ownership): http://www.benton.org/initiatives/ownership.html
- Center for Creative Voices in Media (nonprofit organization dedicated to preserving America's media: the original, independent, and diverse creative voices that enrich our nation's culture and safeguard its democracy): http://www.creativevoices.us/
- Center for Digital Democracy—Media Ownership Action: http://www.democraticmedia.org/issues/mediaownership/index.php
- Center for Media and Democracy (a nonprofit organization serving social change activists, journalists, researchers, policymakers, and the public at large by countering

propaganda, informing and assisting grassroots citizen activism, promoting media literacy, and sponsoring "open content" media): http://www.prwatch.org/cmd
- Creating a Culture of Peace: http://www.omnicenter.org/justicecollection/media.htm
- FreePress (a national nonpartisan organization working to increase informed public participation in crucial media policy debates, and to generate policies that will produce a more competitive and public interest-oriented media system with a strong nonprofit and noncommercial sector): http://www.freepress.net/content/ownership
- Future of Music (not-for-profit collaboration between members of the music, technology, public policy, and intellectual property law communities): http://www.futureofmusic.org/
- Global Media Ownership: http://www.opendemocracy.net/debates/debate.jsp?debateId=24&id=8
- Independent Media Center (network of independent and alternative media organizations and activists): http://www.indymedia.org
- International Freedom of Expression eXchange (a site on media ownership reforms): http://www.ifex.org/es/content/view/full/49805/
- Media Ownership—Get Involved: http://www.mediaownership.org/get_involved.cfm
- Media Reform Information: http://www.corporations.org/media/
- Media Tank: http://www.mediatank.org
- OurMedia (provides free storage and bandwidth for your video files, audio files, photos, text or software, as a way of handling personal copyright and distribution in the digital age): http://www.ourmedia.org
- Prometheus Radio Project (not-for-profit association dedicated to the democratization of the airwaves through the proliferation of noncommercial, community based, micropower radio stations; see their interesting guide to the FCC!): http://www.prometheusradio.org/
- Reclaim the Media (envisions an authentic, just democracy characterized by media systems that inform and empower citizens, reflect diverse cultures, and secure communications rights for all): http://www.reclaimthemedia.org/
- Rocky Mountain Media Watch (Challenges citizens to resist and change the manipulative and toxic formulas of Big Media's news products): http://www.bigmedia.org/
- Take Back the Media: http://www.takebackthemedia.com/owners.html
- The Center for Public Integrity (investigative journalism in the public interest; lobbying, political parties, 527s, etc. See especially Telecommunications and Media Ownership for data on contributions, junkets, lobbying and revolving doors between politics and telecomm industry; also has a cool "media tracker" where you can enter your zip code, and find out about all media coverage and ownership in that area): http://www.publicintegrity.org/telecom/

Note. All Web sites valid as of October 27, 2005. There are, of course, probably many other very useful and informative websites on media ownership. Also, this list avoids blog-like or personal opinion sites.

REFERENCES

Alterman, F. (2003). *What liberal media? The truth about bias and the news.* New York: Basic Books.

Auletta, K. (2005, October 10). Fault line: Money and the future of the *L.A. Times. The New Yorker*, pp. 50-61.

Bagdikian, B. (1983). *The media monopoly.* Boston, MA: Beacon Press.

Bettig, R., & Hall, J. (2003). *Big media, big money: Cultural texts and political economies.* Lanham, MD: Rowman & Littlefield.

Center for Public Integrity. (2005). Media ownership rules. http://www.publicintegrity.org/telecom/default.aspx?act=rules

Croteau, D., & Hoynes, W. (2006). *The business of media* (2nd ed.). Thousand Oaks, CA: Pine Forge Press.

Curran, J. (2000). Rethinking media and democracy. In J. Curran & M. Gurevitch (Eds.), *Mass media and society* (3rd ed., pp. 120-154). London: Arnold.

Einstein, M. (2004). *Media diversity: Economics, ownership, and the FCC.* Mahwah, NJ: Erlbaum.

Entman, R., & Wildman, S. (1992). Reconciling economic and non-economic perspectives on media policy: Transcending the "marketplace of ideas." *Journal of Communication, 42*, 5-19.

Horwitz, R. B. (2005). On media concentration and the diversity question. *The Information Society, 21*(3), 181-204.

Journalism.org. (2003, February). *New federal rules for media ownership: How much does the public know?* http://www.journalism.org/resources/research/reports/fccsurvey/default. asp

McChesney, R. (2004). *The problem of the media: U.S. communication politics in the twenty-first century.* New York: Monthly Review Press.

Napoli, P. (1999). Deconstructing the diversity principle. *Journal of Communication, 49*, 7-34.

Napoli, P. (2001). *Foundations of communications policy: Principles and process in the regulation of electronic media.* Cresskill, NJ: Hampton Press.

Napoli, P. (2003). *Audience economics: Media institutions and the audience marketplace.* New York: Columbia University Press.

Noam, E. (2006). *Media ownership and concentration in America.* New York: Oxford University Press.

Perrow, C. (2002). *Organizing America: Wealth, power, and the origins of corporate capitalism.* Princeton, NJ: Princeton University Press.

Potter, W. J. (2005). *Media literacy* (3rd ed.). Thousand Oaks, CA: Sage.

Starr, P. (2004). *The creation of the media: Political origins of modern communications.* New York: Basic Books.

Thoennes, K. J. (2005). Frankenstein incorporated: The rise of corporate power and personhood in the United States. *Hamline Law Review, 28*, 203-236.

2

MEDIA AND POLITICS IN THE UNITED STATES TODAY

Robert W. McChesney

THREE CRISES OF THE MEDIA

We have a very serious crisis in the United States today. It is a crisis of self-government: the survival of accountable, viable self-governance. Our media are central to understanding the nature of the crisis, in creating and perpetuating the crisis. They are not solely responsible—they only participate in it—but they are a cornerstone in understanding the political crisis in the United States today. There are at least three ways in which media are contributing to this crisis: war coverage, election coverage, and hypercommercialism.

The most important things that media systems do in free societies and democracies, intellectual democracies at least, are to help people participate in two sets of decisions. The first is debating and justifying the decision to go to war. There is no more important decision a government makes than deciding to engage in actions that result in people dying, to put people's lives at risk. The second is coverage of elections, how we select people to have political power or replace them and monitor them. If you flunk those two it does not matter if you get the weather right or the scores right for the ball game. If you do not have a free press, you do not have a free society. And there is a third crisis of our current media system: hypercommercialism, or the conversion of all social communication into commodities.

The War Coverage Crisis

Consider the decision to go to war. In the past century since the United States has become a global power, we have engaged in scores of wars. Most are the quick carpet bomb or troop landing variety we do not even hear about. But seven or eight major foreign wars have involved troops going overseas: from the Spanish–American War to World War I, World War II, Korea, Vietnam, to Central America wars and now the two Iraq wars. And if we look at the record since the Spanish–American War (1898) of the U.S. entrance into these wars and press coverage, it is not a very impressive one (see Solomon, 2005, on the cooptation of media in times of war, especially the Iraq ones). These are the dark pages in U.S. media history. Most discussions pass over these episodes to get to Watergate or coverage of some other impressive area. The reason is very simple. In virtually every major foreign war that the United States has fought since 1898, the government in power at the time the war was being launched, wanted to go to war, but in most cases it didn't believe the people of the country wanted to go to war. So it had to make crucial choices: Does it tell the truth to the people and try and make the case and convince the public to go to war? Or does it try to deceive or underinform them to support the war? Regrettably, almost every the government has opted for Plan B.

Consider Vietnam where we now know from the Pentagon papers and the more recent releases of the LBJ office tapes, of the blatant lying that was done to try to get Americans to support a war that even the leaders at the time knew was fruitless, and which they knew if they told the truth to the American people, the American people would never support the war. That is the legacy that journalists know about; and in good journalism schools they are educated about this dark side.

But along comes the second Iraq war. The government's position, and what was reported, was a series of outrageous claims, now entirely discredited, to justify an invasion of Iraq, which clearly violated international law. And our press coverage was atrocious. And not just some local weekly newspaper, not just some right-wing talk news channel, but the very height of the most prestigious journalism in this country: *The New York Times*, *The Washington Post*.

So our war coverage has not learned anything from the past. If anything, it has gotten worse. Perhaps in Vietnam it was be more understandable for a reporter to get it wrong because there wasn't the same access to information as there is now. When reporters were covering the Iraq story, all they had to do was look at press from any other country, talk to reporters from England or France or Germany, go on the web and Google "WMD" and they would have found plenty of evidence to justify asking more questions before running the story.

There are of course many other examples of this distorted and incomplete reporting. Remember the toppling of Saddam Hussein's statue? People had "risen up in Iraq to welcome us, and they overthrew the statue," but this was an orchestrated media event with very few people actually in the square. Remember the famous "Mission Accomplished," with Bush on the boat in San Diego, flying around in his leather jacket? Remember when they reported catching Saddam Hussein in the hole in the ground? And all of these were played up in our press as if there was absolutely no memory of previous stories.

Consider the initial election in Iraq on January 20, 2005, which was played up again not just on the Fox news channel but in the leading media, *The New York Times*, *Newsweek*, NPR and PBS as a triumphant march of democracy. An election in which most did not even know who they were voting for, in which our leading journalists pontificated and argued that there was 60% to 80% voter turnout when in fact the observers who could make any credible assessment weren't even in the country because it was so dangerous. They were funneling what press releases and government information officers were saying, precisely the mistakes they should have learned from earlier.

From these and many other examples we can sense the depths of the crisis in how the media cover war.

The Election Coverage Crisis

The 2004 U.S. election coverage is a rich subject. It is important to understand that despite the upturn in voting in the 2004 presidential election, in general the voting rate in the United States tends to be abysmally low for a free society, and unacceptably low compared with most of our peer democracies. The number one predictor for whether someone votes is their income; voting is very much class-based in the United States. For example, although there was a big upswing in 18 to 24 year-old voting in the 2004 election compared with the 2000 election—from 36% to 47% (Lopez, Kirby, & Sagoff, 2005)—it was almost entirely in the top third income group of students, and although 59% of this youth group with at least some college voted, only 34% with no college education did. I argue that if one looks at how our system works, it is absolutely rational, especially for lower income and less education people not to vote because of the corruption of our electoral system and the role the media plays in perpetuating that corruption.

And what is the corruption of the system? One can begin with the fact that very few electoral races are contested in this country anymore. In Illinois, there are 21 House of Representative districts. Not one of them was contested because of gerrymandering. (After each Census, voting districts must be redrawn to reflect representative proportions of voters. However, often these districts are "gerrymandered" or redrawn along unusual boundaries to maintain strong majorities of the ruling political party.) Illinois had a Senate seat that was up for grabs last fall. Barack Obama won the seat; but the opponent he ran against wasn't even from Illinois, so there was no reason to vote for the House or Senate in Illinois. Basically, that is how it is state after state. Most races simply aren't being contested because of extensive media campaigns by incumbents, and because gerrymandering eradicates the opposition.

A related problem in election reporting is of course campaign funding. Our system is so drenched in money now that there is little media coverage about it anymore. Democrats and Republicans are both swimming in billions of dollars that pay for these presidential races. More than $1 billion was spent by presidential candidates and national conventions alone in 2004, up 56% from 2000 (Federal Election Commission, 2005). More than $880 million were contributed to the 2004 presidential candidates (Center for Responsive Politics, 2005). For detailed reports on electioneering communications, see the Federal Election Commission (2005b).

And where does the money come from? Overwhelmingly and not surprisingly, more than 80% of individual campaign contributions come from the wealthiest one fourth of 1% of Americans (Miller, 2002). And, of course, extensive funding comes from the telecommunications and media industries (see Center for Public Integrity, 2005, which reports that the communications industry has spent $1.1 billion on lobbying and other influence expenses since 1998 to affect election outcomes and influence legislation before Congress and the White House). For detailed listing of donors (industries, individual donors, Political Action Committees (PAC), soft money, and lobbyists), total raised and spent by House, Senate, and presidential candidates and, spending by advocacy groups (527 committees), see Center for Responsive Politics (2005).

And where does the money go? The more closely contested a race is, the more likely the money goes to paid television advertising. There has been an increase in spending on TV ads of candidates of about 40% every 4 years, going back to the 1980s. Candidates, independent groups, and political parties spent $1.6 billion on television ads in the 100 largest markets in 2004, up from $771 million in 2000 (USA Today, 2004). This enormous amount of money helps to make media companies spectacularly wealthy.

But what about the nutritional content of these ads? Robert Spero, a famous advertising executive (the head of creative for Ogilvy and Mather, one of the leading advertising agencies in the world, back in the 1970s) would go to parties and hear people criticize TV political ads for candidates, saying "it's just like beer advertising." And Spero, being an ad man, became upset because he thought beer advertising had dignity compared with political ads. So he took a year off from his job and did a content analysis of presidential TV political ads from 1952 to 1976. In the resulting book, *The Duping of the American Voter: Dishonesty and Deception in Presidential Television Advertising* (Spero, 1980), he said that if any of those ads were held to the same standards as commercial ads were by the Federal Trade Commission, none would be allowed on the air. They all violated elementary free trade commission standards for advertising.

They also promote cynicism, and that is much of their function: to get people not to vote for their opponent. The goal of political ads is very different from that of a commercial ad. Coke does not have a lot of incentive to run a lot of ads saying, "Pepsi is terrible." Coke really does not benefit if it runs a bunch of ads saying, "The people who work on the Pepsi assembly line use unclean bottles when no-one's looking" because if people do not buy Pepsi, it does not help them. This is not true with candidates. If you get somebody to not vote for your opponent, you win, because even if they do not vote for you, they stay at home. You win.

And what we see from our television media, the place where so many of these ads are placed, is that their response has been to increase the amount of TV advertising while pretty much dropping candidate coverage. There is virtually no coverage of candidates on television anymore. And, tragically, the print coverage that we get, even in our best media, like *The New York Times* or *The Los Angeles Times*, is not about whether political ads are deceptive or inaccurate, such as the anti-Kerry Swift Boat ads (whose claims were not supported by any fact in official Naval reports), for example, but whether people believe them, and whether they are affecting the polls.

A free press would counteract these trends. A free press would discredit the false ads. A free press would say we are going to promote the candidates that do not have a lot of money, who cannot afford to run a lot of ads, and see what they have to say. A free press would try to do everything it could to roll back these anti-democratic tendencies. Instead, our press system does the exact opposite, and we pay the price for it. As discussed below, this is not inherently a failure of the media *per se.* It is a failure of media policy in the United States, and of the conversion of all media content into commercial products.

The Hypercommercialism Crisis

The third crisis or limitation of our press system is hypercommercialism, the commercial colonization of every aspect of our media culture. Nearly all messages and information have been converted either into a direct commodity (such as films, or products spun off from them), an indirect commodity (television and radio, where the programming is designed to attract specific audiences that are valuable to the sponsoring advertisers), or an internal resource of a vertically integrated media conglomerate (where the content can be used multiple times across multiple media, and each company can advertise that content in its other content). Hypercommercialism erodes the traditional barriers between editorial/entertainment and commercial content, or, as it is sometimes put, between church and state. Increasingly, through a number of mechanisms including product placement, the editorial and commercial are merging, with commercial values dominating in a way that was difficult to imagine not too long ago. The purpose of all communication becomes basically to sell, sell, sell. The truth no longer matters. What matters is if the message can convince people to buy whatever it is selling. And that becomes the proof or evidence of truth in a commercial culture.

Concerns about vulgarity and indecency today are the logical result of a concentrated hypercommercial system. You do not hear complaints about vulgarity very often on the local community station. It is commercial values driving network and much cable fare, that foster the most grotesque forms of vulgarity.

So why do we have hypercommercialism, with its associated dreadful journalism? Some people would say, "Gee, well, that is an error. That is not the way the system is supposed to work." However, I argue that that is exactly the way the system is supposed to work—the system is working perfectly. But that is precisely the problem. The system is working very well. The problem with our media system is not that we have bad journalists, they are as good as we have ever had in this country. It is not that we have bad creative talent; it is as good as we have ever had. It is not even that we have bad owners; the owners are largely irrelevant. The owners and CEO of News Corporation, Disney, or TimeWarner are just following the cues of the current media system. They are following what is rational for them to do, what the current market mechanism, shareholders, and media policies reward them for doing. It is not bad people, but it is a bad system, performing fairly efficiently.

FIVE MYTHS ABOUT THE NATURE OF THE MEDIA SYSTEM

The question then naturally arises: Where does the system come from? We are told that this is the natural system, that God wanted it this way, or the founding fathers, or Adam Smith (the author in 1776 of Wealth of Nations, initially titled An Inquiry into the Nature And Causes of the Wealth of Nations, which largely introduced the concept of the invisible hand of the free market), or someone else. Right? In fact, any rigorous look at the history of the founding or our nation, and the origins of U.S. media, shows that is not true at all. There's not something natural about it, it is not a free market system, and it is not the vision of the founding fathers. Indeed, there are at least five major myths have arisen about the nature of the media system.

The First Great Myth: The Current Media System Arose From the Free Market

Our media system is built almost entirely on government policies and subsidies, monopoly licenses, and franchises (McChesney, 1997, 2000, 2004; Starr, 2004). All of our largest media companies are the recipients of enormous governmental largess; the system could not exist without it. Radio and TV monopoly licenses, and cable and satellite franchises, are government granted and protected monopoly privileges, provided at no charge. Copyright is a government-created and -enforced monopoly, to protect content owners (who typically are media conglomerates, who obtain the copyrights in return for distribution contracts from the individual creative artists).

The point here is not to argue against policies and subsidies; they are unavoidable in a press system. But even if you wanted a "free market" system, there would have to be a lot of policies to create. The media system does not develop organically as a free market. Rather, it is all built on policies, subsidies, support, and explicit decisions. So the question is, who are the policies going to be made for, how are they going to be made, and for whose interests are they going to be made?

And here I think it gets sort of interesting. Most people would never have heard of crucial decisions that are routinely made, allocating tens of billions, even hundreds of billions of dollars worth of investment (such as frequency licenses) over time in markets, in core industries, in communication and telecommunication, done behind closed doors. But to understand how the media policymaking process works in the United States in recent decades, think about the great 1974 Academy award-winning film, *The Godfather II.* There is a wonderful scene about half way through that film: It is 1958 in Havana and Hyman Roth, an American gangster, is having a 67th birthday party. Michael Corleone is there, along with other American gangsters, on a patio overlooking Havana. Hyman Roth is dividing up all his holdings in Havana among all the gangsters. As he does this, so we get the hint, he gives them each a slice of his birthday cake, which has an outline of Cuba on it. Everybody is getting a slice of Cuba: one gets the casinos, another the prostitutes. And that is basically how media policy gets done in this country. Very powerful lobbies, meeting behind closed doors, dividing up the media cake with the United States on it, getting the good deals.

However, it is not a conspiracy at all. The media corporations are fighting among themselves, through their lobbyists and associations, to get the biggest slice of the cake (as noted earlier, see Center for Public Integrity, 2005). Yet when they get the benefits of these deals, they hire public relations people to tell us that the success and control of the media companies was won fairly in honest competition in a free market, that the free market philosophy is efficient and is what has made our country great, and government should not regulate it. Just like Hyman Roth, who announced as he was dividing up the country among his gangster friends, "Isn't it great to be in a country with a government that respects private enterprise?"

To give an example of the sort of policy that would never survive any public scrutiny, consider radio broadcasting in the United States in the last decade. In 1996, Congress passed the Telecommunications Act. But before it passed, two paragraphs were inserted into that law by the National Association of Broadcasters that, as far as anyone can tell, were written by the lobbyists themselves. They were never debated in a committee hearing, they were never debated on the floor, never discussed. Those two paragraphs ended the cap on the number of radio stations a single company could own nationally (though there are still limits on the number of radio stations that can be owned in varying sizes of local markets), changing what had been decades of very strict ownership limits that allowed lots of local radio station ownership in this country and highly competitive local radio markets

Since 1996, commercial radio broadcasting in this country has basically been destroyed. Clear Channel, for example, immediately took advantage of this policy change, and now owns just under 1,200 stations, broadcasting in all 50 states, to 110 millions listeners each week. And they are doing what is perfectly rational for them to do in such concentrated markets. They increase advertising and ad rates, and there is more commercialism. Clear Channel's pipes in robot disc jockeys, they are syndicated, they strip out as much localism as possible. Clear Channel's founder and owner told *Fortune* magazine that "We're not in the business of providing news and information. We're not in the business of providing well-researched music. We're simply in the business of selling our customers' products" (cited by Center for Public Integrity, http://www.publicintegrity.org/telecom/analysis/CompanyProfile.aspx?HOID=183). That is rational behavior for them, but there is nothing "free market" about that. That is corrupt policymaking that allows that to happen. If we had a policy of one station per owner, that would not happen. We would still have a competitive, free market. If there were any public involvement in this policy debate over radio, this level of concentrated ownership and commercialism would never exist. Even Senator John McCain (one of the very few elected federal officials who spoke out against some of the greatest frequency giveaways of the Telecommunications Act of 1996)—even Michael Powell, the former head of the Federal Communications Commission (FCC)—have said there is too much concentration of radio.

The Second Great Myth: This System Was Intended by the Founders

The second great myth that protects the concentrated media systems in the United States is that this is the system intended by the founders. According to this argument, the First Amendment basically says that rich people get to make as much money as they want from media, and if you do not like it, too bad. However, it is entire nonsense. In studying the period of the crafting of the First Amendment of the Constitution during the formation of this Republic, what is striking about it is how important press issues were to our founders—especially Thomas Jefferson and James Madison, who wrote at length about them. They were deeply concerned about the importance of having a free press in fostering a free society. They understood that there cannot be a free society without a viable press system and at no time, under any circumstances, did any person in this country and certainly not anyone who crafted our Constitution or any member of Congress for the first 75 or 100 years of this country's history, ever say that freedom of the press meant let's just allow some rich guys make as much money as they want.

In fact, the founders of this country and the first generations of this Republic instituted a series of major policies and subsidies to spawn a free press; otherwise media markets would never have developed to the extent they did (Cook, 2005). The most important of these was the postal subsidy. The U.S. Post Office was by far the largest federal body when it was created; and it was explicitly mentioned in the Constitution. In the 1790s, 70% of the traffic in the Post Office was newspapers, and this rose to more than 90% by the 1830s, so the role of the U.S. Post Office was basically the circulation of our own newspaper system (McChesney, 2004). What newspapers had to pay for postage was the second most important cost after printing newspapers, so a crucial question the founders had to face in 1792 when they set up the U.S. Post Office, as they were required to do by the Constitution, was how much to charge newspapers for being delivered through the mail.

Today, of course, we would expect that there would be someone waving a copy of Adam Smith's *Wealth of Nations*, slamming his fist on the table, saying "These guys have to pay full freight, come on. This is America. No free lunch for these dead beats." In fact, the range of debate in 1792 on what to charge newspapers for distribution extended from one radical position—that newspapers should be heavily subsidized with greatly reduced postage rates, in the interests of democracy, to encourage the distribution of newspapers—to the other position—that all newspapers at all times should be sent for free. To charge anything at all was seen as a form of censorship of ideas because the papers that most needed the subsidy were probably the most dissident papers. The first side won (i.e., heavily subsidized, but not free), but for the next 75 years, the public debate in Congress was routinely about increasing the subsidy for the delivery of newspapers through the U.S. Post Office. For example, abolitionist papers were the ones that needed the subsidies the most; by the 1840s, they had won, so that weekly newspapers were sent for free in their own communities. Thus the poorest, weakest newspapers, including many abolitionist papers, got free distribution in their county.

So it was a political debate, and it was a successful debate. Increasingly, historians are saying this is actually one of the really impressive parts of American history. What we did that distinguishes us from the other countries of the world, was to build a free press. When Alexis de Tocqueville came to the United States in the 1830s, he wrote at length (especially in his book *Democracy in America*) about the print culture in this country—how many more newspapers and books per person there were in here than Britain or France or Canada, and the immersion of print culture and the democratizing effect it had, especially in the northern United States. Such social changes were the result of policies and subsidies that created the print culture, made through enlightened political debate.

In short, freedom of the press is not a right just for those who own it, it is a right for all of us, for a rich diverse press system, for a free society. This truth rejects this second great myth.

The Third Great Myth: Professional Journalism Standards will Protect Us

The third great myth is the idea that professional standards in journalism will protect us from concentrated ownership, hypercommercialism, bias, and cronyism. But this is an important and complex issue (see McChesney, 2004, which devotes two full chapters to this issue).

Professionalism is a relatively recent development in American journalism history. For the first 100 to 120 years of American history, our journalism was stridently partisan. In 1892, for example, if you picked up the newspaper, it would be Republican or Democratic or some branch of Republican or Democratic or some other political party. And if you were to read a Democratic newspaper in the 1892 presidential race, it would be possible that it would not even mention the Republican at all, it would only mention the Democrat. There was little pretense about fairness.

Today (or at least possibly until recently), our response to that system would likely be, "oh my gosh, that was terrible." But maybe not, because the media system worked differently through much of the 19th century; it was a much more competitive market. The city of St. Louis, for example, would have 10 or 15 daily newspapers at any time with a significant range of political opinion, and it was easy enough to start a newspaper. The capital costs were low enough. It wasn't unthinkable to start the 16th paper if you didn't like any of the first 15. By the end of the 19th century, that had pretty much ended. The newspaper economics became driven significantly by the importance of advertising as a source of revenue, which made it virtually impossible to start new newspapers in a given market. So the number of newspapers dropped to one or two in most communities in the country. In that context, having strikingly partisan journalism was unacceptable because there would be few alternative view points, and the newspaper might alienate readers whom the advertisers wanted to reach. Likewise, the commercialization of news, the mad pursuit of profit, undermined the credibility of journalism as a public service; sensationalism or entertainment journalism was the shortest route to a big cash payback. And that was the great crisis of journalism in this country in the progressive era; it spawned professional journalism as the solution. Professional journalism basically meant, "Okay, we are going to have a concentrated press ownership with noncompetitive markets, only one or two companies will control the newspapers in the town, but they will not use that market power to dominate the news; they will have trained professionals that would be responsible for the news. If you're reading a newspaper, you won't know if it is owned by a Republican or a Democrat or a vegetarian. It'll be irrelevant because they are trained professionals doing it." So today we have the notion of objectivity and professionalism in journalism, and that, many feel, should protect us from bias or hypercommercialism fostered by concentrated, commercial media.

Yet there are a number of weaknesses in professional journalism as it developed in the United States. The one that is most striking, but not the only one, is its reliance on official sources. Such sources, in professional canon, are the basis of legitimate political news. But, at its worst, in coverage of war, it means basically newspapers redistribute what people in power debate. And that was new to professional journalism. And it emerged for a number of reasons, including that it is politically very safe, cheap, and easy to do. But professional journalism does not have to be that way. It is not as if the choice is stridently bipartisan journalism or weak professional journalism. In this country in the 1930s, there was a strong movement by our leading journalists of that era, such as Heywood Broun, to challenge that, and say, "Professional journalism should not basically repeat just what people in power say; it should challenge what everyone says." They didn't win that fight, but they were the people who founded the newspaper union, The Newspaper Guild (originally founded in 1933 as The American Newspaper Guild; now a sector of the Communications Workers of America; see http://www.newsguild.org), and that was in their charter. They would not be happy with the journalism that ultimately developed.

Professional journalism in the United States probably reached its peak in the 1960s and 1970s, culminating in the coverage of Watergate. Since the 1970s, the autonomy that journalists were able to develop has been under attack really from two vantage points. First, and most importantly, commercial attack. Increasingly, the big media companies have been able to gobble up newspaper chains, and television networks see doing journalism as bad business. It costs a lot of money, but does not bring a lot of money in. So we have seen much less hard news, much less expensive news, much less investigative news, and we barely have local television news in this country any more. In Denver, Colorado a few years ago, the Rocky Mountain Media Watch conducted content analyses of local TV news (Klite, Bardwell, & Salzman, 1996, 1998, 1999). It was pathetic; it was like the lead story one night was a donkey getting a shampoo. (Klite and colleagues even developed indicators for "mayhem," "fluff," and the "Pavlov index," a measure of highly emotional content used to prepare audiences for advertisements.) In 1998, they petitioned the Federal Trade Commission to deny the re-licensing of four local stations, claiming that their news broadcasts "[are] harmful to citizens . . . are severely unbalanced, with excessive reporting of violent topics and trivial stories and, consequently, inadequate news coverage of a wider range of stories and vital social issues, including local elections, the arts, science, education, the environment, AIDS, children and others. In addition, newscasts present stereotypical and unfavorable depictions of women and minorities" (Rocky Mountain Media Watch, 1998). The point they were making is that there is barely any journalism on television any more, which is the primary source of local news. Most radio journalism basically has ceased to exist in all but a couple of markets and then it is just reading the morning paper.

And most print journalism has suffered, too. One example of the decrepit situation of journalism in the last 20 to 25 years, is the collapse in journalists' morale. Linda Foley, president of The Newspaper Guild, said the number one issue that concerns her membership, way beyond salary, is the commercial encroachment on their work, the inability to do they work they got into the profession to do.

The other blade of the scissor that has attacked journalism has come from the right-wing attack on it, the disciplining of journalism to keep forcing it to the right. This is a central issue in our times, and one I have discussed at length elsewhere (see Nichols & McChesney, 2005). It has the effect of making journalists more likely to deploy a double standard, to tread lightly when covering the activities of right wingers and conservatives. So we cannot rely on the great myth of journalism professionalism.

The Fourth Great Myth: The Media System Gives People What They Want

The fourth great myth that protects media power in this country is the idea that no matter what, even with concentrated markets, the media system gives people what they want. This is especially true for entertainment. If you do not like what the media system does, do not blame the system, blame the audiences who are demanding it. If you want to change the system, change the audiences, do not change the system. It makes a certain amount of sense if you think about it. Television studios and movie companies are out to make profits, they are competitive, they clearly are not going to put out TV shows that people will not watch or movies that they will not buy tickets to.

But on closer inspection, this argument is deeply flawed, self-serving, and even circular. For one thing, it is difficult, perhaps impossible, to use the market to express opposition to the market or market values; it requires political activity to represent those values. For another thing, for markets to give people what they want, basic economic theory states that they have to be competitive markets, so that new companies could enter the markets to provide new choices (i.e., there are no barriers to entry). But media markets are either oligopolistic or semi-monopolistic (and becoming more so owing to lifting of regulatory constraints); they are not competitive. That means audiences do not have consumer sovereignty, a central assumption in free market theory; instead, the media give audiences what they want but within the range of what makes the media company the most money. Audiences develop a taste for it and then start demanding it. Demand does not just create supply; supply to a certain extent creates demand.

If you are never exposed to something of course you will never demand it. The best example of how this process works is with foreign films in the United States. In the early 1970s, roughly 10% of all films shown in American theaters were foreign-language films. All the foreign films would come in with a network of foreign theaters that serviced nothing but foreign films. Seattle, where I lived in the 1970s, had seven, eight, or nine foreign-language theaters. In 1975, Manhattan had 25 theaters that showed nothing but foreign-language films. By the mid-1980s, foreign-language films were down to 2% to 4% theater screen time; by the late 1990s it was down to less than 25% of 1% (McChesney, 2000). Now by the "give the people what they want" theory, you would expect that sometime starting in the late 1970s American people stood up en masse and yelled, "Get those foreign films out of our movie theaters. We are not going near that theater until that foreign stuff is out of there and give us what we want or we are not going back."

In fact, it was not demand creating supply, it was supply creating demand. The mid-1970s saw the rise of the multiplex, the multi-screen theater, which basically drove the single- screen theaters out of business. Instead of a network of independent theaters, there was an internal set of small theaters where there was one projectionist who could handle all the screens, one popcorn maker, one ticket person, and so on. A single-screen theater could not survive against these cost efficiencies and film distribution bargaining power. Foreign film producers, when they came to the United States, increasingly found they had to go to these chains to get their films shown. Suddenly, they had to compete with Hollywood films to get screen space, they had to guarantee to do so much TV advertising (which often was more than the cost of the film), and they could not bundle their film with other films, to lock in a spot on the screen. And some of the films fell away and gradually it just eroded and most theaters gradually stopped showing their films. So in a narrow sense it is true that there is a little demands for foreign language films today; but that is mostly because of the fact that few Americans under 35 would ever think that was a viable option in the first place. The point here is that the media involve a complex relationship between supply and demand. The argument that markets give the people what they want is not meant to shed light on it, but rather to disguise what is actually going on.

Similarly, a core problem with markets is that they generate (or are allowed to generate; again, policies could prevent this) what economists call externalities, where someone other than the producer or consumer pays in some way for the creation of the product; a classic form of externality we all know is pollution. A company makes a product and pollutes the air without paying for that pollution, or paying for the internal machinery to prevent the pollution, someone buys the product at a price for what it costs to make it, and the producer is happy and the consumer is happy. But there is all this pollution in the air that society has to pay for, and it is not factored into the market relationship. That is why it is called a negative externality. Media have enormous externalities, such as the hypercommercialism, terrible journalism, misinformed citizens and thus poor voting and policy decisions, even corruption and poor governance, growing inequality in our society. If the people understood all the costs and negative externalities, and especially if they were forced to pay the true market costs, they would reject much media content and possibly the current media system. It requires policy intervention to force both producers and consumers to pay for the cost of that pollution. We need media policies to address these externalities, to minimize them, and, when there actually are positive externalities, emphasize those. Otherwise, it is easier to believe the great myth that media deliver what people want.

The Fifth Great Myth: Technology Will Set Us Free

The fifth myth is the idea that technology will set us free. There is no need to worry about the media giants, there is no reason to worry about hypercommercialism, and there is no need to worry about the state of journalism. All we need is the Internet, a blog, a mouse, a handheld or digital web camera, or an RSS news feed, and we will have access to all the accurate information and news coverage we could use. We are in the midst of an extraordinary technological revolution and indeed much is changing. But it is clear that the idea that technology alone will allow us to avoid the difficult political and policy fights necessary to develop a democratic media system is simply preposterous. There is no historic or contemporary evidence to support this argument (McChesney, 2004).

Consider just two points about this myth. First of all, the myth asserts that everyone can have a Web site or a blog, and that will counter corporate media control of content. A more formal language used in policy debates is that there is no longer any frequency scarcity, so it does not matter if some companies own most of the traditional media. However, Web sites take time and money to develop and maintain, and expertise to provide accurate content. So the average person's Web site, blog, podcast, or newsletter is neither going to be good, or widely read. The market is an effective censor. To the extent the Internet has emerged as a distinct media sector, largely through search engines and ISPs, it has become the province of the huge telephone and cable monopolies, Google, Yahoo, Microsoft, Time Warner's AOL and Barry Diller's InterActive Corporation. The market has weeded down the crowd pretty quickly.

Second, the Internet is not going to allow new commercially viable content providers to challenge the existing media giants. Not one viable new commercial player has come via the Internet, although there are many claims by the media giants about new competition. A recent report from the Pew Research Center for the People and the Press (2004) notes that those who go online for news at least 3 days a week rose only from 23% to 29% between 2000 and 2004. Only 15% of those who go online at least weekly for news say that their use of other news sources has decreased (71% indicated no change, and 9% reported increases). Furthermore, the top online news sites (in terms of average monthly unique visitors in 2004) were, in order, CNN, Yahoo News, MSNBC (each with about 20 million), AOL News, Gannett, IBS, Knight-Ridder, NYTimes, Tribune, USAToday, and ABCNEWS. Indeed, 17 of the top 25 sites were associated with traditional news companies, who produce offline content for their newspapers, television stations, or magazines (Journalism.org, 2005).

So we have to think creatively about policies to use these technologies to make them actually serve us better if we want to be serious about it. Consider the following example of where policy is so critical even to make these new technologies do what they are capable of doing. With new mesh technology there is a chance to have broadband wireless Internet access anywhere in this country, as a public utility, basically for free. There would be a cost but not to individuals as a per-usage charge; it would be a public utility, much as roads, education, libraries, and so on. There would be much higher bandwidth for uploading and downloading, the uplink would be as broad and thus as fast as the downlink, the more people who use the system the stronger and the less expensive it is, it would be much higher quality and open to everyone automatically, and it would be built into every computer, every device, as a public utility. The cost would be extraordinarily low per person across the population of 280 million people, maybe $50–100 per person per year, if that. In a rational society without vested interests, that is exactly what we would do, 3 or 4 years from now, and soon in the whole world. That would provide all telephone and all electronic communication on a broadband wireless system.

But there is one reason why that might not happen: the phone companies and the cable companies, because their whole survival is premised on us not having that option. They have two wires that you have to go through (and thus pay for) to get broadband Internet access and business in your homes. And they are doing everything in their power with policies to make sure you do not even know about the possibility of broadband wireless Internet.

There are a few crucial areas where this policy fight is going on. In December 2003, Philadelphia passed a law in the city council that they were going to set up one of these wireless networks, make it a public utility, a great wireless meshed system, and everyone would have access to it. The digital divide would be history. Imagine what it would do for commerce as well as for news coverage in a free market. All the businesses that can take advantage of this would drastically lower their communication costs. However, Verizon and Comcast, the two giant telephone and cable companies, are based in Pennsylvania. And they happen to have a certain amount of influence with the governor of Pennsylvania, and they went to the state legislature in Pennsylvania and got a law passed in the state legislature that any community in Pennsylvania that wants to set up its own wireless system for its citizens has to get the approval of either Verizon or Comcast. And the governor signed it.

Fortunately, some local and state governments are fighting to prevent the cable companies and the phone companies from making it impossible to have competitive systems, nonprofit wireless systems owned by the communities. There are other fights in Washington to make this happen. For instance, we need contiguous spectrum, because the meshed wireless broadband system works on a spread spectrum, so it is a much more rational, effective use of spectrum than the current antiquated system. There is enough spectrum to make this policy decision. So counter to the fifth great myth, new technology will not by itself overcome problems of media control. Furthermore, it will take serious debate and media policies to take advantage of some of the potential of new communication technologies.

A CHANGE IN PUBLIC AWARENESS AND OPINION

For decades, it was accepted as absurd to think that the public could possibly challenge this concentrated and controlled media system, respond to these crises, and overcome these myths. People were content. The media lobby was and is superpowerful, investing heavily in influencing the policy process (Center for Public Integrity, 2005). The commercial news media never cover the issue of media, so the public is largely oblivious to their rights and responsibilities. The media system was locked in. Only well after major political reform, along the lines of the New Deal, if even then, could one even consider the idea that media policies could be changed.

And then history does one of those wonderful things. In 2003, the FCC embarked on its review of media ownership rules (mandated by the Telecommunications Act of 1996) with the stated goal of eliminating the rules that limited the numbers of monopoly television licenses that a company could have and that prevented companies from buying up as many as they want and then using their profits to obtain all the other media in the community—cable systems, newspapers, and the like. It would have basically eliminated local media ownership in most cities in this country. It would have created one-newsroom towns, with increased commercial pressures on local news. But, amazingly, 2 or 3 million Americans rose up, across the political spectrum, left and right, and sent letters, e-mails, phone calls, organized hearings around the country, and stopped them, as the U.S. Court of Appeals for the Third Circuit overturned the proposed rule changes (see Media Access Project, 2004).

That outpouring and the court ruling empowered people on the media issue in ways that had not been seen in this country in 75 years. Because as soon as the people understand the media system is not natural, but in fact it is something created in their name but without their informed consent, and is affecting media content to suit the media industry's interests, not the public's interest, it is as though a light goes on. Nothing will ever be the same again. And that is the period we are in now. In January 2003, at a Senate hearing on media ownership and the Senate Commerce Committee, I saw the Republican John Enson from Nevada, popular with the National Association of Broadcasters, stand up at this committee hearing and say something along the lines of, "I'll have you know, in all my time in public life I've never heard any of my constituents show any concern with this media stuff. This is not an issue that anyone cares about." He was probably telling the truth as of January 2003. But now there is not a single member of Congress who has not heard about it or heard about it repeatedly.

The solution to overcoming our current flawed media system and its attendant crises is very simple: We have to popularize the policy debates, draw public attention to them, bring informed public involvement. I think what we would find over time is a type of media system with much more competitive markets, far more local ownership, and a much larger, vibrant, diverse, nonprofit and noncommercial sector. However, if we had those debates and the American people studied that matter and discussed it, and concluded that "No, we want a few companies to own everything, we want to be carpet bombed with ads, we do not like journalism, we love political ads," it would at least been decided on democratically. Regrettably, that attitude is not shared by the media corporations; they are not especially excited about having that debate, and have powerful and well-funded lobbies that repeat these five great myths. So any effort to reform and democratize our media policymaking and media system has to really confront the five myths about the nature of the media system, and confront them directly.

We have seen a fundamental shift following the 2004 election. www.MoveOn.org, and www.TrueMajority.org, each did polls of their memberships and people they worked with, asking, What are the most important issues we have to work on in this country? Where should we put our attention and resources? Number two on each of their lists was *media reform*—ahead of the environment, ahead of the war in Iraq (Nichols & McChesney, 2005). There is an increasing recognition that if we do not work on this issue, all the other issues will not get anywhere. One other indication of how quickly this movement is growing is Free Press (www.Freepress.net), the group that I helped form with John Nichols in 2004, started with one guy sort of sleeping on the floor, in Northampton, Massachusetts. As of mid-2006 the group has 15 organizers and activists and more than 200,000 members working on media ownership, Internet access, public broadcasting, and a number of other core media policy issues.

And we have learned that media reform, rather than being hopelessly impossible, is actually well suited to an incremental strategy. With media reform, one can win tangible victories and they cannot be taken away. For example, there are going to be several hundred low-power FM radio stations in this country in 2006 and 2007 as a result of organizing and activism on the open parts of the FM dial (see FCC, 2005). Media reform efforts can result in changes at the local level. For another example, one of the great secrets of our media system is that all the cable companies get franchises or monopoly rights with municipal governments, and as part of the agreement they are supposed to provide services for the people in the community. Of course, they only do it if they are forced to. If a community organizes, it always wins; the cable companies always concede and give in when they have to do by law. But if no one forces them to do it, they do not. So some communities that have organized, like Grand Rapids Michigan, Ashland, North

Carolina, Chicago, Illinois, or Burlington, Vermont, have spectacular pubic access channels with lots of resources because they get what they are allowed to get by law, they organize, and when the hearings come up they force their city councils to make sure these companies give them what they are entitled to.

So if we organize and commit to media reform actions, we can win major changes in the media system. But you have to organize. And that is what the organization Free Press does. It tries to help communities get together to organize. The struggle is only going to get more intense. The fight, I think, is picking up steam. We are going to see a number of crucial battles, public battles in the coming years. I hope you will go to www.freepress.net, join our movement, stay tuned to it, get on our mailing list, and work to improve our media system, and help solve the crises of our society.

REFERENCES

Center for Public Integrity. (2005). *Networks of influence: The political power of the communications industry.* Washington, DC: Center for Public Integrity. Available at http://www.publicintegrity.org/telecom/

Center for Responsive Politics. (2005). 2004 Presidential election. (Based on Federal Election Commission report released July 11, 2005). http://www.opensecrets.org/presidential/index.asp

Cook, T. E. (Ed.). (2005). *Freeing the presses: The First Amendment in action.* Baton Rouge: Louisiana State University Press.

Federal Communications Commission. (2005). *Low power FM broadcast radio stations.* http://www.fcc.gov/mb/audio/lpfm/

Federal Election Commission. (2005a). Electioneering communications. http://www.fec.gov/finance/disclosure/electioneering.shtml

Federal Election Commission. (2005b). 2004 presidential campaign financial activity summarized. http://www.fec.gov/press/press2005/20050203pressum/20050203pressum.html

Journalism.org. (2005). *The state of the news media: Online.* http://www.stateofthemedia.org/2005/narrative_online_ownership.asp?cat=5&media=3

Klite, P. D., Bardwell, R. A., & Salzman, J. (1996). *Pavlov's TV dogs: A snapshot of local TV News in America 9/20/95.* Denver, CO: Rocky Mountain Media Watch.

Klite, P. D., Bardwell, R. A., & Salzman, J. (1998). Baaad news: Local TV news in America 2/26/97. Denver, CO: Rocky Mountain Media Watch.

Klite, P. D., Bardwell, R. A., & Salzman, J. (1999). *Not in the public interest: A snapshot of local TV news in America 3/11/98.* Denver, CO: Rocky Mountain Media Watch.

Lopez, M., Kirby, E., & Sagoff, J. (2005). The youth vote 2004. CIRCLE (The Center for Information and Research on Civic Learning and Engagement). http://www.civicyouth.org/staff_advisory/index.htm

McChesney, R. (1997). *Corporate media & the threat to democracy.* New York: Seven Stories Press.

McChesney, R. (2000). *Rich media, poor democracy: Communication politics in dubious times.* New York: The New Press.

McChesney, R. (2004). *The problem of the media: U.S. communication politics in the twenty-first century.* New York: Monthly Review Press.

Media Access Project. (2004). *Prometheus Radio Project v. FCC, Case No. 03-3388*, United States Court of Appeals for the Third Circuit, February 2004. http://www.mediaaccess.org/mediaownershipcase.html

Miller, E. S. (2002). *The elephant in the living room.* www.tompaine.com.

Nichols, J., & McChesney, R. (2005). *Tragedy & farce: How the media sell wars, spin elections, and destroy democracy.* New York: The New Press.

The Pew Research Center for the People and the Press. (2004, June 8). *News audiences increasingly politicized; Online news audience larger, more diverse.* http://people-press.org/reports/display.php3?ReportID=215

Rocky Mountain Media Watch. (1998, February 16). *Group challenges FCC licenses of Denver TV stations.* http://www.bigmedia.org/texts2.html#Press%20Release %20FCC

Solomon, N. (2005). *War made easy: How presidents and pundits keep spinning us to death.* New York: Wiley.

Spero, R. (1980). *The duping of the American voter: Dishonesty and deception in presidential television advertising.* New York: Lippincott & Crowell.

Starr, P. (2004). *The creation of the media: Political origins of modern communications.* New York: Basic Books.

USA Today. (2004, November 11). Election ad battle smashes record in 2004. *USA Today* online http://www.usatoday.com/news/washington/2004-11-25-election-ads_x.htm

3

FILTERING THE MEDIA SMOG

Researching the Public's Interest in the Media Environment

Pat Aufderheide

MEDIA SMOG

Our media environment is very noisy, abundant, even polluted. Todd Gitlin (2001) called it "media unlimited." David Shenk (1997) called it "data smog." We have never had more stuff to hear, see, scan, play, select, view. We have never had more channels, and we're about to get so many more—check your cell phone for updates, and plug in that iPod.

The problem of having "Too Much Media" is a good place to start, when we think about media concentration issues. Our own efforts to control our selections, combined with the efforts of large corporations to channel our choices, pose new challenges for people who know that democratic action depends on trustworthy communication.

This data-smog phenomenon is pushed by technological innovation, both in the realms of production and distribution. Digital equipment and the Internet have made us all potential broadcasters. The phenomenon is also pushed forward by the powerful forces of deregulation, which have encouraged many more providers of channels and screens than in the days when we had three networks and one phone company. Our cable company has hundreds of channels, the Internet is littered with blogs and Web sites, we've got video games on screens we couldn't even imagine 5 years ago, there are orders of magnitude more radio stations and TV stations than there were in 1975, here comes digital audio, and oh yes, the cell phone screens. The menu-driven media universe is on its way.

The data-smogginess of the environment is a good place to start considering how to formulate research agendas and problems around the public interest in today's media environment. The once trusty arguments about scarcity and limits do not work in this era of media abundance, as Ithiel de Sola Pool (1983) foresaw with perhaps too much optimism. We are all looking for boundaries to a seemingly limitless amount of media that must be consumed and digested to be understood, which has uncertain reliability and quality. And we lack good enough sorter tools. At the same time, we're deluged by personal messages and spam; we're really sure that our commercial media lack certain kinds of variety, or emphasis; we look at the copycat news channels and sigh; and we all know creators who can't get excellent creative work before large audiences.

MAPPING THE MEDIA: BIG MEDIA AND LITTLE MEDIA

What is the map of media production and distribution morphing into with this explosion of production, this multiplication of screens and channels? One helpful concept in mapping the media today is the dichotomy that Siva Vaidhyanathan (2004) used in *The Anarchist in the Library: How the Clash Between Freedom and Control is Hacking the Real World and Crashing the System*. We could call it *Big Media* (control, centralized, broadcast, commercial) and *Little Media* (anarchy, distributed, networking, individual).

Big Media

Big Media are the mainstream media companies that increasingly are vertically integrated, and that incorporate production, distribution and even theatrical exhibition. How are they dealing with abundance? They are bulking up, both horizontally (buying companies in the same line of business, minimizing competition) and vertically (buying companies along the line of production and distribution for their products, helping to control the process). They are doing so not for synergy, but to hedge against the future, especially against possible competitors, to buy the right to control the pace of change by being one of the prime movers. As well, they want to build brands that restless and distrustful viewers can use as filters for the data smog. Viewers use brands, in other words, to cut through the noise and make a decision more quickly on which information products to pay attention to.

Companies are also addressing the need to attract attention in an ever more crowded environment—one that they relentlessly add to—with program choices: cheaper, nastier, louder, ruder media (whether it's TV shows or games or movies) on the one end, and glossier, more expensive, elaborate, effects- and celebrity-heavy products on the other: Fox's *When Good Pets Go Bad* on one side, and BBC/Discovery's blue-chip nature series *Blue Planet* on the other. Big Media companies are spreading their brands out into new delivery modes, and so increasingly staking out positions on information and communications policy issues as arcane and wide-ranging as digital rights management (controlling digital products by putting digital locks on them) and spectrum reallocation (the federal government's ongoing process of reassigning spectrum to mass media, telecommunications, and policing services as digital innovation changes needs). They know that the old business models are not working well now and their challenge is to control user choice as much as they can.

They face threats not only from each other and from commercial upstarts, but also from the digital universe that began to bloom with email, took off with the Web and is sailing into the future with peer-to-peer software. It is a much bigger phenomenon than the "piracy" label that (Motion Picture Association of America and Recording Industry Association of America (RIAA) executives keep using for it. It is Vaidhyanathan's anarchy, or Little Media.

Little Media

Little Media lives on distributed networking. It is decentralized, ubiquitous, constantly proliferating. Anyone there can become a producer and a distributor. Little Media can be a grassroots, friends-and-family kind of place, where what lures eyeballs is human connection, the trust that the sender or the maker is your friend, your family, someone like you. Growth occurs through the expansion of self-generating user networks—sometimes so rapidly that people use phrases such as "viral marketing." This is the world that MoveOn (www.moveon.org) negotiates so well, the one that indymedia (www.indymedia.org) enthusiasts inhabit, the world of bloggers and IMers and downloaders and podcasters. It is chaotic and unvetted, and it is as easy for bad information to spread as good.

In Little Media, business models are only just now emerging, but they are very interesting. They are tied to the ability of networks to form and spread, as Robert Greenwald discovered when MoveOn distributed DVDs of his 2002–2004 series of anti-Bush documentary films, and he was able to pay back the original investment within a month of Internet-only sales. This was in stark contrast to the reluctance of any major distributor to carry any critical documentary into theaters (with the exception of the celebrity-clout Fahrenheit 911—but even that was rejected by its original distributor) in the election season. They create what *Wired* magazine is happy to call "the long tail," or the ability to market with very low costs to very small markets for a very long time, and thus get out of the search for large markets to make a profit, which has always been the business of Big Media..

CROSSING THE BORDERS BETWEEN BIG MEDIA AND LITTLE MEDIA

Little Media continues to be anarchic and unpredictable, open to new entrants, fertile ground for new makers, a zone of experiment and even of challenge to Big Media. But there are plenty of Big Media outposts in Little Media. In fact, 13 of the 20 top news and entertainment sites on the Internet are run by major TV and newspaper companies, according to the media reform group Free Press (www.freepress.net). The most downloaded podcasts from Apple's iTunes in the first weeks after it launched its search service included "Queer Guy Hip Tips," from the TV show, and ABC News (Athavaley, 2005). And there are crossover points between Big and Little Media, like the DVD rental company Netflix (www.netflix.com), which harnesses the power of the Internet to find tiny markets and serve them well with Big Media products.

Both Big Media's explosion of channels and services over the last two decades and the proliferation of Little Media over the last decade have led deregulation enthusiasts to argue that we do not need to worry about media concentration any more. That was the Federal Communication Commission's position in 2003, when it attempted to further loosen limits on media concentration. After great public outcry, the issue was taken up in court by public interest groups, and the court disagreed with the FCC. The court basically told the FCC that a major TV station and a blog were not comparable news sources, and to come back with a better argument for raising ceilings. This docket was just one piece of evidence that we need more and better research into today's media environment.

We can see both as researchers and as citizens that innovation is creating new opportunities, and even changing business practice. But Big Media remain the platforms through which people learn what they think about what other people think.

Consider how Big and Little Media interacted at the time of the invasion of Iraq. Recall that during the Iraq invasion, polls were running 70% in support of the president's decision to go to war with Iraq. That 30 opposing percent was an awkward number for mass media, where majority rules. TV news managers were extremely concerned about how to cover the rampup to the invasion. Frank Magid, a research firm, told TV stations the week after the war started that covering antiwar protests could, according to a survey, affect ratings. Alienating the

70% would not be good. Stations immediately took the advice. Anchors were uniformly supportive of the war, and station Web sites were heavy on the "support your troops" web links (Farhi, 2003). Of course you could go to the BBC on the web (http://www.bbc.co.uk); and a lot of people did. But they were pretty sure their neighbors were watching local Fox News.

Down there in Little Media, the Beastie Boys wrote "In a World Gone Mad," and let people download it for free (if people could find the Web site). But it didn't make news in Big Media, where people were finding out what other people thought. Blogs bloomed everywhere, but they mostly added attitude. Hard information was largely drawn from a few sources—CNN, MSNBC, Fox, and so on. As the Pew American Life and Internet Project reported, when the more than half of American Internet users who used the Internet to get information about the war went to their URLs, they most often turned to U.S. TV news networks' sites (Rainie, Fox, & Fallows, 2003). E-mails ricocheted around the world, carrying slideshows and jokes and URLs and petitions. The petitions were interesting because several of them were either useless or outdated, but nothing seemed to stop them circulating even so. On the Internet, unreliable data travels as quickly as reliable data.

These were phenomena we watched, in a small demonstration project at our Center for Social Media, where we tracked how people were using Little Media to express anti-war perspectives. You can see the results at our Web site; it's a project called War Beyond the Box (http://centerforsocialmedia.org/warbeyondbox/index.htm). What we saw was Big Media having an enormous ability to influence polls and what the pollsters call public opinion, and Little Media mobilizing large numbers of people who felt solidarity in their grievances and the silencing of their views in Big Media. We saw a heightening of already polarized views, a heightening of resentment, grievance, cynicism, and mistrust.

Media Abundance and the Public Interest

We know far too little about user practices in coping with media abundance. User choice tells us what tools people are using to take the shortcuts they have to. It will show us which filters—what means of selecting from the welter of material—people turn to when they know they need to know something. And that will tell us how the various strategies companies now use to narrow user choice work, and how Little and Big Media interact. It all goes back to media's powerful shaping role in social and political identity, and to the notion that there is a public interest in media. The public interest has been a highly elastic term, often bent to wishes of special interests. But as a key phrase in the Communications Act of 1934, it has reminded policymakers in every era that communications is not merely a business but a creator of culture (Aufderheide, 1999). And in a democratic society, it is either helping or not to create a functioning public.

What is ultimately at stake is not the quality of our media but the quality of public life. The great American philosopher John Dewey described the public as the body of people who have to deal with the shared consequences of private, corporate or governmental actions, because those actions affect the society as a whole (Dewey, 1927). And of course, public life has intimate ties with the media environment. When people pay attention to media, they invest in the importance of that perspective, argument, information. As the communication scholar Jim Carey (1989) says, "Reality is a scarce resource." People fight over it. Media is about the most expensive real estate of all, the real estate inside your head. It's about how you understand the world. That's what Roger Ailes, the dark right-wing genius behind Fox, really understands very well. Behind the "we report, you decide" façade is a pronounced ideological mission to change the inside of people's heads, so that a right-wing agenda becomes their "new normal."

Our past invocations of the public interest in media have not been terribly strong. In the early days of the FCC, in the 1930s, the public interest was basically defined as the government asking stations to which they licensed broadcast spectrum (a public resource), Are you commercializing the spectrum or just wasting it? Since then, however, government policy has consistently turned communication into property (Streeter, 1996), and the construction of communications policy has been a never-ending wrangle of commercial interests in the name of the public (Starr, 2004).

As mass media consolidated, and as civil rights movements grew, organized public stakeholders demanded something in return to public life from the mass media—a kind of noblesse oblige from the media barons, who held a tight grip on the few media outlets. The public got a few things, including: strong encouragement for public affairs, kids' shows, and community-oriented programs; requirements to keep program records open to the public; requirements to check in with community leaders to find out how local folk saw the TV or radio station's performance. Localism, public affairs, kids' shows: each of these were rough markers for responding to the needs of public culture, of a public sphere (Aufderheide, 1999). That kind of noblesse oblige—what was called "serving in the public interest"—is over. For decades, the public interest has been defined at the federal government level as pretty much "what the public is interested in." That is, the "public" has been transformed into the "individual consumer." This transformation of the definition for public interest was the result of the great deregulatory revolution charted so well by Robert Horwitz (1989). Deregulation was a major impetus to create both much more competition for eyeballs and ears—all those screens and speakers—and much more media concentration.

Our problem of coping with too much media of uncertain quality, relevance, and reliability feeds back into the old problem of how to be members of a public, how active members of a democracy can coordinate to bring outside pressure to bear on organized systems that channel power.

Media Control and Ownership Issues

The next big reorganization of media is around the control of funnels and filters. That is where the policy action is going to be. There are many ownership issues here, and they go far, far beyond the question of who owns MGM today. They are about who controls the menu of options for the many choices we make about media now. For instance:

- Who controls the broadband pipes and spectrum that deliver digital data to your screen and upload your digital work? Not you the end user. At the moment, it is probably a cable or phone company, and they probably have restrictions on how much digital material you upload. Some cable companies have restricted customers to 10 minutes or so of upload, and most broadband providers offer much faster download than upload speeds. Why? Because the programmers want to stay in charge, and because they want you to consume material much more than you want to produce and share it. Those companies, both cable and phone, also have successfully resisted letting any other companies rent their broadband service to offer competition.
- Who controls the electromagnetic spectrum that is getting you wireless connections? Most spectrum is leased by Big Media and Big Communications users. Wifi connections run over public spectrum, but wi-fi's tiny amount of bandwidth is cramping what could be a truly public new communications network. In order to make available more spectrum to the general public, so that we can use small sender-receivers in our computers to build a grassroots network of communication, the FCC will have to take spectrum away from powerful vested interests.
- Who controls the software that runs the menu and selection services for program services? Large companies such as News Corp. have staked claims to menu services, and the industry is in a race to see who can develop the most effective search tools. Whoever owns the search and selection process owns the viewer's choice.
- Who controls the machines on which you view material? Not you, even if you bought it. Companies are designing those machines to refuse to copy digital material. That would criminalize a lot of behavior that comes naturally to digital technology, and it could squash innovation a lot quicker than it will save a dying business model.
- Who controls the digital file? Not you the receiver. The digital file is a lot easier to copy than a book or a CD, as the RIAA (http://www.riaa.com) wants to remind you every day. Or, if it were passworded, it could be a lot harder. Digital Rights Management, which Microsoft and its image archive company Corbis (http://pro.corbis.com) have embraced and promoted their version vigorously, "locks up" digital files. So maybe you get to read your ebook three times before paying again; and forget about sharing it with your book group.

Or maybe you watch a movie on your machine but can't go back and show your friend the really funny part. Under the Digital Millennium Copyright Act, it is also now illegal to try to crack open a locked digital file.

- Who controls the intellectual property in the file? Whoever else, not you. Copyright has never been longer—currently, 70 years plus the life of the author—and it could get longer. Extra-long copyrights are a boon to people with an extremely valuable copyright (like the words to "Happy Birthday," or "The Lion Sleeps Tonight," or the character of Mickey Mouse). But for most people, locking up material for a century or so just means that it does not get to be part of popular or public culture except on the terms of the copyright holder. (The copyright holder is rarely the creator, since to get distribution most creators sell their copyrights to these same media corporations.)

These are all ownership issues. Each of them deals with the control challenges posed by the hyperproductive and porous media environment. Every one of them is in contest at the moment.

RESEARCH NEEDS ABOUT FUNNELS AND FILTERS

Research that explores emerging structures of both persuasion and control of the end user will be eye-opening. We need to know more about corporate practices as new technologies create new opportunities for control and surveillance, and it would help if the FCC would require more transparency in reporting from media corporations. We need to know more about emerging niche markets for media, and how those define new stakeholders and how they serve some interests and not others. We need to know more about information cycles—how and how quickly information moves through the society.

It is not enough to be assured that information is out there if anyone wants to inform themselves. The dynamics of this media environment are far too powerful to be reassured by that informed-citizen piety. Michael Schudson (1998) wisely pointed out that the informed-citizen argument is a way of transferring down to individual the responsibility that all too often individuals cannot actually exercise. Being an informed citizen is an exhausting and unrewarding full-time job, and most of us already have one of those.

Finally, it will be important to look carefully at the filters that are designed to help the public become the public it can be. That is the arena of public media—public radio, public TV, public access cable, DBS channels that were set aside specifically to serve the public interest, media arts centers, computer technology centers, media associated with schools and universities, low-power radio, and Web-based public media initiatives. This has been a dowdy little neighborhood of the media environment. But maybe it is time to take another look at what works in those zones. Public media could have unique advantages in a media environment of unmanageable abundance. Most of it is still local, which means we can use technologies of communication to reinforce our inner circle of public life, right here at home. Historically, these zones encourage—not always, but sometimes—respect for diverse voices and diverse points of view. Public media are likely candidates for providing honest-broker services, acting like public parks in this polluted environment.

Big Media filters our choices through branding (which is a form of persuasion), through manipulation of our access, and through manipulation of our usage choices (both of which are forms of control). Big Media directs our interests at what vendors and advertisers are interested in.

Little Media funnels our choices through trust networks, some of which really aren't trustworthy for the jobs we need to do as our public selves, especially in terms of getting accurate information about what people we don't agree with think. Between the two, we are pulled between profit, partisanships and passions. Public media could play a role of actively mobilizing us as members of the public, vetting information, sifting grassroots expression, analyzing media strategies, investigating long term trends, seeking out innovative cultural trends—the kinds of things public media do now, in odd and glorious moments.

We need research that asks questions about data smog, the interpenetration of Big and Little Media, the subtle and pervasive imposition of control upon anarchy, about the role of honest brokers and the nature of public media matters, and the forms and implications of funnels and filters. If our media systems cannot actively serve democratic culture and help the public find itself, then they have failed the basic public service test. And then a media system becomes toxic to democracy.

REFERENCES

Athavaley, A. (2005, July 8). Mainstream media is tuning in to 'Podcasting'. *Washington Post*, pp. 1, 7.

Aufderheide, P. (1999). *Communications policy and the public interest: The Telecommunications Act of 1996*. New York: Guilford Press.

Carey, J. W. (1989). *Communication as culture: Essays on media and society*. Boston: Unwin Hyman.

Dewey, J. (1927). *The public and its problems*. New York: H. Holt.

Farhi, P. (2003, March 28). For broadcast media, patriotism pays: Consultants tell radio, tv clients that protest coverage drives off viewers. *Washington Post*, p. 1.

Gitlin, T. (2001). *Media unlimited: How the torrent of images and sounds overwhelms our lives*. New York: Metropolitan Books.

Horwitz, R. B. (1989). *The irony of regulatory reform: The deregulation of American telecommunications*. New York: Oxford University Press.

Pool, I. d. S. (1983). *Technologies of freedom.* Cambridge, MA: Belknap Press.

Rainie, L., Fox, S., & Fallows, D. (2003). The Internet and the Iraq war: How online Americans have used the Internet to learn war news, understand events, and promote their views. *Pew Internet and American Life Project.* Washington, DC, Pew Internet and American Life Project: 11. Available at http://www.pewinternet.org/pdfs/PIP_Iraq_War_Report.pdf.

Schudson, M. (1998). *The good citizen: A history of American civic life.* New York: Martin Kessler Books.

Shenk, D. (1997). *Data smog: Surviving the information glut.* San Francisco, CA: Harper Edge.

Starr, P. (2004). *The creation of the media: Political origins of modern communications.* New York: Basic Books.

Streeter, T. (1996). *Selling the air: A critique of the policy of commercial broadcasting in the United States.* Chicago: University of Chicago Press.

Vaidhyanathan, S. (2004). *The anarchist in the library: How the clash between freedom and control is hacking the real world and crashing the system.* New York: Basic Books.

II

HISTORICAL AND POLITICAL ISSUES

4

THE CLAIMS OF COPYRIGHT

Public Purposes and Private Property

Mark Rose

This volume is concerned with media ownership and specifically with the increasing concentration of media of various kinds in vertical conglomerates that many fear represent a threat to creativity, competition, and democratic discourse. These developments in the media industries necessarily involve copyright, the body of law that regulates what is now, rather abstractly and bloodlessly, called "content" —that is to say, the texts and images through which works of art, stories, analyses, and information of all kinds circulate as commodities.

The structure of the media industries has been changing (see chapters by Epstein, Holt, and Noam, this volume). Likewise, copyright law has been changing, often under direct pressure from the media industries. Attention has been focused lately on two areas of copyright law in particular. The first area is the way in which the digital environment affects the traditional contours of copyright (see e.g., Lessig, 1999; Litman, 2001). What, for example, does "copying" mean when nearly every action one takes on a computer, including reading or storing a document, involves making a copy? Copyright is about regulation and the media industries are understandably concerned about how to regulate the circulation of the commodities they sell—music, texts, and movies—in the digital environment. Others, however, are concerned with the way the imposition of legal principles developed in the context of print culture may represent a real threat to the circulation of information and knowledge when aggressively imposed on a digital culture.

The second area of copyright law on which attention has recently been focused is the length of the term of protection that copyright provides. The Sonny Bono Copyright Term Extension Act of 1998 (CTEA) extended the basic copyright term by 20 years to the life of the author plus 70 years. For works made for hire, the act established a term of 95 years from the date of publication or 120 years from the date of creation, whichever is shorter. In practice, then, many works will now remain protected for more than a century after they have been created. Even without taking media consolidation into account, many critics of the CTEA consider that the extended length of copyright protection now represents a serious threat to cultural vitality. These issues recently figured in the important case of *Eldred v. Ashcroft* (538 U.S. 916; 2003) in which the U.S. Supreme Court rejected a challenge to the CTEA on constitutional grounds.

But length of term is only one aspect of copyright doctrine. In fact the threat results from the combination of the depth of modern copyright protection together with the length of term. Until well into the 19th-century, copyright was a very limited monopoly, merely the right to print or reprint a specific text. It did not include the right to prevent others from making new versions of a text—translations or abridgements—or from making other kinds of adaptations such as an entirely new story employing, say, the characters from *Star Trek*. Even in the current period of increasing media consolidation, the extended terms might be innocuous if we still had the copyright doctrine that was envisioned in the early Republic when the first U.S. Copyright Act was enacted in 1790.

A CONCEPTUAL TENSION UNDERLYING COPYRIGHT

In order to understand the nature of the present moment in media development, then, it is important to understand the history of copyright. My purpose is to provide a sketch of this history, starting with the English experience in the 17th and 18th centuries and then moving to the American experience after the Revolution. More specifically, I suggest that the history of copyright reveals a fundamental conceptual tension, one that takes the form of opposing claims. Thus, on the one hand, a claim is made that copyrights are properties like any other properties and that the essential purpose of copyright is to protect private property. This is a position that is taken today by, among others, the media industries whose representatives speak of the circulation of illicit copies of music and movies in digital form as "theft." On the other hand, an equally compelling claim is made that the "real" purpose of copyright is not to protect property but to promote the public good. After all, note the opponents of term extensions and expanded protection, the U.S. Constitution grants Congress the power to enact copyright laws not to protect property but "to promote the progress of science and the useful arts"—that is, to advance the public good.

Presented starkly as a conflict between private property and public interest, it is evident that this conceptual tension reenacts an opposition that has energized Western culture from the 17th century to the present. Which principle is logically prior, the private or the public? Which has the final claim, the individual or society? This is not, I think, a question to be answered because to do so is to court absolutism in one form or another. On the one hand lies the chaotic absolutism of the extreme libertarian for whom individual rights always trump social claims. On the other lies the repressive absolutism of the totalitarian for whom communal claims, hypostasized as the state or the will of the people, always trump individual interests.

Private and public are in fact a pair of mutually defining concepts, neither of which can be understood except in relation to the other. Furthermore, in their modern form, private and public are historically located notions, concepts born in the 17th century in the process of the great social transformation from a traditional hierarchical society based on rank and nobility to a marketplace society based on the principle of commercial exchanges between equal individuals. It is precisely in this period, in late 17th century England, that the modern concept of copyright as the right of an author to control the printing of his or her work emerges.

EMERGENCE OF COPYRIGHT IN ENGLAND

Before the precipitation of the modern dialectic of private and public, the circulation of printed texts, like much else, was tightly regulated. There were two dimensions to the regulation of printed texts: the policing of content and the orderly management of the printing industry. In England, content was regulated by licensing laws which were intended to prevent seditious or blasphemous works from being printed. The enforcement of these laws was possible because of the existence of a crown-chartered guild of printers and booksellers, the Stationers' Company, which was also responsible for the orderly management of the commercial aspects of the trade (Blagden, 1960). No one but a member of the company was allowed to print and no one but a member of the company could register a claim to the exclusive right to print a particular title. This the member did by entering the name of the work to be published in the guild register book, claiming the title as his "copy." In return for monopoly rights on printing, the guild served as the agent of the crown in ensuring the regulation of content.

The early regime of press regulation broke down in 1695 when for political reasons related to the emergence of a constitutional monarchy, the licensing laws were allowed to lapse (Astbury, 1978). This led to a period of chaos in the printing trade and to futile guild petitions for the reinstatement of licensing. But eventually it became possible to see that the specifically commercial issues with which the printers and booksellers were primarily concerned might be separated from issues related to censorship, and in 1710 the Statute of Anne (8 Anne ch. 19), the world's first copyright statute, was enacted (on the early history of copyright in Britain, see generally Deazley, 2004; Patterson, 1968; Rose, 1993).

This law codified the traditional guild practices whereby exclusive printing rights were established by entry of the title in the company register. But whereas under the early regime only members of the Stationers' Company could own copyrights—outside of the context of the guild the notion of copyright had no meaning—the statute established authors as the first owners of their copyrights, rights which they then were able to sell to the booksellers. Moreover, whereas under the guild regulations copyright was perpetual, the Statute of Anne specified a limited term of protection of 14 years from the date of publication with the possibility of a second term of protection if the author were still living at the end of the first.

Was the Stationers' Company of the 16th and 17th centuries a private or a public entity? Insofar as it was an independent organization with its own rules and regulations, the Stationers' Company might be considered private. But insofar as it was the creation of the crown and its agent in maintaining order in the realm, the company might well be considered an organ of the state. In fact the modern concepts of public and private do not quite apply to such an entity. But when the essential guild practices were codified as the Statute of Anne, the shape of a new system began to emerge. Now the guild took on the appearance of a private organization, a commercial collective that dealt in commodities regulated by public statute. Authors, too, acquired a new legal standing, one defined by the modern dialectic of private and public. Now for the first time the author was a legally recognized proprietary figure, the first creator and owner of a property produced in private and then brought into the public marketplace for sale.

The Statute of Anne incorporated the dialectic of private and public in another way as well. Before the statute, individual booksellers' copyright claims were perpetual. Once entered in the guild register, a title belonged to a bookseller forever and might be bequeathed to his heirs or sold like any other property. Moreover, old books were protected no less than new, and many of the Latin and Greek classics, including Aesop, Cicero, Ovid, Terence, and Virgil, were owned by the Stationers' Company itself, which could then assign individual titles to whichever guild members it chose. Before the statute, then, when the book trade was regulated in all its aspects, the public domain—the commons of freely circulating texts open to anyone to print or reprint—simply did not exist. By setting limits on the term of protection, the Statute of Anne actually created the public domain. Now individual titles were treated as special limited properties that at the end of the term passed from private to public.

The booksellers and printers who had lobbied for the Statute of Anne had couched their pleas in terms of their need for parliamentary protection of their properties. But, significantly, the statute was not titled, as it might have been, simply an act for "securing the property of books." In fact the title is worth quoting in full: "An act for the encouragement of learning, by vesting the copies of printed books in the authors or purchasers of such copies, during the times therein mentioned." As this suggests, the drafters of the statute were concerned both with property rights and with the broader civic good, and the regime that the statute inaugurated was one in which, at least in principle, private rights were balanced against public purposes. The crucial element in furthering parliament's civic agenda was of course the provision for limited terms.

At first the great booksellers of London, the guild members who controlled nearly all the copyrights of value, simply ignored the matter of the term limits. But when the statutory terms began to expire, and when parliament refused to extend them, the booksellers developed a new strategy to protect their expiring copyrights. Copyrights, they maintained, were and had always been private property that was no different in principle from property in a house or land. The limited copyright provided by the statute was merely supplementary to an underlying common-law right of property that authors acquired through their labors. This common-law right was transferred to the bookseller by deed when the author sold the copyright and, like any other property right, it lasted forever.

The booksellers' claims provoked a long legal struggle over the "question of literary property," as it was called, that resulted first in the landmark decision of *Millar v. Taylor* (4 Burr 2303; 1769) in which the Court of King's Bench, the highest common-law court in England, determined that there was indeed a common-law right of literary property and that this right lasted forever. Five years later, however, in *Donaldson v. Becket* (4 Burr 2408; 1774), the House of Lords, acting as the supreme court of Great Britain, reversed the decision in *Millar*. This decision led to an explosion of reprinting activity as classic texts such as Shakespeare, Milton, and Locke, long considered the private properties of individual booksellers, became available for anyone to print in cheap editions. The Statute of Anne created the public domain in principle, but it was the *Donaldson* decision sixty-four years later that made the public domain a fact.

Following its usual practice in such matters, the House of Lords merely voted on the *Donaldson* matter: No formal opinion was written. Therefore it is difficult to say on what grounds or principles the Lords acted. Nonetheless, a legal consensus eventually formed that what the Lords held was that there was indeed a common-law right but that it was taken away by the statute and that therefore the copyright term was limited to whatever parliament specified. The *Donaldson* decision thus in effect affirmed the conception of a balance between private and public claims implicit in the Statute of Anne.

In the years after *Donaldson* the English book trade adapted to the consequences of the Lords' decision (Belanger, 1978). The question now was no longer whether the copyright term should be limited but how long protection should last—that is, exactly what kind of balance between private rights and public claims should be struck. Interestingly, now the principal agitators for a longer term were authors rather than booksellers, a fact that reflected both the changes that had occurred in the trade and the increasing professionalization of authorship. At the time of the *Donaldson* decision, Samuel Johnson remarked that a single term of the author's lifetime plus thirty years would be "sufficient to reward the writer without any loss to the publick" (Johnson, 1774, p. 445).

But many prominent authors, including Robert Southey and William Wordsworth, remained at least in theory copyright absolutists who insisted that copyrights should last forever. They were championed in Parliament by Thomas Talfourd, himself an author and friend of Wordsworth. Talfourd, who understood the need for compromise, led a campaign to extend copyright to the life of the author plus sixty years. Talfourd and Wordsworth were opposed, however, by Thomas Babington Macaulay who spoke for the public interest, famously describing copyright as "a tax on readers for the purpose of giving a bounty to writers" (Macaulay, 1841, p. 201). Copyright, Macaulay said, is essentially a monopoly created for a public purpose and therefore an evil enacted to produce a good. How long should the monopoly last? Not "a day longer than is necessary for the purpose of securing the good" (Macaulay, 1841, p. 199). The result of the authors' movement was the new compromise reached in the Copyright Act of 1842 (5 & 6 Vict. c. 45). This adjusted the balance somewhat further toward private interests by setting the term as the life of the author plus 7 years or a total of 42 years, whichever was longer (on copyright reform in early 19th century England see Feather, 1994, and Seville, 1999).

DEVELOPMENT OF COPYRIGHT IN THE UNITED STATES

Let us turn now to the story of copyright as it developed in the United States. The *Donaldson* case was decided 2 years before the American Revolution. At this time, copyright was not a significant issue in the American colonies, which did not have either authors or a book trade on a scale comparable to that in England (Tebel, 1972). After the Revolution, however, a number of American authors, notably the scholar and lexicographer Noah Webster, lobbied for American copyright legislation. Starting with Connecticut in 1783, most of the American states passed copyright enactments modeled on the Statute of Anne, but a web of separate and varying state laws was clearly inadequate since an author would have to register his work separately in each of the states (Patterson, 1968, and on American copyright generally Bugbee, 1967). Thus, one of the provisions adopted by the Constitutional Convention in 1787 was the section granting Congress the power "to promote the progress of science and the useful arts by securing, for limited times, to authors and inventors, the exclusive right to their respective writings and discoveries" (U.S. Const. Art I, § 8).

This clause is explicit in stating that the basic purpose of copyright is the furthering of the public interest, the promulgation of knowledge, rather than the protection of property. Nonetheless, the means by which this public purpose was to be accomplished was by granting private rights. Writing in *The Federalist* to explain the proposed Constitution, James Madison (1788) remarked:

> The utility of this power will scarcely be questioned. The copyright of authors has been solemnly adjudged, in Great Britain, to be a right of common law. The right to useful inventions seems with equal reason to belong to the inventors. The public good fully coincides in both cases with the claims of individuals. The States cannot separately make effectual provision for either of the cases. And most of them have anticipated the decision of this point, by laws passed at the instance of Congress. (p. 267)

Significantly, Madison saw no conflict between the public and the private dimensions of copyright, and, indeed, at this early point in the development of the republic none was apparent.

The first U.S. copyright law, passed by Congress on May 31, 1790, was titled, "An act for the encouragement of learning, by securing the copies of maps, charts, and books, to the authors and proprietors of such copies, during the times therein mentioned" (1 Stat. 124). The title of the U.S. act was strikingly similar to that of the Statute of Anne, even to the point of using the term "copy," which by 1790 was a somewhat archaic word for copyright, and it followed the provisions of the English act very closely in most respects, including the setting of two fourteen-year terms of protection. Thus, like the English statute, and like the empowering clause in the Constitution, the U.S. Statute of 1790 presented itself as seeking to achieve public purposes—the encouragement of learning—through granting limited private rights.

As already noted, the circumstances of the book trade in the United States were very different from those in Britain. Whereas English publishing was centralized in London, American publishing was more dispersed. Furthermore, newspapers, pamphlets, and other ephemera played a much larger role proportionally in the American trade than they did in England. There was also a cultural difference insofar as print played a special role in the civic imagination of the young republic, which emphasized what Meredith McGill (2003) has called a republican "culture of reprinting." This stressed the importance of the circulation of texts as the sign and guarantor of liberty as opposed to private rights. Of course, the republican ideological position reflected a printing trade that to a considerable degree relied for "content" on English works, including popular novelists such as Scott and, later, Dickens, that were not protected under U.S. copyright because only American citizens or residents were eligible for protection. Civic-minded as it might have been, the republican position was also the object of bitter and largely justified complaints by English writers.

Naturally the claims of authors were also part of American copyright discourse. In his comments in *The Federalist*, Madison recalled the English case of *Millar v. Taylor* when he noted that before the revolution the copyright of authors in their works had been decided in Britain to be a common law right. And Noah Webster (1788), writing 2 years before passage of the U.S. Copyright Act, expressed his conviction that the common law right would also be acknowledged by American courts. Was there an underlying common law right of literary property, and, if so, did it persist even after the passage of the Copyright Act of 1790? *Donaldson v. Becket* answered the question for Britain, but the Statute of Anne had never been in effect in the colonies and *Donaldson* did not resolve the matter for the United States. The issue was decided, however, in the foundational case of *Wheaton v. Peters* (33 U.S. 591; 1834), the first copyright matter to be decided by the Supreme Court.

Wheaton v. Peters was a curiously self-reflexive and ingrown matter. Henry Wheaton was the author of a complete series of U.S. Supreme Court reports from 1816 through 1826. He was succeeded in office by Richard Peters who published a volume of cumulative condensed reports containing Wheaton's reports. Wheaton claimed both that he had a common law right in the text of his reports and that his right was protected under the statute. Peters, on the other hand, denied that law reports could be the subject of copyright and rejected Wheaton's statutory claim on the grounds that Wheaton had not fulfilled the requirements for securing a copyright which included the delivery of a deposit copy of the work to the Secretary of State. By a five to two majority, the Supreme Court ruled that there was no federal common law right and that Wheaton could only assert a claim under the statute. Therefore, the case turned on whether Wheaton had in fact fulfilled all the statutory requirements. In his capacity as court reporter, Wheaton had been required to deliver eighty copies of each volume of his reports to the Department of State. But did the delivery of these eighty copies constituted fulfillment of the statutory requirement? No, said Justice McLean, writing for the court, every single condition specified under the statute was essential to securing a copyright. "The eighty volumes were delivered for a different purpose; and cannot excuse the deposite [sic] of the one volume as specially required" (33 U.S. 667).

Wheaton, which is regarded as the foundational U.S. case with respect to copyright, thus established that the only protection available to an author for a published work was the statute. It also spoke for a limited, formalistic conception of copyright. Copyright, the *Wheaton* court in effect declared, was a state grant made for public purposes, not a private property right. Moreover, in order to secure a copyright, an author had to comply punctiliously with every one of the statutory formalities which included advertising the work in newspapers as well sending copies to Washington for deposit. This strictly limited conception of copyright found expression in later cases as well. Thus in the famous case of *Stowe v. Thomas* (23 F. Cas. 201; 1853), a circuit court declared that a German translation of Harriet Beecher Stowe's *Uncle Tom's Cabin*, prepared for the Pennsylvania market, was a new work and not an infringement of Stowe's copyright. Likewise in *Holmes v. Hurst* (174 U.S. 82; 1899), the Supreme Court found that because of a technicality having to do with the serial publication of Dr. Oliver Wendell Holmes's *The Autocrat of the Breakfast Table* in the *Atlantic Monthly*, an unauthorized reprint taken from the magazine and not from the book version was not an infringement.

Although *Wheaton*, like *Donaldson* in Britain, established that copyright protection after publication was limited to the statutory terms, it by no means ended the debate over the nature and purposes of copyright. Eaton S. Drone, for example, the author of the most authoritative American copyright treatise of the 19th century, insisted that literary property is property like any other and maintained that both *Donaldson* and *Wheaton* had been incorrectly decided and that copyright should last forever. "The law which puts an arbitrary terminus on the ownership of literary property," wrote Drone (1879), "is the same in principle with one that would abridge the farmer's right to his orchards and grainfields. If there were the remotest danger that this principle would ever be applied to material possessions, every English tongue would clamor for a new Magna Charta" (p. 51). Likewise, Mark Twain maintained that copyrights should last forever. Testifying before Congress in 1906 on the proposed extension of copyright protection to the life of the author plus 50 years, Twain (1906) acknowledged that copyright was limited, but claimed that he did not understand why.

> I understand, I am aware, that copyright must have a term, must have a limit, because that is required by the Constitution of the United States, which sets aside the earlier constitution, which we call the Decalogue. The Decalogue says that you shall not take away from any man his property. I do not like to use the harsher term, "Thou shalt not steal." But the laws of England and America do take away property from the owner.... I am quite unable to guess why there should be a limit to the possession of the product of a man's labor. There is no limit to real estate. (p. 110)

And a century later, Jack Valenti, the former president of the Motion Picture Association of America, expressed similar sentiments when he testified before Congress that "Creative property owners must be accorded the same rights and protections resident in all other property owners in the nation" (cited in Lessig, 2004, p. 117). Like Twain, Valenti acknowledged that the Constitution does not allow perpetual copyright. But in 1998, at the time of the latest extension of the copyright term, Valenti is reported to have proposed that copyright should last forever less 1 day (144 Cong. Rec. H9952).

ENDURING TENSIONS AND RECENT CHANGES

The booksellers of the 18th century, Wordsworth, Talfourd, Drone, Twain, and Valenti, and of course, countless others as well, have insisted that copyright is property. This position has been countered by the House of Lords in the *Donaldson* decision in 1774, by the *Wheaton* Court in 1834, by Lord Macaulay in 1841, and recently and emphatically again by the U.S. Supreme Court in *Feist v. Rural Telephone* (499 U.S. 340; 1991) when it cited the Constitutional Clause and reiterated that the "primary objective of copyright is not to reward the labor of authors, but 'to promote the Progress of Science and useful Arts.'" The point that I have been making so far is that the tension between the private and public claims are fundamental and unresolvable. In the 18th century, this tension manifested itself in the struggle over whether copyright was perpetual or limited in term. But when that issue was resolved in *Donaldson* the essential debate continued, both in England and in this country, in the form of claims like that of Wordsworth and Twain that copyright really ought to be perpetual and that therefore, at the least, the term should be extended.

As I noted earlier, however, length of term is only one aspect of copyright doctrine. The last 200 years have seen other changes as well. One change has been the lateral expansion of copyright to cover more and more kinds of productions. The Statute of Anne protected only "books and other writings." The U.S. Statute of 1790 protected maps and charts as well, and later American legislation extended protection to musical compositions, prints, engravings, the performance of dramatic compositions, photographs, paintings, drawings, chromolithographs, statues, fine art designs and models, technical drawings, motion pictures, and sound recordings. The current U.S. statute broadly provides protection for all "original works of authorship fixed in any tangible medium of expression" (17 U..S. Code § 102), a definition that applies to computer programs as well as to other kinds of productions that were inconceivable when the first copyright statutes were enacted. "Original" in the context of copyright doctrine means only that the work was independently created rather than copied, and "authorship," too, is a technical term of very broad application.

Another change has been the increasing depth of protection. As I noted earlier, copyright originally was simply the right to print or reprint a specific text. Based on a Lockean theory of property—the author's claim was founded on the labor of producing a work—copyright, did not cover abridgements, adaptations, or as, we have seen in the case of *Stowe v. Thomas*, translations, all of which required new efforts. Today, however, copyright protects not just the right to print but the right to make copies of any kind, including for example photocopies and, in principle, manuscript transcriptions. Today, too, copyright protects the separate elements of a work, not just the work as a whole. Thus employing a substantially similar version of a single scene adapted from a play constitutes infringement. Moreover, today copyright protects the right to prepare derivative works such as a sequel to a story or a screenplay adapted from a novel. Naturally, the derivative nature of a work will not always be acknowledged and it is common, especially in the movie industry, for allegations of infringement to arise in cases where there is no literal similarity between the works in question. We have evidently come a long way from the limited right with which copyright began.

How did the limited copyright of the Statute of Anne and the U.S. Statute of 1790 become the deep copyright doctrine of today? What became of the civic-minded, formalistic, and parsimonious copyright doctrine of the *Wheaton* Court? The U.S. Supreme Court may continue to insist that the primary objective of copyright is the public good but in practice the balance between public and private claims has shifted enormously since the early 19th century. How has this happened? As I have suggested elsewhere, part of the answer may be rhetorical. Copyright lawyers have historically had a much easier time thinking about copyright in terms of property rights than in thinking about how to formulate the claims of civil society, and that orientation has been more and more emphasized in the last 100 years or so. The discourse of public rights is weak compared to the well-developed discourse of property and this is one of the reasons that the legislature has repeatedly been sympathetic to claims for the extension of protection (Rose, 2003).

But there are broader considerations to be taken into account as well. As the concern with "the encouragement of learning" and the "progress of science" suggests, early copyright legislation is an Enlightenment product steeped in the values of the late 17th and 18th centuries. Copyright forms in the context of the long social process that Jürgen Habermas (1989) famously identified as the emergence of the "public sphere," the civic arena distinguished from the family on the one hand and the state on the other. (1942). But, as Habermas has indicated, in the context of the advanced capitalism of the 19th and 20th centuries a structural transformation occurs, a "hollowing out" of the public sphere by new social, political, and economic forces so that what remains of civic society is similar to the earlier formation in name only. An adequate discussion of the transformation of U.S. copyright doctrine in the period since *Wheaton*, then, would have to take account of much of the social, cultural, and economic history of the 19th and 20th centuries, including the shift from an agrarian and agricultural to an urban and business-centered society, the emergence of the large corporation as the typical mode of business organization, and to rise to dominance in the 20th centuries of the American mass culture industries.

Focusing more narrowly on the transformation of legal doctrine, it is possible to locate the early signs of a doctrinal shift in practically the same moment as *Wheaton*, which I take to be the high-water mark of the classical republican approach to copyright as a civic instrument. Decided in 1841, *Folsom v. Marsh* (9 F. Cas. 342; 1841) concerned the charge that Charles W. Upham's *The Life of Washington in the Form of an Autobiography* infringed the copyright on Jared Sparks' authoritative 12-volume collection of *The Writings of George Washington.* In fact, Upham's *Life* consisted in substantial part of letters and other documents printed in Sparks' collection. The defendants asserted that these were public documents that belonged to the United States at large and that in any case one author had the right to quote, select, and abridge the writings of another in the process of composing of a new work.

The case turned on the abridgement question. After determining that the plaintiffs did indeed have a valid copyright in the Washington material, Justice Joseph Story, one of the most influential figures in 19th-century American jurisprudence, considered the question of the author's use of the materials. Story acknowledged that Upham's biography of Washington was "an exceedingly valuable book" and that Upham had taken only so much material as had suited his limited purpose as a biographer. But that Upham had taken only a selection from Sparks's collection did not resolve the matter.

> It is certainly not necessary, to constitute an invasion of copyright, that the whole of a work should be copied, or even a large portion of it, in form or substance. If so much is taken, that the value of the original is sensibly diminished, or the labors of the original author are substantially to an injurious extent appropriated by another, that is sufficient, in point of law, to constitute a piracy pro tanto. (9 F. Cas. 348)

Story expressed regret at having to reach a conclusion that would interfere with what he considered the "very meritorious labors" of the defendants but nonetheless he found that Upham's *Life* was an infringement.

Folsom v. Marsh was explicitly concerned with the question of when one author might justifiably use materials created by another. In his decision Story suggested principles for the analysis of difficult issues of this sort, recommending that courts "look to the nature and objects of the selections made, the quantity and value of the materials used, and the degree in which the use may prejudice the, sale, or diminish the profits, or supersede the objects, of the original work" (9 F. Cas. 348). Thus, Story articulated the principles that have since become known as "fair use," and *Folsom* has gone down in legal history as the case that first defined fair use doctrine. This reputation would seem to imply that *Folsom* represented a loosening of copyright doctrine, a more liberal attitude toward the use of existing works. But in fact the opposite is the case.

Indeed, as L. Ray Patterson (1998) identified, *Folsom* contained the seeds of an enormous doctrinal transformation. In this case, Story focused not so much on the contributions of the defendants, their "meritorious labors," as on what the Upham took from Sparks and on the degree to which the biography may have diminished the commercial value of the original work. The doctrinal effect was thus to shift copyright from being in principle a part of public interest law, a limited statutory monopoly granted for public purposes, to becoming in practice what the booksellers of the eighteenth century and such authors as Noah Webster and William Wordsworth claimed it should be, a part of property law. As Oren Bracha (2005) suggested, in this doctrinal shift can perhaps be found the architecture of the legal channel through which have flowed many of the commercial and ideological forces that have transformed the limited copyright envisioned in the Constitution, the Statute of 1790, and *Wheaton v. Peters* into the deep and expansive copyright of today.

COPYRIGHT AND MEDIA CONCENTRATION

Let us return now to the subject of the increasing concentration of media of various kinds in vertical conglomerates. The consequence of such a concentration is that a single conglomerate can control multiple kinds of media outlets—magazines, book publishers, radio and television networks, cable outlets, music and movie production companies—and can manipulate the relations between them to achieve economies and maximum impact. Thus, for example, vertically integrated media corporations can arrange for the internal transfer of rights between units at much less than market value. (The economy in this case is likely to be achieved at the expense of an individual author.) Thus, too, a campaign to promote a movie can be coordinated through multiple media that may appear to be independent but in fact are not. Moreover, a movie can be designed to exploit materials and personalities already controlled by the conglomerate—say a character from a prior movie, a novel published by the conglomerate's publishing arm, music from its record company, personalities from its network—to achieve an extraordinary degree of synergy in the quest for market domination. And all of this synergetic activity may have immediate political consequence as well, as, for example in the promotion of a movie that tacitly reinforces or challenges some aspect of social philosophy or foreign policy.

But the issue of domination and control is even more profound than at first appears because, as the history of copyright reveals, one form of vertical integration, the media conglomerate, is in fact being layered onto another. Whereas copyright originally protected only the right to print a particular text, today it protects not just the right to one work but a right to any further works that might possibly be generated. Thus, the owner of a copyright controls a right that in itself may be considered vertical. And when a media conglomerate owns such a work, it possesses a degree of control over "content" without precedent in the modern era. Indeed, if there is a precedent it is probably to be found in the pre-modern era, in the absolutist regimes of the 16th and early 17th century before the precipitation of the modern dialectic of private and public when the circulation of texts was strictly regulated both for content and to prevent undue competition within the trade.

At least one other historical observation is in order. In the 18th century, the long struggle over whether copyright would be perpetual or limited was fought in the name of the author's right. But in fact authors were little concerned with copyright and it was the booksellers, particularly the great booksellers of London, the men who owned nearly all the most valuable copyrights, who pressed the argument for perpetual property rights. In the 19th century, when the circumstances of authorship had changed, it was the authors themselves who pressed for extended terms. Today, however, it is once again the publishers—that is, the movie and music companies—that are most prominent in pressing for extended terms and other measures. From one point of view, it may seem that history has come full circle. But it is also apparent that the media conglomerates are not individuals like booksellers in the 18th century.

Part of the process that Habermas identified as the "hollowing out" of the public sphere is associated with the rise of the giant corporations, the "incorporation of America" as Alan Trachtenberg (1980) termed it, that occurs in the latter half of the 19th century. Prior to the late 19th century, most businesses were either individually owned or were partnerships between individuals. The rise of the corporate form made possible the mustering of vast capital resources and enabled the development of a national market in the period after the Civil War. The first giant corporations were formed to build the railroads that spanned the American continent, but the process of organizational transformation and capital consolidation that began with the railroads spread to manufacturing and other industries, including of course the media (see generally Perrow, 2002). The concentration of media organizations into vast conglomerates can be seen as a continuation of this long, on-going process (see generally Starr, 2004). In both cases, the federal government played a major role in subsidizing the emergence of corporations in general and media industries in particular.

In the 19th century, American corporations managed to establish themselves legally as artificial persons enjoying nearly all the rights of natural persons, including the "due process" and "equal protection" provisions of the Fourteenth Amendment, designed to secure the civil rights of individuals (Thoennes, 2005). Unlike natural persons, corporations are immortal. Their identities persist even though the natural persons whom they employ in capacities from board members to clerks change over time. But like individuals corporations can sue and be sued, can own property, can enter into contracts, and can assert, among other civil rights, the private individual's right to free speech. Are the giant media corporations private or public entities? Legally they are private, but in their vastness and their domination of the circulation of cultural and informational products of all kinds, they plainly have a public dimension as well. In fact the very concepts of public and private do not quite apply to these entities, as those concepts did not quite apply to the Stationers' Company of the 16th and 17th centuries. Chartered guilds like the Stationers' Company were the creatures of a time before the precipitation of the modern dialectic of private and public. The media conglomerates collapse that dialectic. And yet, through the instrument of copyright, a legal construct that emerged precisely in the tension between public and private, and that expanded greatly as private claims were pressed with increasing intensity in the 19th and 20th centuries, the media conglomerates now control much of the public discourse of the present day.

CONCLUSION

The history of Anglo-American copyright reveals a continuing tension between the claim that the rationale for copyright is the promotion of the general good and the claim that it is really about the protection of private property. At the same time copyright has developed from a limited monopoly right to print a particular text to a "vertical" right to control not only one work but any further works that may possibly be generated. When a vertically integrated media conglomerate — an organization that for many purposes asserts the legal status of a private person—owns copyrights today, it possesses a degree of control over public discourse that is probably without precedent since the absolutist regimes of the 16th and 17th centuries.

REFERENCES

Astbury, R. (1978). The renewal of the licensing act in 1693 and its lapse in 1695. *Library, 33*, 296-322.

Belanger, T. (1978). From bookseller to publisher: Changes in the London book trade, 1750-1850. In R. G.Landon (Ed.), *Bookselling and book buying* (pp. 7-16). Chicago: American Library Association.

Blagden, C. (1960). *The stationers' company: A history, 1403-1959.* London: Allen.

Bracha, O. (2005). *Owning ideas: A history of Anglo-American intellectual property.* Unpublished SJD dissertation, Harvard Law School.

Bugbee, B. W. (1967). *Genesis of American patent and copyright law.* Washington, DC: Public Affairs Press.

Deazley, R. (2004). *On the origin of the right to copy.* Oxford, UK: Hart Publishing.

Drone, E. S. (1879). *A treatise on the law of property in intellectual productions in Great Britain and the United States.* Boston: Little, Brown.

Feather, J. (1994). *Publishing, piracy and politics: An historical study of copyright in Britain.* London, UK: Mansell.

Habermas, J. (1989). *The structural transformation of the public sphere: An inquiry into a category of bourgeois society* (T. Burger, Trans.). Cambridge, MA: MIT Press.

Johnson, S. (1774). Letter to [William Strahan?], March 7, 1774. In G. B. Hill (Ed.), *Johnsonian Miscellanies* (Vol. 2, pp. 442-446). Oxford, UK: Clarendon Press.

Lessig, L. (1999). *Code and other laws of cyberspace.* New York: Basic Books.

Lessig, L. (2004). *Free culture: How big media uses technology and the law to lock down culture and control creativity.* New York: Penguin Press.

Litman, J. (2001). *Digital copyright.* Amherst, NY: Prometheus Books.

Macaulay, T. B. (1841). A speech delivered in the house of commons on the 5th of February, 1841. In *The life and works of Lord Macaulay* (Vol. 8, pp. 195-216). London: Longmans.

Madison, J. (1788). The federalist no. 43. In H.C. Lodge (Ed.), *The federalist.* New York: G.P. Putnam's.

McGill, M.L. (2003). *American literature and the culture of reprinting, 1834-1853.* Philadelphia: University of Pennsylvania Press.

Patterson, L. R. (1968). *Copyright in historical perspective.* Nashville, TN: Vanderbilt University Press.

Patterson, L. R. (1998). The worst intellectual property opinion ever written: *Folsom v. Marsh* and its legacy. *Journal of Intellectual Property Law, 5,* 431-452.

Perrow, C. (2002). *Organizing America: Wealth, power, and the origins of corporate capitalism.* Princeton, NJ: Princeton University Press.

Rose, M. (1993). *Authors and owners: The invention of copyright.* Cambridge, MA: Harvard University Press.

Rose, M. (2003). Nine-tenths of the law: The English copyright debates and the rhetoric of the public domain. *Law and Contemporary Problems, 66,* 75-87.

Seville, C. (1999). *Literary copyright right reform in early Victorian England: The framing of the 1842 copyright act.* Cambridge, UK: Cambridge University Press.

Starr, P. (2004). *The creation of the media: Political origins of modern communications.* New York: Basic Books.

Tebel, J. (1972). *A history of book publishing in the United States. Vol. 1: The creation of an industry.* New York: R.R. Bowker.

Thoennes, K. J. (2005). Frankenstein incorporated: The rise of corporate power and personhood in the United States. *Hamline Law Review, 28,* 203-236.

Trachtenberg, A. (1982). *The incorporation of America: Culture and society in the gilded age.* New York: Hill & Wang.

Twain, M. (1906). Copyright in perpetuity [testimony before congress in 1906]. *The Green Bag,* 2nd ser, 6 (2002), 109-115.

[Webster, N.] (1788, February) Acknowledgements. *The American magazine, 130.*

5

NETWORKING AND BROADCASTING IN CRISIS

Or, How Do We Own Computable Culture?

William B. Warner

OWNING MEDIA CULTURE

In recent years, a "neo-liberal" regime of deregulation has been promoted by the U.S. government as a way to clear the way for individuals to engage in productive new forms of wealth accumulation, thereby allowing increasing numbers to become part of "an ownership society" (George W. Bush, Second Inaugural Address, January 20, 2005). However, several trends suggest that the expanding size and influence of big media corporations means that increasingly they will own and control our media culture. There are two traditional factors driving this shift: a) corporate copyright owners have won a steady expansion of the breadth, depth, and length of copyright, and b) media corporations are seeking to protect themselves from risk by increasing in size, and a relaxed enforcement of the antitrust laws is allowing them to do so.[1] However, the issue of media ownership has been complicated by a third important factor besides the absolutist expansion of copyright or a tendency toward corporate giantism. I am referring to the development, in the last decades of the 20th century, of the networked computer, and especially by its most popular form, the Internet. This chapter explores the forces and issues that have come into play around the urgent question: How should we institute the networked computer as part of our media sphere? How, for example, are we to balance a copyright holder's interest in controlling the circulation of their products on the network against the user's wish to protect computing as a remarkable new technology of inscription? At the end of this chapter, I offer a pragmatic approach to these questions.

When Napster developed a networking technology so that millions of worldwide users could share music online, the big music companies not only sued Napster for violation of their copyright in court, but the subsequent struggle between big media companies and youthful computer users exposed a divergence of views on what the networked computer is and should be. Although for many of the creators and users of the networked computer, it is the most valuable cultural technology since the invention of printing, for big media producers the networked computer has emerged as a menace, an instrument of thieves, and a destroyer of business models. So the media content industry is working—through Congress, through the courts, and through technology agreements—to change the architecture of the networked computer, away from the open source principles inscribed in the Internet protocols, so that in the future individuals will not be able to copy and distribute legally protected media. Viewed historically, the extreme positions and alarms sounded by partisans of both the traditional media companies and the partisans of "new media" are a necessary part of the struggle around the institutionalization of the network computer, a struggle that is legal, economic, and cultural, and will have important implications for all of us.

NETWORKING AND BROADCASTING

The currently unfolding legal and social struggle around institutionalizing the networked computer is the latest episode in a long American history of the rivalry and symbiosis between two modes of communication, which I term *networking* and *broadcasting*. Table 5.1 summarizes the main comparisons between networking and broadcasting, with more detailed discussion in this and the next two sections.

The early meanings of these words are a way to recover the distinctness and strangeness of each of these communication acts. Long before networks were fashioned from wires, computers, and humans, a network was the "open-laced fabric" used to fashion nets to trap fish or birds, the "lines or channels that cross and interconnect" (as in a train network); or "a complex, interconnected group or system" (as in an espionage network) (*American Heritage Dictionary* 1992)]. Long before broadcast was applied to radio, to "broad-cast" was to sow seeds widely, rather than

TABLE 5.1. *Comparing Two Modes of Communication: Networking and Broadcasting*

Networking	Broadcasting
Definitions	
• "An open-laced fabric" used to fashion nets to trap fish or birds • "A complex, interconnected group or system" (as in an espionage network) • "An extended group of people with similar interests or concerns who interact and remain in informal contact for purposes of mutual assistance," (American Heritage Dictionary)	• To sow seeds widely, rather than placing them in the narrow grooves of a furrow • To communicate so that one "scatter[s] widely abroad." [OED]. • To propose "to bring [something] to public notice." Near synonyms—announcing, advertising, and publishing (American Heritage)
Modes of Communication	

• One-to-one communication
• Topology
  • horizontal
  • symmetrical
  • decentralized and egalitarian
• Access is easy and "cheap"
• users control content
• extemporaneous, informal style
• less regulation
	• One-to-many communication
	• Topology
	  • vertical
	  • asymmetrical
	  • centralized and hierarchical
	• Access is difficult and expensive
	• Broadcaster controls content
	• Generic conventions and correct usage
	• More regulation

Media Forms	
• Correspondence by post • Telegraph • Telephone • E-mail • Peer-to-peer file sharing • Blogs	• Publication in print • Film • Radio and tv broadcasting • Spam • Web pages • Pod casting
Social Forms of the Long 18th century (1660-1830)	
• The club • The political caucus • The committee (e.g., the political committees of correspondence) • The learned societies (e.g., The Royal Society)	• The public (as audience) • The author (©) • The "work" (of art) • Criticism (of literature, art, culture) • Celebrity (e.g., Byron) • Fandom

Ethical Imperatives and Law	
• The taboo on reading other people's mail	• The law of copyright
• A new vigilance about computer viruses	• The taboo and laws against plagiarism
• Vilifying the hacker	

Characteristic Utopian Ideals for Communication	
• If we would all link together • The end of politics and difference = "the ideology of communication"	• Since the *whole* nation is watching, it becomes *one* • Hitler's Nuremberg rallies • Frank Capra: *Meet John Doe* • Cronkite: "And that's the way it is."

placing them in the narrow grooves of a furrow; and then metaphorically, it is applied to any communication that one "scatter[s] widely abroad" (OED, online). Each supports a distinct social practice, a different dream for communication: while networking brings together "an extended group of people with similar interests or concerns who interact and remain in informal contact for purposes of mutual assistance," broadcasting, like its near synonyms—announcing, advertising, and publishing—proposes "to bring [something] to public notice" (*American Heritage Dictionary*, 1992).

How does one compare these two general modes of communication? Here I use fairly conventional terms from communication studies to account for the difference between networking and broadcasting. The contrast is useful precisely because it is rather abstract and formal. As a mode of communication networking involves one-to-one communication; its topologies, or the shapes of communication flow, are horizontal, symmetrical (i.e., network users all tend to be doing the same sorts of things), decentralized (the job of making meaning is widely distributed), and thus (at least potentially) egalitarian. Crucially, access to networks is usually relatively easy and "cheap." Most crucially, networks involve user control of content. Networks promote an extemporaneous informal style, and less regulation (than broadcasting).

Broadcasting features one-to-many communication; its topology is invariably represented as vertical, asymmetrical (broadcasters are doing something very different than those receiving the broadcast), centralized, and thus, implicitly hierarchical. Several traits of broadcasting are implicated in one another: Access is difficult and expensive, the broadcaster enjoys control of content, and broadcasters tend to develop fixed generic conventions and correct usage. Historically, broadcasting has brought more regulation than networking. Sometimes, this regulation is represented as a technological necessity. The notion that networking is more egalitarian or democratic than broadcasting pivots on two terms of this contrast:

Because network users are both the receivers and senders of messages, networking puts its users in charge of content formation; and, second, access to networks, especially in the United States, has been comparatively easy and cheap.[2]

There are many examples, taken from the long history of media culture, of the dominant affiliation of various forms of communication in the modern period, broadly defined as 1700 to the present. The first networking took place through private conversation, and my first example of networking can been seen as an extension of conversation: the manuscript letter mailed between correspondents.[3] This form of networking was given vast expansion during the 18th century through the institution of a postal system that was public, regular, and national. In many ways the emergence of the modern postal system anticipates the effects later ascribed to the Internet: By enabling global flows of information, the postal network had a transformative effect on its users' sense of and relation to geography, economy, and politics. In the 19th century, networking through writing and voice is given electronically enhanced form through the telegraph and the telephone. Twentieth-century computer networks have provided powerful new modes of networking like e-mail and peer-to-peer file sharing.

The earliest broadcasting was one person addressing a group. The importance that Athenian democracy and the Roman republic accorded the training of young men in rhetoric is an early recognition of the power of broadcasting as a means to influence public opinion. The development of moveable type in the 15th century greatly expanded the broadcasting potential of writing. Martin Luther's unauthorized, vernacular translation of the Latin Bible into German could not have achieved its challenge to Papal authority without Gutenberg's invention. It is difficult to imagine the development of the most popular broadcasting forms of the 17th and 18th centuries—the broadside ballad, the newspaper, and the novel—without the expansion of print. In the 19th and 20th centuries, the new technologies of film, radio, and television gave broadcasting unprecedented cultural centrality. At the same time, because of the complex *way* in which these new broadcasting technologies became major new media institutions, broadcasting in America became associated with large, vertically integrated, and often monopolistic corporations—the movie studios and the broadcast networks. Late in the 20th century, computer networks provided a much more decentralized way to broadcast content through web pages, spam, or "webcasting."

My contrast of these media forms, as modes of networking and broadcasting, leads to one of the crucial points of this chapter. Although I have listed various media forms on one side or the other of the tables, broadcasting and networking do not align, in any simple way, with a particular medium or technology. Any given media technology has a plasticity and lack of essence, that allows it to be used for either networking or broadcasting or both. What turns out to be crucial to the way a media technology takes root in culture are the decisions made, during the early stages of a development of a new media technology, about how it is to be used. These decisions often have a highly arbitrary and contingent quality;

they depend on the collision of (a) the desires and values of users, (b) the market, (c) the law, and finally the (d) inherent potentialities of new technology.[4]

Thus, although radio appears on the broadcasting side of the comparison, histories of radio show that it did not have to turn out that way. The Marconi Company promoted radio as a system for secure, long distance, point-to-point communication, in short as wireless telephony. Radio development in America received early support from the U.S. Navy as a way to enhance ship-to-ship and ship-to-shore communication. But amateur radio enthusiasts also played an important role in early American radio: ham operators formed into clubs in which young men communicated with each other in an informal one-to-one, one-to-few style of networking. Only after World War I, when the control of radio that the U.S. Navy had wrested from private companies was transferred to the Radio trust (RCA, GE, Westinghouse, and AT&T), did the previously informal practice of broadcasting emerge as a profitable business.[5] Through the radio conferences sponsored by Secretary of Commerce Herbert Hoover (1922-1925), an early champion of the "partnership" of business and government, as well as through the Radio Act of 1927, the federal government gave decisive priority to radio broadcasting. This position was consolidated through the establishment of the Federal Communications Commission (FCC) in 1934.

Radio offers an obvious 20th century example of the rivalry between the networking and broadcasting potentials of a medium. From the point of view of the creative early adopters of radio, who fashioned a network out of a two-way technology that required technical skill, the commercialization of radio as broadcast medium was a loss; a network of early users was "captured" by and for broadcast radio. Radio thereby lost much of its earlier participatory and egalitarian character.[6] Of course, radio's early adopters could continue to use their ham radios on more remote frequencies; but the amateur was no longer central to radio. However, from the point of view of the triumphant media companies—NBC, CBS, and ABC (each of which took "broadcasting" as its middle name)—the development of the broadcasting potential of radio was a natural evolution *and* an unqualified technical and cultural triumph: It created a national media culture for all citizens to share, irrespective of their region or level of literacy. It led, for example, to new, aesthetically valuable media forms, such as the live wartime broadcasts of Edward R. Murrow, Orson Welles' ingenious parody of the radio news broadcast in *The War of the Worlds*, and to sitcoms like *I Love Lucy*, that would later migrate from radio to the other great broadcasting medium of the 20th-century, television. The rise of radio and television broadcasting also changed the cultural idea of "broadcasting" in a fashion that is consequential for this chapter: In the public mind, broadcasting came to mean "real-time broadcasting," and for government regulators, broadcasting was represented by a certain group of media companies (NBC, CBS, ABC) that had succeeded in gaining FCC licensed control of what was once said to belong to the people: the airwaves. Now they *were* the nation's "broadcasters." Any future changes in media policy would have to be negotiated with these powerful "stakeholders."

THE SOCIAL LIFE OF COMMUNICATION

Communication is not just about getting a message from Point A to Point B. Communication emerges from, and then helps to support, complex social forms. Many of the social forms associated with networking emerged in the long 18th century (1660 to 1830): political caucus, the club, the committee, and the learned society (see Table 5.4). The political caucus and the committee becomes the matrix for new highly participatory forms of political organization, and the learned society lies behind the development of the modern disciplines like science. The emergent postal service was crucial to the networking at a distance necessary for political committees and learned societies. Broadcasting, through print publication, helped to enable the emergence of the "public" (in the very modern sense that there is a collective, anonymous audience out there to receive what is published for them), the author (as the originator and owner of the work), the "work" (as a unique expression of an author), criticism (as an institution to assess and interpret works), the celebrity (the "star" whom everyone knows), and finally, fandom (fanatical fans, who experience the celebrity as the condensation of value). Most historians of print culture agree that these distinctly modern ideas and practices would not have emerged in the way they did without print-enabled broadcasting (Chartier, 1994; Eisenstein, 1980; Johns, 1998; McLuhan, 1961).

Because networking and broadcasting are complex social activities, societies have elaborated a complex group of codes, procedures, and laws to shape and sustain each. Although the ways of shaping human communication are variable and complex, they can be as rigid and public as the law or as informal as a personal vow to keep the secrets of a club or society, or the plea to be a good communication listserv citizen. Sometimes, problematic communication behavior is discouraged by a change in the communication infrastructure. Thus, the eaves dropping that was common in the early days of telephony—when live operators connected phone users and many had "party lines"—became impossible with the introduction of automated switching and private phone lines for residential customers. On the networking side, the rise of the post encouraged the development of a taboo on reading other people's mail. In the United States, this taboo was encoded into law. Telecom laws have been developed to guard against pornographic use of telephony, and the U.S. government's recent "do not call list" is a popular way to diminish intrusive commercial phone calls. In 2006, we seem to be witnessing the emergence of a new vigilance about computer viruses and a concomitant criminalization of the hacker, although these taboos are only now being encoded into law (Lessig, 1999). The early decades of the networked computer has allowed us to grasp the complexity of the process by which ethically charged communications protocols are socially negotiated. Thus, listservs and on-line environments developed communitarian ways to discourage "flaming," whereas system administrators are charged with the responsibility of warning against the inappropriate uses of corporate and university networks.

On the broadcasting side, the development of the modern market in print depended on the modern law of copyright, first enacted in England, as the law of Queen Anne, in 1710. The idea that the author was the creator and owner of the book, which was published under his or her name, helped to promote the negative characterization of copying without attribution as plagiarism. But the idea of plagiarism seems to emerge with the idea of the author as the owner of the work (Rose, 1995). Scholars have noted that medieval and renaissance writers, like Chaucer and Shakespeare, who borrowed from other writers freely and without attribution, would have had a very difficult time comprehending the modern idea of plagiarism. Newspaper broadcasting has been shaped by intra-institutional enforcement mechanisms, such as the codes that define the professional conduct of a journalist who would cover sensitive national security issues. By contrast, the FCC has developed legal sanctions to protect licensed users of the radio spectrum against unlicensed broadcasts. The music industry's effort to suppress peer-to-peer file sharing has involved an explicitly double pronged strategy: to pass laws to punish file sharing, but also to "educate" users so they accept that peer to peer "sharing" is really "theft" of their copyrighted material.

NETWORKING VERSUS BROADCASTING

This brief, historically inflected contrast of networking and broadcasting suggests that our ongoing institutionalization of the networked computer has activated the antagonism between these two modes of communication, the values they encourage, the social practices they sponsor (e.g., as user or consumer). This antagonism has a practical and an ideological aspect. As a practical matter, it is often difficult to say what the networked computer is, and therefore, what ethical norms or legal rubric should be applied to its use. When Congress passed *The Communications Decency Act* in 1996, lawmakers relied on the well-established legal precedents in telecommunications and broadcasting law to justify restrictions on indecent expression over the Internet. In overturning that law, in Reno v. ACLU, the Supreme Court interpreted the Internet as a medium for publication: "The Web is thus comparable, from the readers' viewpoint, to both a vast library.... From the publishers' point of view, it constitutes a vast platform from which to address and hear from a world-wide audience." Once the Internet was understood as a form of publication and speech, it was a logical next step to overturn the Communications Decency Act as an unconstitutional infringement on First Amendment rights to freedom of speech (Kang, 2001).

Although the online computer was understood by its inventors as a form of networking, the success of the Napster peer-to-peer file-sharing system showed the big media companies that computer-mediated networking could have enormous practical implications for their businesses. Thus, in the typology of media forms listed in Table 5.3, notice that the four important Internet applications fall on *both* sides of the networking–broadcasting divide. E-mail and peer-to-peer file sharing can be understood as forms of networking that build on the posted letter or earlier systems for sharing data (the encyclopedia, the catalogue, the library). However, web pages and spam, because they can take one message to many at the same time, can be understood as a computer-enabled broadcasting.[7] "Web-casting" of audio and video feeds develops a still more exact simulation of radio and television broadcasting.

The divergence of opinion on how we should understand the peer-to-peer computing—as "sharing" through a network, or as *de facto* broadcasting—has profound implications for the question of ownership. Because the paradoxical form of property called "copyright" was developed so that authors could sustain property rights in an object, even as they broadcasted it to others, it seems appropriate to understand copyright and the "author" as characteristic inventions of broadcasting (Rose, 1995). But what happens when copyrighted objects are freely shared on networks? Although we sometimes have copyright in what we transmit through networking (e.g., in the personal letters we write), we are not used to thinking of what is shared on networks (phone conversations, telegraphs, e-mails, real-time chat) as owned by anyone.[8] I think this helps explain the recent social acceptability of peer-to-peer file sharing. Although peer-to-peer file sharing, when aggregated through millions of simultaneous users, can produce the effects of broadcasting, a user experiences it as having the intimacy, autonomy, and informality associated with networking. However, there are signs that this may be changing. Although free downloading of copyrighted music flourished in the years after Napster (1999), the availability and popularity of MP3s for sale by Apple (through its software and hardware pair, i-tunes and the i-pod in 2001), and legal action by copyright holders led by the Recording Industry Association of America (RIAA; 2003) may be casting free downloading of copyrighted content into disrepute.[9]

However, the recent antagonism between networking and broadcasting cannot be reduced to a merely practical struggle to define the proper uses of a new technology of communication. Nor should it be reduced to a battle between two huge sectors of the information economy—the traditional broadcast industries (music, film, radio, and television) and the relatively new networking companies (computer hardware and software, networking). Rather, the antagonism between networking and broadcasting so evident in today's courtrooms was anticipated by the claims made for the networked computer by its inventors and early promoters. In fact, the evangelical early promoters of the "network society" saw computerized networks as a media infrastructure so flexible, smart, and transformative, that it could absorb the traditional broadcasting media of print, radio and television. In his recent lectures on the early days of computer innovation, Alan Kay claimed that the print revolution offered an historical model for the sort of transformation of culture and knowledge the first computer engineers sought to advance. Only, as he put it modestly, "this time we'll do it right."[10] Or, as

Netscape's first annual report declares, the "nternet changes everything" (Netscape annual report, 1996). This grandiose, and quasi-imperial project—by which old media will be rendered obsolete by new media—was sustained by a "cyberlibertarian" ethos of hyperindividualism, bad attitude, and the widely circulated slogan, "information wants to be free." In this notorious slogan, "information" is personified as an agency with a quasi-divine trans-historical autonomy and power.[11] Sean Fanning's development of the Napster peer-to-peer file-sharing system came from this libertarian culture of hacking and open-source computing. His software offered a way to "rip" or "liberate" songs from their corporate cocoon (the CD), so that, as binary digits on the computer hard drive, they could be shared across the network, and then remixed into a medium given final form by the user.

Media scholars have contributed to this highly polarized, and sometimes tendentious, understanding of the relationship of new computerized media to "old" media. In a fashion reminiscent of Marshall McLuhan's (1967) prophecy that modern electronic communication would transform the world into "a global village," early media theorists of the networked computer proclaimed "the second media age" (Mark Poster), where there would be a fundamentally new organization of communication, knowledge, subjectivity, and community. In *Communication Theory: Media, Technology and Society*, David Holmes (2005) debunked the claims that networked computing would create a new kind of interactivity and a new kind of online community. Holmes argued that broadcasting continues to repair the social wounds of modern urban life and ground what social integration we have. He applied many of the arguments developed for computerized networking to earlier forms of electronic broadcasting, especially TV broadcasting. He insisted that "the broadcast communication environment still frames our individual lives" (p. 17); "only broadcasting has the power to be "live" (p. 106); "broadcast is the only medium that can be performative" (p. 108), and thereby attract a mass audience (p. 110). According to Holmes, broadcasting can create communities through shared spectacles (like the coronation of Queen Elizabeth II) and through "identification, recognition, interactivity" (between audience and the broadcasting medium; Holmes, 2005, p. 210). For Holmes broadcasting gets the palm as the most influential and important form of media, and by the end of the book, as that which can underwrite a desirable social integration.[12]

Looked at through the prism of the long history of media, the Internet hype of the late 20th century resembles the enthusiastic early days of the appearance of a new medium of communication, especially that associated with the telegraph (Czitrom) and the telephone (Marvin, 1990). At the center of this neo-liberal utopian fantasy is a notion of the power of the Internet to link all to all, in freedom and community. The fantasy is liberal because its transformation of culture is fueled by the agency of diverse individuals, who freely chose the form and content of their communication.

But broadcasting incites its own dream for social transformation through communication. The collectivist fantasy incited by broadcasting at first appears to resemble that incited by networking, but is actually quite distinct. Broadcasters promise to make the nation one by linking them as viewers/listeners of one iconic object. Broadcasting's motivating political ideology may be republican or fascist. Think of the iconic function of Caesar's image on coins and statues, of Washington's portraits, or of Hitler's persona at the huge rallies at Nuremberg, where the roar of hundreds of thousands are mingled with his voice and broadcast around the world (by radio, news reels, or newspaper photographs). Frank Capra's film (1941) *Meet John Doe* offers a critique of the fascistic potential of broadcasting, while at the same time vividly formulating American's distrust of, but fascination with, the power of big media. The desire of big media to unify America around one version of reality is given an avuncular tone in Walter Cronkite's famous sign-off for the CBS evening news for most of the 1950s and 1960s: "And that's the way it is." Thus, both networking and broadcasting offer different forms of what Armand Mattelart (1996) called the "ideology of communication" that is "at the heart of representations of modernity" (p.xi)—the idea that by communicating humans can overcome distance, difference, and politics itself.

THE BIAS OF AMERICAN COMMUNICATION; OR, WHAT WE CAN LEARN FROM 18TH CENTURY AMERICA TO SHAPE MEDIA POLICY FOR THE 21ST CENTURY?

At least since Alexis de Tocqueville's *Democracy in America*, historians of communication have detected the following bias or tendency in American communication, especially when it is compared with its European antecedents: "greater openness and transparency in the public sphere; higher levels of commercialization; greater decentralization; more rapid extension and ubiquitous penetration of communication networks; and greater receptivity to new products and technologies." (Starr, 2004, p. 14). This long-standing bias in American communication expresses itself in an odd ambivalence about communication: first, Americans have evinced a relentless optimism about the positive potential of communication technologies, but, at the same time, we harbor a profound distrust of central control of public media.[13]

Where did this bias first come from? And how did it get coded into American culture and law? Elsewhere, I argue that this communications bias was consolidated in the decade before the American Revolution (Warner, 2005). Here, I offer an exceedingly condensed overview of the Anglophone media sphere as it existed just before the revolution, as well as a brief characterization of those acts of networking crucial to the organization of the Whig resistance to Britain. Most of us are familiar with the ideological, political, and military struggles that led to American independence from Britain. But too little attention has been paid to the communications innovations that American revolutionaries developed to challenge British policy toward the colonies. To run the empire, the British administration used letters written to the Royal governors of each colonies as well as the print broadcasting system of the day—the newspaper of record, the *London Gazette*, distributed by Royal Post to 50 newspapers of Great Britain and the 42 newspapers of British America—to disseminate decisions, laws, and opinions from London, throughout Great Britain, and by fast, regularly scheduled packet ships to the colonies of British America.

When the Boston Whig revolutionaries set out to contest imperial decisions, they had recourse to networking. Samuel Adams worked through the Boston town meeting to establish a committee of correspondence to interlink the towns of Massachusetts into resistance against British administrative measures. Like the Internet, these committees become a very flexible new interface for broadening political participation. Although these pre-revolutionary acts of resistance were grounded in the networking provided by political clubs, town meetings, and a postal system open to all, to reach "the public" they also needed access to print. So, on November 20, 1772, the Boston town meeting ordered their committee of correspondence to print and deliver their letter to the 260 towns and districts of Massachusetts, requesting their opinion on recent British measures, in the form of a pamphlet called *The Votes and Proceedings of the Town of Boston* (see Fig. 5.1).

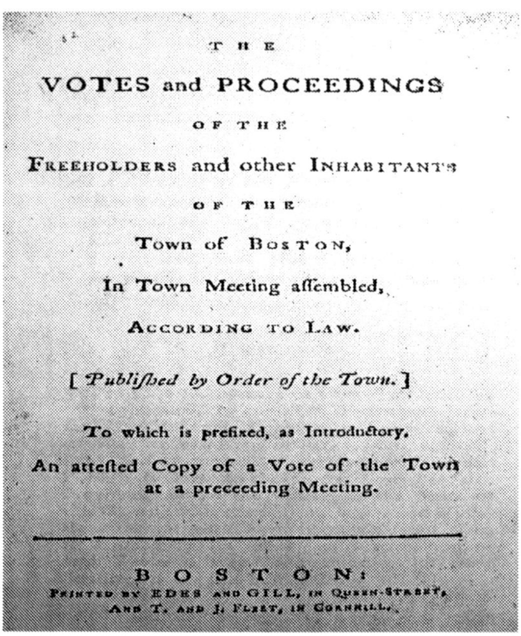

THE

VOTES and PROCEEDINGS

OF THE

FREEHOLDERS and other INHABITANTS

OF THE

Town of BOSTON,

In Town Meeting aſſembled,

ACCORDING TO LAW.

[*Publiſhed by Order of the Town.*]

To which is prefixed, as Introductory,
An atteſted Copy of a Vote of the Town
at a preceeding Meeting.

BOSTON:
PRINTED BY EDES AND GILL, IN QUEEN-STREET,
AND T. AND J. FLEET, IN CORNHILL.

FIGURE 5.1. The Boston Pamphlet, November 1772.

When the Boston committee of correspondence received supportive written responses from the other towns, the Boston Whigs published these responses in newspapers that were sympathetic to their cause, like the *Boston Gazette.* In addition, the news of the organization of the Massachusetts committees of correspondence spread through the newspapers of the colonies as well as Britain. On March 12, 1773, Whig leaders in Virginia, after reading newspaper excerpts of the writings of the Boston committee, proposed an intercolonial network of committees of correspondence, and in the year that followed, as there is an intensification of political crisis around Boston's destruction of the tea, a network of committees of correspondence emerged by 1774 to receive and articulate the political authority that was flowing away from British institutions (like the Royal Governors and councils, the courts, the post). At the end of the First Continental Congress (October 1774), the Congress formed an Association that called for the formation, throughout the colonies, of committees to enforce trade embargoes, muster the militia, and (eventually) pursue a political revolution.

It was crucial to the success of this networking operation that the committees had access to the broadcasting medium of the newspapers of British America. This access depended on an information architecture that was decentralized and resistant to government control.[14] As Table 5.2 suggests, the Boston committee's political initiatives could enter this media sphere because of the distinct features of these newspapers. The 42 papers of British America were locally owned and run by a printer-editors; several used this slogan to assert their independence and general appeal: "Open to all parties, but influenced by none." These papers were loosely connected with each other by the practice of franking free copies through the post to each other. This gave them some of the properties of the wire services (like Reuters, UPI, and AP) later developed in the 19th century. Since newspapers were regarded as ephemera and there was no copyright on a paper's content, the content was free to other newspapers in both senses of the word—it cost nothing and it was available for recopying. Finally, because of the failure of earlier prosecutions for seditious libel in both Britain an America, the colonial press was also free from political coercion, producing a media sphere that was "censorship resistant." This proved crucially important once the revolution began in earnest.

TABLE 5.2. *Traits of the Anglophone Newspapers of the British Empire*

- Locally owned and edited: "*Open to all Parties, but influenced by None*"
- Free mailing ("franking") of free copies of newspapers sent to other newspaper editors
- No effective ©: 18 newspapers develop into an information commons for free copying
- Failure of prosecutions for seditious libel in both American and Britain
- *Outcome*: the British Atlantic media sphere proves to be 'censorship resistant'

When, after 1789, the American founders set about to shape the legal underpinnings of the media policy for the new republic, they carried a memory of the pre-revolutionary practices of networking and broadcasting, and valued them as engines of popular sovereignty. So the founders developed communication protocols that privileged what was free, public, and open, and wove them into the First Amendment, the first Copyright act, and the first Postal Bill. In other words, the successful struggle against the British state encouraged the generation of the revolution to set up protections for precisely the sort of de-centralized communication, networking, and mobilization that had been essential to the revolution's challenge to British rule.

With the First Amendment, the founders developed a powerful formula for protecting freedom of speech and the press as constitutive elements of the new American political culture. I quote, preserving the three-part division observed in legal studies:

1. Congress shall make no law respecting an establishment of religion, or prohibiting the free exercise thereof;
2. or abridging the freedom of speech, or of the press;
3. or the right of the people peaceably to assemble, and to petition the Government for a redress of grievances.

Here freedom of the press is embedded within four other expressive rights and freedoms: to worship, to speak, to assemble, and to petition the government. These range from the most personal and private (religious worship) to the most public and political (to assemble and petition government for redress). The antithetical structure of the first clause (on religion) anticipates the double stance of the new government toward media: it desists from the "establishment" of any official government newspaper (like the London Gazette), and it refrains from preventing others from the "free exercise" of their media freedoms. Because the political critique of British policy by the press was indispensable for mobilizing pre-revolutionary opposition, here freedom of the press is linked to explicitly oppositional forms of public expression: peaceful assembly and the petition for redress of grievances. Finally, the freedom of speech and of the press is protected from legislative control with a verbal formula that gets its power from a double negative: "Congress shall make *no* law . . . *abridging* the freedom of speech, or of the press" (italics added). This double negative hollows out a theoretical space for freedom of speech and the press prior to, and immune from, the law-making powers of Congress.

The founders also took steps to guard speech and invention against undue commercial control. To encourage the unencumbered circulation of information, the first American copyright bill, passed in 1790, adapted the limited copyright and patent law that then prevailed in Britain and America (a 14-year term, renewable once). By adapting to the legal framework of the new republic British copyright law, as limited in the then-recent ruling of the House of Lords in *Donaldson v. Becket* (1774), the founders assured that writing and inventions would pass quickly into the public domain and become free to users. This restricted understanding of copyright was consistent with the early republic's privileging of the public good. This bias can be detected in a Benjamin Franklin declining the Pennsylvania legislature's offer of a 10-year patent to his stove: "as we enjoy great advantages from the inventions of others, we should be glad of an opportunity to serve others by any invention of ours, and this we should do freely and generously" (Morgan, 2002, p. 23).

Finally, the Postal Act of 1792 (Table 5.3) encouraged the development and circulation of American newspapers as a means to link far-flung states into one print media sphere: first, by setting postal rates so that personal and commercial letters provided a substantial subsidy to newspapers, so that heavily traveled eastern routes supported new western routes, and so that the American postal system functioned as a subsidized public service rather than a for-profit business.[15] In each of these laws, the founders chose protocols for the flow of media that favored an information culture that was public, open, and free. Why did they do this? Such a media policy would allow them to network and broadcast to fight tyranny should it emerge within the new republic.

Equally important is what the early framers of American communication policy *did not do*: They did not set up a federally financed and government sponsored newspaper (like Britain's London Gazette) or a national university. When new communication technologies emerged in the 19th and 20th centuries—the telegraph, the telephone, the radio, and television—the early American reticence about centralized control of information continued to discourage government ownership of each new communication system, even though the federal government was often an indispensable sponsor of early research. Space does not permit me to discuss the most perverse effect of this decentralized, anti-government ownership media policy: in the 20th century the federal government (through the formation of the FCC) collaborated with the largest corporations in gradually delivering de facto control and ownership of the public airwaves into the hands of a few big media corporations.[16]

TABLE 5.3. *First Postal Act: 1792*

- Heavily traveled eastern routes support new western routes
- Private letters subsidize newspapers
- The postal system is a public service rather than a private business

FREE CULTURE?

How can we carry the early American media policy into current debates around copyright, media ownership, media diversity, and freedom of expression? Lawrence Lessig, a law professor at Stanford, developed ways to insert the networked computer into the policy and legal frameworks developed for earlier media forms. In 2004, Lessig published a book entitled *Free Culture.* In that title "free" is both an adjective modifying the noun "culture" and an imperative verb: free culture and free culture! Appropriately enough, the book is available for free on the Internet: http://free-culture.org/freecontent. Lessig gives a vigorous and thoughtful discussion of the broad stakes in the copyright war unleashed by Napster (Table 5.4). What Lessig demonstrated are the powerful negative synergies between the three quite distinct developments to which I quickly referred at the beginning of this chapter: the concentration of media ownership; the steady expansion of the term and scope of copyright; and finally, the evolution of the computer into a "universal media machine," the instrument for producing, distributing, and reading/watching/listening to/consuming cultural products (Manovich, 2001). Lessig demonstrated the very problematic aggregate effects of these trends: Because big media owns more of culture for longer times, and because big media is simply bigger, big media has an incentive to enforce copyright by using the networked computer to monitor and control how we use cultural products. They can do this through legally mandated hardware and software design, continuous surveillance of the computer user, encryption that is illegal to circumvent, and restrictive licensing.[17]

Some ask, what's so bad about this? Here is one answer. If the Internet protocols were recoded to make the networked computer a copyright enforcement device, it could transform our media sphere. Within such a system of networked media distribution, you might lose the uses traditionally enjoyed with books, tapes,

TABLE 5.4. *Negative Synergies: Lawrence Lessig's Free Culture*

- Media concentration aided by
- Relaxation of FCC ownership rules
- Vertical integration of super-big media
- Steady extension of the term and scope of copyright
- The networked computer evolves into the universal media machine (replacing paper, screens, radio, TV, . . . etc.) [Lev Manovich]
- Possible result: not "free" culture, with unregulated zones, but a seamless media sphere of regulation, "permissions," and control: *The Matrix*

videos, and CDs: These include (a) repeat cost-free reading/listening, (b) sharing with friends, (c) annotating and marking up, (d) excerpting for use in writing, (e) archival copying, and, (f) right of "first sale." In short, what is at risk is what Lessig call free culture: not culture that is "for free"—after all, you've often paid for it—but culture that is "free" to use in most of the ways you wish, because your use has never before been regulated by license or controlled by technology. In place of our old free culture, we might have a seamless culture of regulation and "permissions," with precious left outside. Such a future for culture has certain uncanny resemblance to life in 1999 film *The Matrix*: in that film, culture, as well as reality itself, is not something each person has; it is a system of domination devised to control humans.

While the stark moral polarities of Lessig's argument, between "free culture" on the one hand, and a "regulated" or "permission" culture on the other, are valuable for mobilizing sentiment against the expansion of copyright, Lessig's recourse to the most loaded terms of our political culture—"free" and "freedom"—obscures rather than clarifies the fault lines exposed by the copyright wars. I argue that we are witnessing the collision of two distinct communication cultures, one associated with broadcasting and the other with networking. Over the course of the 20th century, the scope and power of broadcasting was expanded by film, radio, and television, that of networking through telephony and the networked computer. But, as suggested, this is not just a clash of technologies. It is also a contest of social and aesthetic values.

The copyright wars should not force us to choose between networking and broadcasting, or to subordinate one to the other. Instead, we need a media policy that will allow both to flourish. The broadcaster's commitment to the integrity of copyright is grounded in more than economic interest; it is also part of affirming the value of the author(s) as the responsible creator(s) of a well-made work. Full technical control of a broadcast helps ensure the temporal and formal coherence of the broadcast, assuring that it reaches the public in one form, which is crucial if it is to become part of a shared culture. From Milton's *Paradise Lost* to Lang's *Metropolis*, the products of this cultural system have flowed to, and enriched the culture of, the public.

But networking has much to contribute to creative innovation. While for the broadcaster, networking computing threatens to dim the aura of the work and obscure the creator's design, from the vantage-point of network user, even peer-to-peer file sharing can be the pathway to advanced practices of inscription. Now the user can break out of the temporal constraints of real time broadcasting and the black boxes of analog media, and lift an image from the film, record the voice of the actor, and take a video clip, and this user-fabricator can then remix them to their own liking (Manovich, 2001). In the last 40 years, artists and scholars have shown how the mobility and intelligence of computer-enhanced networking can be used to advance knowledge and art.[18]

TWO PRINCIPLES THAT SHOULD GUIDE THE INSTITUTIONALIZATION OF THE NETWORKED COMPUTER

Media Ecology

Media and legal scholars have suggested the usefulness of applying an ecological analogy to the debates around copyright and the networked computer: Just as unbridled human development of forests, wetlands, and waterways threatens the species diversity necessary to planetary life, so unbridled media concentration and copyright expansion threatens the cultural diversity on which human culture depends.[19] Media concentration of the sort pursued by the ten great world media conglomerates not only hastens the day of a world media mono-culture; it also extends the power of what might be called a regime of unconscious censorship: What gets censored out of the media system are all of those creative projects and new technologies that big media never thought to support, because one's business model made it appear to be unprofitable, or a direct threat to profits. Thus, the film and television companies did not invent the VCR and the music companies did not develop the i-pod/i-tunes system for the distribution of music over computer networks.

Table 5.5 suggests the components for a media policy that will enable the networked computer to figure as part of a diverse and sustainable media ecology. Such a media policy begins by recognizing that both broadcasting and networking are necessary and valuable modes of communication. It is therefore crucial that we don't let one mode of communication co-opt or disable the other. So networks should not be able to turn all broadcasting into freeware. If this were to happen, creators would no longer be able to make a living wage off of book or song writing, and movie makers could not afford to produce something as ambitious and expensive as the modern feature film. Conversely, we should not let entrenched broadcasting industries lock-down networks. The Digital Millennium Copyright

TABLE 5.5. *Toward a Policy Promoting a Rich and Diverse Media Ecology*

- Both broadcasting and networking are necessary and valuable for culture
- Do not let one disable or co-opt the other
- Broadcasting is not equal to freeware, and
- No lockdown of networks (DMCA, DRM)
- Broadcasting benefits from open networks
- Networking is nourished by broadcasting (e.g., Trek fanzines and conventions)

Act's anti-circumvention provisions criminalize what Ed Felten has dubbed the "freedom to tinker," understood as the "freedom to understand, discuss, repair, and modify the technological devices you own."[20] Various "digital rights management" schemes disable the user agency associated with traditional literacy—copying, excerpting, marking up, and so on[21]—as I described them earlier.

Open networks, which enable copying and sharing, are essential to media production. This was true as early as the Renaissance, when literate gentlemen and ladies kept commonplace books, in which readers copied passages from books, so they could consult them for edification, and in composing their own texts or speeches.[22] But copying is no less essential for our own time. For example, special effects labs that work for the movie studios need access to old content to make new content. Finally, networking is nourished by broadcasting. Film scholar Constance Penley (1997) has demonstrated the way broadcasting (of the Star Trek television show) incited the development of fan networks, trek conventions, and "fanzines," which build on the original television show in ingenious (and sometimes outrageous ways), and this distributed (copyright infringing) production, in turn nourished the production and success of the Star Trek films.[23]

A Technology of Inscription

Although the networked computer has profound implications for earlier forms of broadcasting (from print to film, radio, and TV) and earlier forms of networking (most crucially, postal mail and telephony), a policy that will nurture the full cultural potential of the networked computer will attend to the computer's provenance. The networked computer emerged from the university and defense department labs—that is, sites for research and knowledge production. This helps explain the design goals of the open-source protocols that were developed for the Internet and the World Wide Web: In this setting, listening and spectatorship (of the sort sponsored by radio and television broadcasting) are less important than writing, ownership is less important than innovation, and sharing is crucial to collaboration (Abbate, 2000; Berners-Lee, 2000). Above all, early researchers developed the network computer as a technology for writing (and quickly sharing) code, natural language, and anything (like algorithms or images) that could be constructed of either or both.[24]

Understanding the networked computer as a technology of inscription is important for understanding the popularity of computer technologies like peer-to-peer file sharing. From my informal questioning of avid Napster users, I have concluded that the appeal of Napster, and its successors, Morpheus and Kazaa, did not simply consist in the music being free. Radio has always provided an abundance of free music. In order to make MP3 files an object for storage and manipulation the user must incur the costs of a computer, a network connection, and (often) a CD burner as well. What was most attractive to Napster adepts is the way this new technology changes the music listener into a user who can both "read" and "write" music as code. Napster mobilizes the speed, flexibility, and agile immateriality of digital code, to give users a new power to appropriate music and make it their own.[25] Peer-to-peer file sharing is one in a long line of technologies of inscription central to the development of the Internet. Thus, if one surveys the history of the Internet, one repeatedly finds that the surprise breakthrough successes have been those that enabled users to win agency through technologies of inscription: email on the Arpanet; listservs on the Internet; web page composition on the WWW; real-time chat on AOL; and most recently, "instant messaging" (IM) on both computer and digital cell phone.[26]

The recent history of the networked computer can help guide us in shaping media policy (Table 5.6). Because the networked computer has emerged as one of the main ways we read and write, anything we do to modify the networked computer needs to pass this litmus test: "***Does this change protect and extend the full powers of reading and writing as they have operated within culture?***"

America's long experience suggests that a robust media ecology requires the following elements. First, we should strive to sustain broad media ownership and protections for decentralized artisan producers, so they can continue to exercise substantial control of the production process. This will require sustaining a copyright that is limited in its breadth, depth, and term length, perhaps by rolling back the recent extensions of copyright, introducing a renewal term, or redefining what constitutes a "derivative," a "copy," "publication," and "fair use"(Lessig, 2004; Rose, this volume). Only with a limited copyright will cultural products return to the public domain out of which they were created, and only by curbing those who demand an absolutist interpretation of copyright can we sustain a decentralized and innovative "culture of the remix" (Manovich, 2001). Because copyright enforcement has evolved into our most active censorship regime, we need to limit it in order to sustain robust protections for freedom of expression.[27]

Finally, the infrastructure necessary for networked computing should be managed, to the extent that it is feasible, as a public resource rather than private preserve. Universal postal service, the public library, and the Internet protocols

Table 5.6. *Media Policy for the Networked Computer*

Litmus Test Question:
Does this policy protect and extend the full powers of reading and writing as they have operated in culture?

- Broad ownership and protections for decentralized artisan producers
- Limited copyright (so that cultural products can return to the public domain)
- Robust protections for freedom of expression
- Media infrastructure managed as a public resource rather than private preserve: e.g., postal service; public library; Internet protocols (TCP/IP)

(TCP/IP) are American wonders: they are instances when this country chose to expand media infrastructure as a public resource: the post office helped knit the early republic together; the public library gave broad information access to literate American citizens (including many new citizens); and the Internet (TCP/IP) and web (http, html) protocols and markup languages are relatively simple software programs that computers use to connect to the Internet and the web. Although the Internet protocols were developed with US government funding, in an act of startling generosity and wisdom, these protocols were released to the world as "open source software"—code published without copyright, without charge, and thus open to use by all. In doing so, our own epoch countersigned the values used to fashion the media policy of the early republic, where the circulation of information and knowledge was supposed to serve a public good (rather than private gain), circulate freely (so it costs as little as it can, and is protected from censorship of various sorts), and, where access for both readers and writers is as open as possible. Will these values guide our institution of the networked computer?

NOTES

1. On the extension of copyright, see Lessig (2004) and Rose (this volume). From the rather vast literature on the increase on media concentration, see, for example, Chomsky and Hermann (1988), McChesney (2004), and Miller (2002).
2. Note that such a contrast, in order to be broadly useful, excludes a set of middle possibilities—one-to-few communication (the office memo, the end-of-the-year "what's our family been doing" letter), as well as few-to-few communication (e.g., CB radio, listservs, etc.). In each case, one can fairly ask whether these are forms of augmented networking, or narrow-band broadcasting. The coming of computer-mediated communication has incited study of the different media attributes of new and old media. See, for example, Rice (1987), who showed that that attributes of different media are variable and context specific.
3. For most of history, postal systems served empires and states (from the Persian court to the Roman Empire to the Holy Roman Empire to the Courts of Louis the XI). But in 1600, the viceroy of the Netherlands finally authorized the Taxis mail carriers to charge for carrying private letters (Siegert, 1999).
4. Lawrence Lessig (1999) offered the most influential discussion of the way current negotiations around the future of the networked computer are inflected by architecture, as defined by software code and hardware design, by the market, as it dictates what capital will fund and consumers will buy, by the law, through decisions by Congress, the courts, and regulative agencies like the FCC, and finally, by ethical norms, as they inflect the actual practice of media users.
5. For an authoritative history of early radio, see Douglas (1987).
6. For a history that argues the crucial early role of the ham radio operator, see Czitrom (1982).

7. However, even this topology is open to debate. The big music companies that sued Napster see peer-to-peer file sharing not as a kind of networking between individuals, but as a system of "darknet" broadcasting. "The idea of the darknet is based upon three assumptions: 1: Any widely distributed object will be available to a fraction of users in a form that permits copying. 2: Users will copy objects if it is possible and interesting to do so. 3: Users are connected by high-bandwidth channels. The *darknet* is the distribution network that emerges from the injection of objects according to assumption 1 and the distribution of those objects according to assumptions 2 and 3" (Biddle, 2002, p. 2; see also Lasica, 2005).
8. For a discussion of the case that established the author's owner over a private letter (*Pope v. Curll*) see Rose (1995).
9. Although the Pew Internet and American Life surveys report that the percent of Internet users who say they download music files in general has increased from 19% in February 2004 to 22% in January 2005, it is still 10% lower than reported in October 2002. File downloaders are more likely to say they use online music services than say they use p2p services. A PIP and comScore Media Metrix report noted by the Pew site reports that 14% of Internet users say they no longer download music files, and 20% of those who continue to do so say they do so less often because of the RIAA lawsuits. Music and Video Downloading Moves Beyond P2P (2005); http://www.pewinternet.org/PPF/r/153/report_display.asp; 14% of Internet users say they no longer download music files: Data memo from PIP and comScore Media Metrix (2005); http://www.pewinternet.org/PPF/r/124/report_display.asp; Sharp decline in music file swappers: Data memo from PIP and comScore Media Metrix (2005). http://www.pewinternet.org/PPF/r/109/report_display.asp
10. The computer revolution hasn't happened yet. Lecture given on March 8, 2002 at the UC/Santa Barbara conference, *Interfacing Knowledge*. See http://dc-mrg.english.ucsb.edu/conference.asp
11. In a rather devastating and systematic critique of the structuring myths of cyberlibertarian ideology, Liu (2004) showed how a series of partly contradictory programmatic positions—free speech and strong privacy, strong encryption and free access to public information, pro-consumption yet anti-production, pro-entrepreneur but anticopyright and antiregulation—founder through a refusal of the central negotiation of modern society: the political negotiations between the claims of the individual and the community. Liu argued that the political postures of cyberlibertarians do not arise from the traditional locus of sovereignty—the people—"but in fact they arise *ex* nihilo in the process of the restructuring the People's authority as follows: Information → People → Network. In place of God, let the transcendental sanction of freedom be a free-floating, abstract yet also strangely personified Information (as in the expression, 'information wants to be free'" [p. 245]).
12. From the vantage point of this article, there are several fundamental problems with David Holmes' analysis. First, he does not understand networking and broadcasting as "modes" of communication, but as "mediums" [*sic*] (Holmes, 2005, p. 14), a term and concept he never explained. This terminological gap in his history means that the broader possibilities of these two modes of communicating, sometimes implicitly glimpsed in his reference to Raymond Williams' work of television, resolves into a more simplistic opposition between the television (as the chief modern representative of broadcasting) and the Internet (as the most influential popular form of networking).

But even within the terms of this argument, there are problems with Holmes' subordination of networking to broadcasting. First, Holmes' pervasively modernist bias—he is concerned almost exclusively with 20th century media and theorists—doesn't allow him to understand the long history of both networking and broadcasting as I have sketched it here. Second, Holmes's analysis of the Internet is framed and evaluated by the values—visibility, real-time "live-ness," mass communication, influence, and identification—commonly associated with television, rather than values—like agency, compositional control, equality, affiliation and friendship, trust, and mobilization—more commonly associated with networking. Finally, Holmes' restrictive understanding of networking (as computer networking), and the special privilege he accords the socially binding, ritual function of television, means that Holmes grossly underestimates the idea I have developed here: the cultural centrality of networking as a strong and valuable complementary "other" of broadcasting.

13. Sometimes both of these attitudes can be found in one media text. La Capra's *Meet John Doe* develops a sustained critique of the fascistic dangers to American democracy of big media. The capitalist villain, E.B. Norton, makes clever use of newspapers and radio to build the "John Doe Clubs" into a movement that could win him the White House. However, although this film warns of the dangers of two forms of broadcasting, it also makes use of film to allow its viewers to achieve an empathic identification with "humanity" through the celebrity icon as ordinary man, Gary Cooper/John Doe.
14. I discuss the early American newspaper as an information architecture at much greater length in Warner (2005).
15. Thus, newspapers provide only 3% of the revenue but provide 70% of the weight. See John (1995, especially chapter 2, The Communications Revolution); also McChesney (2004, chapter 1).
16. For an indispensable overview of the way policy debates have unfolded over the long history of American media policy, and of how an early public spirited media policy gave way to one rules by the imperatives of large commercial enterprises, see McChesney (2004).
17. The Digital Millenium Copyright Act (1998), "digital rights management" schemes, and obligatory licensing, by arming content producers with the power to prevent all copying, circumvents the traditional balance built into copyright law between the claims of copyright holders and the public interest in the circulation of cultural objects in the public domain (Rose, this volume).
18. For an overview of 40 years of computer art, see Paul (2003) and Wardrip-Fruin and Monfort (2003).
19. The problem of media diversity was implicitly addressed in Newton Minnow's famous May 9, 1961 speech: He challenged assembled television executives to spend one entire day watching television: "I can assure you that you will observe a vast wasteland" (Minnow, 1961). Historians of television have documented the factors that turned television into something barren and monotonous. With three dominant networks chasing the same large, general audience and responding to the same commercial imperatives, programming quickly gets channeled into a narrow set of genres and popular formulas. Within this market-based entertainment system, one of the few ways to expand television is by increasing the number and diversity of channels. This only happened with the rise of cable television brought a vast increase in the number of channels, as well as a much more liberal regulatory environment. Does anyone who

remembers the heyday of NBC, CBS, and ABC, think that these channels could have invented MTV, CNN, or the programming of cable channels like HBO (*Sopranos, Sex in the City*, etc.)?

20. See Professor Ed Felten's Freedom to Tinker website: http://www.freedom-to-tinker.com/.
21. For a full overview of "digital rights management" see the Berkeley Law School Conference: *The law & technology of DRM.* http://www.law.berkeley.edu/institutes/bolt/drm/
22. Rhetorical study during the Renaissance was, "inculcated by means of commonplace notebooks, . . . Students compiled these notebooks in the course of their readings in order to create a stock of ideas for their own speeches and compositions" (Hobart & Schiffman, 1998, p. 99).
23. The UCSB scholar Constance Penley (1997) showed, in her studies of the *Star Trek* fanzines and conventions, how networking and content sharing help to seed the creation of new science fiction.
24. The provenance of the Internet in acts of inscription is being recovered by such recent university-based research initiatives as the University of California's Transliteracies project, which focuses on the adaptation of textual and reading practices to digital, networked multimedia, and vice versa (http://transliteracies.english.ucsb.edu/category/research-project/).
25. For a long time, British cultural studies has argued that individuals and groups are not passively programmed into a mass by the culture industry; instead they have developed cunning ways to appropriate cultural products to their own subcultures (Hall, 1996). Napster offered its users active powers of appropriation only dreamed of by British cultural studies.
26. As noted earlier, this shift of the media user from listener/spectator/reader to writer has been oversold as an unprecedented "liberation" of the consumer. But, one does not need to subscribe to that narrative to grasp the expanded possibilities for writing brought by the networked computer through hypertext writing, web editing, Flash composition, and so on.
27. Freedom of expression in the classroom often requires unending citation and discussion of copyrighted material. In the pre-digital, pre-networked classroom this was easy to do, and was widely accepted as part of fair use doctrine. But, when content is digitized for placement on-line, it can quickly be construed as an instance of copyright infringement.

REFERENCES

Abbate, J. (2000). *Inventing the Internet.* Cambridge, MA: MIT Press.

American Heritage dictionary of the English language. (1992). Third edition. Boston: Houghton Mifflin.

Berners-Lee, T. (2000). *The origin, design and ultimate destiny of the World Wide Web.* New York: Collins Press.

Biddle, P., England, P., Peinado, M., & Willman, B. (2002). *The darknet and the future of content distribution.* http://crypto.stanford.edu/DRM2002/darknet5.doc

Boyle, J. (2003) The second enclosure movement and the construction of the public domain. *Law and Contemporary Problems, 66*(1&2), 33-74.
Chartier, R. (1994). *The order of the book.* Stanford, CA: Stanford University Press.
Chomsky, N., & Herman, E. (1988). *Manufacturing consent.* New York: Pantheon Books.
Czitrom, D. (1982). *Media and the American mind: From Morse to McLuhan* (pp. 60-90). Chapel Hill: University of North Carolina Press.
Douglas, S. (1987). *Inventing American broadcasting, 1899-1922.* Baltimore, MD: Johns Hopkins University Press.
Eisenstein, E. (1980). *The printing press as an agent of change.* Cambridge, UK: Cambridge University of Press.
Hall, S. (1996). *Stuart Hall: Critical dialogues in cultural studies.* London, UK: Routledge.
Hobart, M., & Schiffman, Z. (1998). *Information ages: Literacy, numeracy, and the computer revolution.* Baltimore, MD: the Johns Hopkins University.
Holmes, D. (2005). *Communication theory: Media, technology, society.* London: Sage.
John, R. (1995). *Spreading the news: The American postal system from Franklin to Morse.* Cambridge, MA: Harvard University Press.
Johns, A. (1998). *The nature of the book: Print and knowledge in the making.* Chicago, IL: University of Chicago Press.
Kang, J. (2001). *Communications law and policy: Cases and materials.* Gaithersburg, MD: Aspen Law & Business.
Lasica, J. (2005). *Darknet: Hollywood's war again the digital generation.* Hoboken, NJ: Wiley.
Lessig, L. (1999). *Code and other laws of cyberspace.* New York: Basic Books.
Lessig, L. (2004). *Free culture: How big media uses technology and the law to lock down culture and control creativity.* New York: Penguin Press.
Liu, A. (2004). *The laws of cool.* Chicago, IL: University of Chicago Press.
Manovich, L. (2001). *The language of new media.* Cambridge, MA: MIT Press.
Marvin, C. (1990). *When old technologies were new: Thinking about electric communication in the late nineteenth century.* Oxford, UK: Oxford University Press.
Mattelart, A. (1996). *The invention of communication* (S. Emanuel, Trans.). Minneapolis: University of Minnesota Press.
McChesney, R. (2004). *The problem of the media: U.S. communication politics in the 21st century.* New York: Monthly Review Press.
McLuhan, M. (1962). *The Gutenberg Galaxy: The making of typographical man.* Toronto: University of Toronto Press.
McLuhan, M. (1967). *The medium is the massage: An inventory of effects.* New York: Bantam Books.
Miller, M. C. (2002, January 7). What's wrong with this picture? *The Nation.* Chart of "The Big Ten" media conglomerates at http://www.thenation.com/special/bigten.html.
Minnow, N. (1961). (http://www.janda.org/b20/News%20articles/vastwastland.htm)
Morgan, E. (2002). *Benjamin Franklin.* New Haven, CT: Yale University Press.
Netscape Annual Report. (1966). Netscape Corporation.
Oxford English dictionary online. http://www.oed.com
Paul, C. (2003). *Digital art.* London: Thames & Hudson.
Penley, C. (1997). *Nasa/Trek: Popular science and sex in America.* New York: Versa.
Rice, R. (1987). Computer-mediated communication and organizational innovation. *Journal of Communication, 37*(4), 65-94.
Rose, M. (1995). *Authors and owners: The invention of copyright.* Cambridge, MA: Harvard University Press. http://dc-mrg.english.ucsb.edu/conference.asp

Siegert, B. (1999). *Relays: Literature as an epoch of the postal system.* (K Repp, Trans.). Stanford: Stanford University Press.

Starr, P. (2004). *The creation of the media.* New York: Basic Books.

Wardrip-Fruin, N., & Monfort, N. (2003). *The new media reader.* Cambridge, MA: MIT Press.

Warner, W. (2005) Communicating liberty: The newspaper of the British Empire as a matrix for the American Revolution. *ELH*, 72(2), 339-361.

6

THE AGE OF THE CONGLOMERATE, OR HOW SIX COMPANIES ATE THE NEW HOLLYWOOD

Jennifer Holt

THE END OF THE MEDIA CONGLOMERATE?

Each year, the world's most influential media moguls, technology wizards, and important investors fuel up the corporate jets and gather in Sun Valley, Idaho at Herb Allen's annual retreat for the movers and shakers of the entertainment industry. It is an intriguing mix of family vacation and the art of the Fortune 500 deal, mostly off limits to journalists and media. In the summer of 2005, however, Sumner Redstone had an announcement to make to those reporters who were gathered in the resort's lobby. The Viacom CEO heralded the end of an era that has been building for the last 20 years when he proclaimed, "The age of the conglomerate is over" (Sutel, 2005).

This announcement came as his company, one of the largest entertainment conglomerates in the world, was about to split into two: one half focusing on its growing cable channels that include MTV, VH1, and Nickelodeon as well as the Paramount Studios, the other half structured around Viacom's broadcast properties with the centerpiece being the CBS network, acquired in a $35 billion deal just

6 years prior, along with Infinity Broadcasting and dozens of television and radio stations. The impetus, of course, was the judgment of Wall Street and the heavy weight of unrealized synergies, a sluggish advertising market, and the declining value of its radio and billboard properties that are dragging the company's fortunes down.

It also comes as some of the most highly touted deals of the last 10 years have already proven to be disasters by most standards. Disney's $18+ billion deal to acquire ABC has not exactly produced the dividends and synergistic opportunities that architects of the merger had hoped for: ABC experienced a dramatic rating slide in prime-time and news programming shortly after the takeover and languished in third place in the ratings race for years, and a precipitous slide into fourth place by 2001. Numerous high-profile management shake-ups and personnel crises have also plagued ABC, and the network is presently being propped up by the surprise hits of *Desperate Housewives* and *Lost,* while deeper problems linger and creative methods to fuse their various brands have been decidedly lacking. After all, there are only so many trips that families on ABC's Extreme Makeover: Home Edition can take to Disneyland.

Another grave disappointment was also the biggest media merger in history: the union of AOL and Time Warner. This was hailed as the moment of transformation for entertainment in the Internet era when it was announced in early 2000; as the 21st century began, so did a new age in entertainment, according to insiders and analysts alike. The deal was touted as the definitive marriage of old and new media—and "a stunning affirmation of the new reign of the internet" (Usborne & Wilson, 2000, p. 1)—that was to kick off a revolution in the industry. Unfortunately, the plunge in stock prices, culture clashes between the two companies, and corporate infighting revealed something quite different: the deal has been a catastrophe by all accounts. Aside from the collapse in the company's value (estimated at $184 billion when the merger was first announced and now hovering around $73 billion), revenues and subscriber numbers at the internet division have dropped significantly, more than $500 million has been paid out to settle securities fraud charges, and the highly anticipated cross-promotion opportunities between the old and the new media have failed to deliver on their lofty promises. By the conglomerate's second birthday, four books on the fiasco were already in the works, it had lost two thirds of its stock value, thousands were laid off, and 1 year later it posted an annual loss of nearly $100 billion, the largest ever in corporate history. In a desperate attempt at an extreme image makeover for investors, the Time-Warner board voted to drop the AOL from the company's name in September 2003, but stock prices are presently a long way from recovery.[1]

The age of the conglomerate has surely seen better days.

However, this era is far from over. In fact, in many ways, it appears that the party was just getting started. This ever-shrinking universe of titans that control today's entertainment has continued to consolidate and thrive, as have their respective industries. Five conglomerates (Time Warner, News Corp., Disney,

Viacom, NBC Universal) produced 80% of the prime-time network series for the 2005 fall season (Lowry, 2005). Two of their companies (Warner Bros. and Fox) are currently responsible for two thirds of the prime-time schedules in broadcast. Six studios, all with yearly production budgets that hover around the $1 billion mark, often account for 90% of the box office, and four corporations control 90% of what is seen on television. Films that are made purely to exploit a conglomerate's holdings and distribution pipelines are still alive and well. And, although Viacom may be moving toward a more streamlined model, many of the company's chief rivals are expanding. A consortium led by Sony purchased MGM in 2004, Hong Kong Disneyland opened in September 2005, and News Corporation continues to spread outward in all directions including cyberspace, acquiring the popular Web site Myspace for $580 million in cash the same month as Redstone's announcement.

REAGAN-ERA ROOTS

For these opportunities and revenue streams, the moguls of today have Ronald Reagan to primarily thank, as his policies were the spark that jump started the new age of the conglomerate 25 years ago. Since then, the industry has developed into a small handful of tightly diversified empires built entirely around entertainment-related properties, as opposed to the old conglomerate era of the late 1960s characterized by a small handful of film studios under the same corporate roof as funeral homes, parking lots and auto parts companies. It was President Reagan's firm commitment to deregulation and, more importantly, the reinterpretation of anti-trust during his administration by his Department of Justice that ushered in this formation of the New Hollywood media conglomerate.

These two highly politicized elements of presidential policy were the key ingredients that combined to propel the entertainment industry into a new paradigm of political economy. At the outset of the Reagan era, Hollywood's various businesses were largely separate entities; movie studios, broadcast networks and cable companies were disconnected from one another structurally and strategically. In the two decades that followed, cable, film, and broadcast were transformed from largely separate industries into businesses feeding off of one another to survive as collective holdings in a small handful of vertically and horizontally integrated media empires.

There are many important factors that have contributed to the current political economy and the new "Age of the Conglomerate"—primarily the development of new technologies, the larger industrial/economic trend towards globalization, escalating costs and the attendant drive to find new efficiencies of scale, and even creative thinking by very well-financed moguls—these have all created change in the corporate blueprint of entertainment. However, it was ultimately the manner in which deregulation was implemented and antitrust was reinterpreted during the Reagan Administration that allowed for a new archetype to emerge, thus setting the stage for the wholesale industrial consolidation that was to follow. Consequently, when tracing the circumstances and conditions that have rendered the current political economy possible, profitable and even necessary in the New Hollywood, the most productive lens to look through is the one focused on the arenas of regulation and policy.

ANTITRUST AND THE ENTERTAINMENT CONGLOMERATE

From the passing of the Sherman Act in 1890 to the end of the Reagan era in 1990, the first 100 years of antitrust enforcement can be characterized by distinct cycles in government attitudes toward big business, largely determined by the prevailing economic and political climate. The Sherman Act—sometimes referred to as "the Magna Carta of the antitrust movement" (Geisst, 2000)—established federal antitrust policy as a way to prevent companies from engaging in unfair business practices or undermining competition in their industry. It was borne out of the Industrial Revolution and the public outcry over the "trusts" and monopolists of the day that were largely controlling the economy, including Carnegie and his interests in steel and railroads, Rockefeller's Standard Oil Company, Jay Gould and J.P. Morgan being among the most famous. The Sherman Act was passed in order to counter the growth of these trusts and limit their chokehold on prices. It was amended by the Clayton Act nearly 25 years later in 1914 to address growing horizontal integration (mergers between two firms in the same industry), which was also stifling competition. After this, there was no new antitrust legislation passed until Congress approved the Celler-Kefauver Amendment to the Clayton Act in 1950. This amendment was designed to address the growing fear of "increasing concentration of industrial assets into the hands of a few" (Fox, 1986, p. 8). It essentially gave merger law some more substance and "bite" to carry over into the second half of the 20th Century.

However, antitrust laws "are among the least precise statutes enacted by Congress" and have therefore developed and changed with the legally determined definitions of what constitutes competition, markets, and monopoly (cited in Conrad, 1997, p. 7). In fact, antitrust is one of the most politicized forms of legislation, and has even been theorized as a "subcategory of ideology . . . a microcosm in which larger movements of our society are reflected and perhaps, in some small but significant way, reinforced or generated" (Bork, 1993, p. 10). Consequently, as leading antitrust scholar Herbert Hovenkamp (1985) noted, "Almost every political generation has abandoned the policy of its predecessors in favor of something new" (p. 1). Still, antitrust regulation remains as a way for government to threaten big businesses that exhibit monopolistic tendencies with legislation that could force changes in their structure or conduct.

Accordingly, the application of antitrust statutes has undergone numerous cycles and changes since the drafting of antitrust's foundational principles. The Progressive era (roughly from the 1890s to the early 1920s) saw inquiries into monopoly control launched in broad and comprehensive attacks on trusts and cartels in every industry from motion pictures (Edison's Motion Picture Patents Company also known as "the Trust") to petroleum (Rockefeller's Standard Oil) and tobacco (Duke's American Tobacco). There was a period of practical nonenforcement around the Great Depression in order to prop business back up and speed economic recovery. However, Thurman Arnold, the Antitrust Division's Chief during Franklin Roosevelt's second term, soon expanded the division's budget and staff and began vigorously attacking big business for "not being responsive to the needs of the individual consumer" in the late 1930s (Gersst, 2000). The conglomerate craze of the 1960s and the birth of the multinational corporation brought even more aggressive assaults on big business supported by the predominantly liberal school of antitrust enforcement and policy developed by the Warren Court (1953–1969).

During this time, federal officials were quite hostile to mergers and often embraced the principles embodied by the Harvard School, which dominated the courts from the 1930s through the 1960s. The Harvard School focused on detailed case studies and industry-specific analysis to formulate policy. It was an empirical approach to antitrust and relatively aggressive in its approach to stemming industry concentration, viewing vertical integration (ownership of all phases of a business from production to distribution and sale) as harmful to the competitive health of an industry. The idiosyncratic, extremely particularized approach embodied by the New Deal, which created an enormous and varied anti-trust apparatus, was inspired by Harvard School principles. During the 1930s and 1940s, for example, government enforcement agencies such as the Department [DOJ] of Justice and the Federal Trade Commission [FTC] conducted studies to determine particular anti-trust policies and guidelines for each industry. Market-specific agencies were created (i.e., the Federal Communications Commission [FCC] for communication) and staffed by experts in the field in order to regulate each different business sector. The underlying notion was that industries are inherently different from one another and they need special rules exclusively applicable to their unique circumstances.

The Chicago School of antitrust policy that informed and guided the Reagan Administration was just the opposite; this approach applies a single set of operational principles to a wide variety of industrial markets and contexts instead of considering them case by case.[2] Chicago theorists assert that markets operate the most efficiently (and most competitively) when left alone, and their approach results in minimal antitrust enforcement and is extremely tolerant of mergers and industry consolidation. The main argument of the Chicago School is that efficiency should be the only goal of the Sherman Act, and this efficiency will only develop when the market is left alone; antitrust enforcement merely interferes with that process, according to this philosophy (see Conrad, 1997; Hovenkamp, 1985). This neoclassical economic approach became increasingly influential on policy and regulation in the 1970s.

Consequently, this period saw a much more relaxed approach to antitrust enforcement while the country also experienced an economic decline and shifting political winds. The Chicago School continued to hold sway into the 1980s with President Reagan. Almost immediately, Reagan established his commitment to deregulation and minimizing the role of government as he embraced the changes that had begun brewing under Presidents Nixon, Ford, and Carter, and significantly extended them into most realms of industry. Antitrust analysts have described this period as being driven by "free market law and economics scholars" who provided "the theoretical underpinning for the policy of an Administration that promised business to get government 'off its back'" (Conrad, 1997, p. 10). In the end, it would be these politicized changes in attitude that would also prove to be the core of synergy between politics and the entertainment industry during the Reagan era.

Since the days of the Reagan Administration, the DOJ has consistently regarded consolidation in the media as benign—even beneficial—to surviving in the competitive global marketplace. Thus, as the slippery business of defining a monopoly has changed alongside economic and political ideology, the entertainment industry has been both beneficiary and victim of these various interpretations. Since the 1980s, however, this industry has often found itself on the winning end.

INITIAL CONFLICTS ON THE PATH TO CONSOLIDATION

President Reagan's economic policy in particular was extremely tolerant of mergers, acquisitions, consolidation and conglomerate growth. The view of his antitrust enforcers was that vertical integration was often harmless; it was perceived as a strategy promoting economic efficiency as opposed to encouraging anticompetitive behavior or restraint of trade as charged by opponents of this approach. It has even been said that William Baxter, President Reagan's first assistant attorney general in charge of antitrust, changed both "the vocabulary and the central paradigm of antitrust" (Fox & Sullivan, 1987) as he instituted change that was driven by the deregulatory and free-market philosophies of the Chicago School. As the scope of acceptable business practices expanded and the parameters of antitrust laws enforced by Baxter narrowed, the entertainment industry crawled towards the mountaintop of vertical and horizontal integration and left most regulatory constraints on growth and consolidation by the wayside.[3]

Still, although a discernible cycle certainly took hold during this time, it is important to note that it was not always a smooth or obvious path toward unbridled consolidation between and among various media industries. Indeed, the manner in which antitrust policy was applied and interpreted was quite contradictory and uneven, especially at the beginning of the Reagan Administration when the DOJ intervened repeatedly in the affairs of the film industry as they attempted to encroach on the growing business of cable television. In 1980, for example, the studios appealed to the DOJ to investigate HBO's allegedly monopolistic business practices. HBO and its complete market dominance troubled the studios and they were incensed by the service's monopoly and monopsony power in the network program service market.[4] According to industry analyst Harold Vogel, (1998) by 1981, HBO had even surpassed the large theater chains and emerged as the studios' largest customer, licensing more than $130 million that year alone. However, the studios were met with quite a surprise when the DOJ not only refused to stem the development of HBO, but it turned around and sued the studios instead (Paramount, Universal, 20th Century Fox, and Columbia) for their attempt to create Premiere, a major cable distributor. In the end, the studios were forced to abandon their venture and wait a few more years before getting involved in cable distribution.

As the drive to consolidate across industries heated up and the film studios saw the profit potential in the burgeoning pay-TV market, they attempted to enter cable ownership through another door; but again the DOJ stepped in and intervened. When Paramount and Universal tried to purchase Showtime from Viacom and merge it with The Movie Channel in 1983, the DOJ objected, claiming that this combination would create vertical integration and barriers to entry in the emerging market. This attempted merger of HBO's two primary competitors in pay-TV was another failure to integrate, so the major studios remained on the outside of the cable revolution looking in.

What was at stake here was more than merely business expansion or industry growth; in fact, the very definitions of various entertainment "markets" were being formulated for purposes of antitrust regulation. The two markets of pay-TV and theatrical film were regarded as distinct and separate by the courts in the initial confrontations between the studios and the cable industry. They were not seen as being part of the same "business of entertainment" but instead as discrete and independent industries from one another. These definitions and decisions would shape the formation of home video and pay-cable by forcing their separation from the major film studios at the outset and allowing companies such as HBO to profit from their perceived status as Hollywood "outsiders." In turn, the structural transformation of the global entertainment industry would be initially controlled, and the drive toward conglomeration somewhat restrained.[5]

The manner in which the DOJ handled these early cases also signaled that the preliminary tenor of policy and regulation in this tumultuous, critical period of cable development would be one that was protective of already-established interests and players in cable television. Primarily, HBO and its parent company, Time,

Inc. would be the beneficiaries of this official disposition. After all, in a period of 3 years, the DOJ pursued antitrust action twice against the studios in their attempts to establish pay-TV channels/services that would compete with the dominance of HBO.

These early conflicts over the major studios' attempts to create or control a competitive cable distribution service additionally illustrate the pervasive resistance and suspicion with which the government still viewed Hollywood in relation to matters of industrial structure and antitrust. The lingering suspicion of the film studios, given their long history of anticompetitive behavior (indeed, the major studios were sued for major anti-trust violations and forced to restructure by the DOJ *twice* in the industry's first 50 years), certainly prevented the conglomerate era from taking off immediately. By aggressively and actively policing studio involvement in the cable marketplace, the DOJ would thus forestall the consolidation of the major film, television, and cable companies for the time being.

Interestingly, this activity was taking place in the context of the breakup of the government-sanctioned monopoly enjoyed by AT&T for 70 years. In 1982, an antitrust lawsuit that the DOJ had begun 8 years earlier finally resulted in the dismantling of the world's largest and most successful telecommunications network. The company's size and power fell out of sync with the prevailing public and political sentiments regarding monopoly and government-protected big business in the wake of the Vietnam War and the Watergate scandal. Consequently, AT&T and its 22 local subsidiaries were restructured into one long-distance company (AT&T) and 7 regional local carriers (what became known as the "Baby Bells") effective January 1, 1984. Once in office, the Reagan Administration certainly put immense pressure on the DOJ to drop the AT&T suit, but there was already too much momentum at this point, not to mention the time and resources invested, to justify simply dropping the case. Even during his Presidential campaign in 1980, Reagan had said of the AT&T case, "the government's suit was a mistake," (Horowitz, 1989, p. 239) However, these events would largely mark regulation's last hurrah. On the same day that the Reagan Administration finished with AT&T, they also settled their longstanding antitrust suit against IBM which began in 1969 as an investigation into the company's alleged attempts to monopolize the business computer market. Essentially, William Baxter dropped the IBM case after 13 years because it no longer had merit, according to the new Administration. And thus began the DOJ's change in attitude to one of great tolerance toward matters of concentration, vertical/horizontal integration and antitrust.

CHANGES IN ATTITUDE AND OWNERSHIP

This shift in the federal government's position on vertical and horizontal integration actually began with the formation of Tri-Star in 1983, a studio uniting cable, broadcast and film interests with the blessing of the DOJ. This partnership of HBO, Columbia Pictures, and CBS created an instant vertically integrated entertainment company—a brand new Hollywood major studio with $400 million in funds from well-established players across the media spectrum. Many elements of Tri-Star's business plan served as a model for a new generation of Hollywood studios and moguls. The company itself was a monument to the necessity of exploiting new and varied forms of distribution, and also a demonstration of how important cable holdings would be to the entertainment complex of the New Hollywood.

As Stephen Prince (2002) also noted, the birth of the studio heralded "the inexorable movement throughout the decade toward greater concentrations of media ownership and control of interlocking markets" (p. 31). Indeed, it further indicated a paradigm shift in the philosophy of the DOJ in relation to antitrust enforcement in entertainment. In fact, the greatest legacy of Tri-Star endures in the manner that the studio ushered in a new era for antitrust enforcement in entertainment. The days of aggressive regulation were over. After Tri-Star was established, the DOJ did not stop any vertical, horizontal or collaborative mergers in the entertainment industry for the rest of the decade. Instead, consolidation and "synergy" would quickly become an accepted reality of doing business and the government assumed their new position as "casual observer" of the marketplace.

Changes in ownership were just the beginning; the real transformation would come when the FCC attended to the key piece of regulation that lingered—the Financial Interest and Syndication Rules ("fin-syn") which were put forth by the FCC in 1970. It was fin-syn alone that could open the gates to entirely new dimensions of industrial structure, efficiency, and synergy in Hollywood. Fin-syn essentially dictated that the broadcast networks could not own the programming that they aired or syndicated, forcing the television industry to look to the Hollywood studios for product. The rules pushed the networks out of the syndication market and also separated production/distribution from exhibition in a sense, forcing the networks to choose: they could either earn their money from producing programming or from broadcasting programming. As they were set up to be broadcast networks, not production networks, their choice was relatively simple.[6]

Initially, the rules were enacted to loosen the grip of network power and expand the market for independent producers.[7] They had also been formulated as a response to insistent lobbying from the film studios, which were having severe troubles of their own in the late 1960s and were growing increasingly dependent on their revenues from television production. By 1970, the networks held financial and syndication concessions in roughly 98% of their programming; (Mac Donald, 1994) taking this revenue away was a devastating financial blow to broadcasters and a huge boon to Hollywood producers. However, the power and the true legacy of fin-syn was to separate the *markets* and *industrial structures* of broadcast and filmed entertainment, not just the products. This government-enforced boundary that divided the ownership of film and television ensured that these media could not merge while fin-syn remained on the books. Consequently, it would not be until the mid-1990s when the rules were fully repealed that the industry at large would be able to take full advantage of the strategic opportunities afforded by conglomeration.[8]

By 1984, the Hollywood studios had begun inching towards a new conglomerate era by capitalizing on the growing tolerance of the DOJ and enacting new strategic alliances with the cable industry. As the majors amassed ownership stakes in the growing medium, Fox and MGM/UA remained notable exceptions. However, this would soon change and these two companies would play a pivotal role in creating the new conglomerate era. Ted Turner and Rupert Murdoch—both industry outsiders—would use the film assets of the two companies to reinvent what "television holdings" could mean for the studios as they increased their own resources, integrated these media and modernized the relationship between film and TV for the New Hollywood . . . all while evading the long arm of fin-syn, which seemed to have everyone else shackled to outmoded blueprints of corporate design.

MURDOCH AND TURNER: NEW DESIGNS ON THE OLD ECONOMY

As the decade wore on, horizontal and vertical integration became key strategies in cable's growth and consolidation . . . and business was booming. Program distributors were also suppliers. Station owners were increasingly controlled by multiple system operators (MSDs), essentially companies that owned more than one cable provider. The efficiencies of teaming production/distribution/exhibition were readily available to this burgeoning industry and cable companies of all kinds were taking advantage of those opportunities. The cable industry was thus free to grow and/or consolidate in a variety of ways that incorporated synergy into their plans and their approach often seemed to brazenly defy antitrust policy as it was applied to film and television, highlighting the discrepancy among regulatory standards in the various industries. Cable systems, for example, could own financial stakes in cable channels, which many of them did; John Malone's TCI (the nation's largest cable provider in 1982 with more than 2 million customers [Parsons & Frieden, 1998]) was heavily invested in cable programming, as were Warner Communications, Inc. and ATC (owned by Time, Inc.), among others. This stake that the cable companies (which would eventually be called MSOs) had in the programming on their "networks" was precisely the type of vertical integration that fin-syn and the consent decrees were designed to prevent in broadcasting.

Things did not look so rosy for broadcasters, however. As cable began to explode and deregulation took hold, the broadcast networks were at risk of losing their privileged position along with the government protection they had enjoyed in the marketplace since radio. Broadcasters could own the cable programming services (or what are commonly referred to as cable channels, i.e., ESPN) but they were prohibited from owning the cable delivery *systems* (or provider/distributors, commonly referred to as MSOs, i.e., TCI), which was where the power and control were centered because the MSOs determined, among many other things, which networks would be on a cable system and which would not. Furthermore, as Douglas Gomery (Compaine & Gomery, 2000) noted, "there were no restrictions on combining program ownership and cable ownership" (p. 207). The MSOs became an extremely powerful consortium of large corporations with formidable lobbying power in Congress able to function as monopolies and sole distributors of cable programming in local markets. As a result, by the mid-1980s the "Big Three" were left watching helplessly as the competition was allowed to expand their reach and power by means that would have been illegal if the networks had done it themselves. In much the same way that the major film studios were excluded from ownership when television took over, the networks were shut out of the distribution game once cable took off.[9]

What's more, at this point the solvency of the broadcast television industry was being assaulted on all fronts: Competition from new technologies and new regulatory philosophies endangered their survival in the marketplace. Network profits had been declining since 1980 due to increased programming costs (for both news and entertainment) and declining advertising revenues (thanks to cable, a recession, and rising interest rates). The Big Three networks saw their cumulative profits cut in half in the 4 years between 1984 ($800 million) and 1988 ($400 million Auletta, 1992). Audiences were diminishing as cable slowly began draining network viewership and the networks all began instituting massive layoffs amidst desperate budget crises. The three networks went from claiming 90% of the nightly audience in the 1970s to less than 60% in a span of just 15 years. (Auletta,1992). These threats brought hungry corporate raiders sniffing about in search of a good deal, and the walls were closing in by 1985. These dangers were all intimately connected, and ABC Chairman Leonard Goldenson presciently complained in late 1984, "the FCC has so deregulated the industry that broadcasters are no longer insulated from unfriendly takeovers" (cited in Auletta, 1992, p. 38).

In fact, deregulation proceeded swiftly with Mark Fowler at the helm of the FCC beginning in 1981 and, with the increase in ownership limits in 1985, station sales went through the roof and all three major networks were taken over within 2 years.[10] The huge corporations that had been circling around the industry's declining market values and underperforming assets were also able to take advantage of lowered interest rates to finance their acquisitions. Cap Cities took control of ABC, GE bought RCA (parent company of NBC,) and real estate magnate/chairman of Loew's Corporation Laurence Tisch gained control of CBS out from under the network's founder William Paley.[11] This was the first change in ownership ever for CBS and NBC, and the first time for ABC since the network's merger with United Paramount Theaters in 1953.

Despite the massive deregulation taking place, there was a very important set of rules that cable systems still had to abide by: the FCC's "Must Carry" rules instituted for the first time in 1972. These rules mandated that the cable systems carry all local television stations, regardless of whether the station had low ratings or duplicated another network/channel that was already on the system, in order to ensure the survival of the local stations and the broadcast networks. While cable owners chafed against these rules and fought them in court, their existence allowed for the proliferation of independent and new UHF stations, which would have otherwise been lost due to their higher position (frequency) on the dial (channels 14–69) and inferior signal strength. Now, their reach and coverage was practically equalized by this new technology and the UHF stations were able to attract significant audiences with distribution over cable and satellite. It was a dramatic change in fortune for those on the UHF dial.

When Ted Turner and Rupert Murdoch entered the arena, the "must carry" rules suddenly had new meaning. The vision of these media entrepreneurs bolstered by the mandate and opportunity created by the must carry rules combined to redesign the paradigm for vertical integration in the New Hollywood. Consequently, these two men were able to use UHF stations to do exactly what the networks were prohibited from doing: enjoying a financial stake in the programming that they distributed and exhibited. This lack of uniform regulation across media industries was the linchpin in cable's ability to expand during this time, and also the reason Turner and Murdoch's empires took off in the mid-1980s. Both relative outsiders before that time, they would bring a new vision to entertainment industry partnerships and consolidation, combining film and television, content and distribution, into conglomerates that were expanding in previously unthinkable directions.

Turner began by buying a destitute UHF station, using RCA's Satcom I satellite to bounce his signals to the rest of the country and, with it, created the first Superstation on cable. His initial strategy was to amass programming assets that could exhibited on his station and replayed often, lowering operating costs and fashioning himself on a vertically integrated model from the start. Turner always focused on owning content, and even utilized the Atlanta Braves and Atlanta Hawks as inventory to plug into his distribution networks, using underperforming sports teams as relatively cheap programming. He would not be the first entertainment enterprise to own a sports team (CBS owned the New York Yankees from 1964 to 1973) but he was the first to hold all of the *rights* to his own team.

Turner's investment philosophy became the mandate of the new entertainment conglomerates. "You need to control everything," he said. Often citing Standard Oil as a model, since Rockefeller owned the oil fields, gas stations, pipelines, trucks and distribution networks, Turner very clearly saw from the start that "controlling everything" is the path to power. "You want to have a hospital and a funeral home, so when the people die in the hospital, you move them right over to the funeral home next door. When they're born, you got 'em. When they're sick, you got 'em. When they die, you got 'em . . . the game's over when they break you up" (cited in Auletta, 2001, p. 154).

His launch of CNN in 1980 was a big step in gaining more control, soon adding Headline News and, after a failed bid to takeover CBS, continued by buying up MGM/UA (but only retaining the lucrative film library—which would soon be the basis for his cable network Turner Network Television and ultimately Turner Classic Movies), the Hanna-Barbera library, which was then used to launch his Cartoon Network, and numerous other media properties. His approach to building his vertically integrated company was unique because of his primary emphasis on content as opposed to distribution. His vision was pieced together from the bottom up, acquiring programming *first* and then developing the networks, instead of the other way around. Also, the configurations of most of the other major entertainment companies in the 1980s were an amalgamation of distribution networks, creative production arms, and exhibition outlets in various media. Turner's vision was more focused, streamlined and tightly diversified than that. He stuck to film libraries and cable programming/networks, eschewing most other avenues of ownership in the entertainment industries.

Turner and this new vision of vertical integration also became the future of cable and represented a giant leap towards the new conglomerate era. He took a bankrupt UHF station and transformed it into a company that now stands as the heart of the AOL-Time Warner global entertainment conglomerate. He did this by unconventional means that combined media in ways that had never been done before . . . all which became standard industry practice just 10 years later. There was only one of his contemporaries who could match his creativity in constructing entertainment synergies while evading federal regulations: Rupert Murdoch and News Corporation.

Murdoch was busy designing his own aggressive blueprint for vertical integration in American media *and* he was taking it global at the same time. With his aggressive acquisition strategy, Murdoch forced the reconsideration of decades' worth of antitrust policy and broadcast regulations. He began rapidly building his American assets in the 1980s, making an unsuccessful attempt to buy Showtime in August 1983 and launching another losing bid when he tried to take over Warner Communications in 1984.[12] After that, he went on a buying binge in late 1984 that "stunned media industry executives, some of whom worried that their companies might be next on his shopping list." (Moore, 1987, p. 90). Murdoch finally made his mark on Hollywood and the American media landscape 1 year later when his News Corporation bought 20th Century Fox from Martin Davis in 1985.

Barely 6 weeks after his Fox purchase was official, Murdoch made another deal, this one triple the size: on May 6, 1985, even before the Fox purchase was completed, Murdoch announced a plan to buy the Metromedia stations for $1.85 billion. These were popular, well-placed, independent, successful properties in

Washington, DC, New York City, Chicago, Los Angeles, Houston, and Dallas and would form the core of the fourth broadcasting network. It was a way to both expand his holdings and capitalize on synergy, as Murdoch could use the holdings of 20th Century Fox and at the same time, siphon off some of the estimated $8.3 billion in advertising revenue that was going to the other three networks.

With this, Murdoch created the first rival to ABC, CBS, and NBC since the demise of the DuMont network 30 years earlier.[13] More importantly, he managed to structurally integrate the film and broadcast industries for the first time, creating a blueprint for entertainment conglomerates that would endure into the new millennium. It was just like the studio era but bigger, as Murdoch could funnel his guaranteed supply of entertainment into the lucrative arms of theater exhibition *and* television programming.

The FCC's definition of a "network" was key to Murdoch's successful exploitation of the 20th Century Fox library and the integration of his resources as program supplier and exhibitor. For the purposes of fin-syn the commission defined a network as "providing 15 or more hours per week to at least 25 affiliates in 10 or more states" ("Charting," 1986). Murdoch started with just 8 hours a week, and with this minimal schedule, he was able to avoid the "network" label and evade the regulatory policies that confined the behavior (and profits) of the networks.

In May 1990 the network was granted a 1-year waiver in order to encourage and promote competition in the marketplace. This allowed Fox to expand its programming schedule to 18.5 hours and pose a greater competitive threat to the "Big three" without being subject to the limitations of fin-syn. In fact, Fox never had to abide by the restrictions of fin-syn because it was given another exemption in 1993 when moving to a full prime-time programming schedule because of its status as an "emerging network." (A user's guide," 1993, p. 34). This special dispensation allowed Murdoch to program 22 prime-time hours a week like the other networks, yet he was still free to continue his lucrative syndication operation without having to choose between maintaining that income stream and expanding his network programming operation.

It took less than 1 year from his arrival in Hollywood for Murdoch to become a major player in American film and television and just two deals for him to integrate these two entities that the government had labored to keep separate for the previous 40 years. Since then, Murdoch has gone on to set the standard for the vertically integrated global media giant and is widely regarded as the shrewdest strategist in the entertainment industry. Ultimately, Murdoch's impact in the Reagan era transformed the way that business was done in television. He revalued syndicated programming, restructured the broadcast industry, instituted new competitive measures in the relationship between cable and the networks, and redefined the medium for advertisers, audiences, industry players, and former competitors in entertainment media.

As Albert Moran (1996) noted, Murdoch's purchase of Fox in 1985 was more than just acquiring product for his media outlets in the United States, Europe, and

Asia; it was also a move towards solidifying the global character of the New Hollywood media conglomerate. Moran suggested that in buying Fox, Murdoch "caused an abandonment of the strategy of disaggregation in favour of a new round of integration along vertical as well as horizontal lines" (p. 13). These lines were also stretched across the globe by the end of the decade, with his properties in Australia, Europe and North America. Now, Murdoch has holdings everywhere except Antarctica. He has an interest in more than 100 newspapers, some of the most profitable magazines in the world, book publishing, broadcasting, cable networks, film, Internet providers, and satellite distribution systems that reach all over the planet.

In the end, Turner and Murdoch both used film to reinvent the political economy of television, increasing the value of both industries in the process and redefining the potential of synergy in entertainment. Their prescience and trailblazing inspired the birth of two more studio-owned broadcast networks—Paramount's UPN and Time-Warner's WB Network. However, these networks would have to wait until 1995 and the final repeal of fin-syn before they would be on the air. The fact that fin-syn remained in place ensured that although the industrial economy was shifting in the late 1980s, the real seismic activity was yet to come. Once the Financial Interest and Syndication Rules were repealed in 1995, the earthquake would soon follow, as explained later. This was manifested in mergers, takeovers, and new studio-based networks that were all conceived and propelled by the notion of synergy, uniting all aspects of the entertainment industry under one corporate roof for the new millennium.

OLD HOLLYWOOD/NEW HOLLYWOOD

At the same time that Turner and Murdoch were uniting film, cable, and broadcast properties, the Hollywood studios were revisiting their golden years and once again buying up prime theaters all over the country. The five vertically integrated major studios had been forced to sell off their theater chains when they lost a Supreme Court case in 1948 (*U.S. v. Paramount Pictures, Inc. et al.*). The resulting consent decrees—known as the *Paramount* Decrees since Paramount was the first defendant named in the case—would essentially prevent vertical integration in the motion picture industry for the next four decades. That is, until a very quiet memo was released by the DOJ in 1985, which declared that while the government would not formally terminate the *Paramount* consent decrees, the DOJ (1985) would support such legal motions in cases where the action was "in the public interest (p. 2). Basically, antitrust chief J. Paul McGrath was unwilling to expend the resources necessary to terminate the decrees in the judicial system but he supported their dissolution. This statement extinguished the authority of the decrees, if not legally then practically; by removing the looming menace of any government lawsuit, the DOJ eliminated the only compelling reason to comply with the *Paramount* decrees.

It was at this point that the essence of *Paramount* was quietly transformed from an iron fist wielded by the federal government into a façade embraced by the studios; that façade has effectively and simultaneously solidified their power while keeping potential government scrutiny at bay. Consequently, the continued existence of the *Paramount* decrees on the books after 1985 gave the appearance that the film industry was still being actively regulated by the DOJ, when in practice the DOJ was actually supporting the film industry's bid for renewed consolidation and autonomy.

Consequently, 1986 was a watershed year in the march toward a new conglomerate era; even *Variety* noted it as a moment of "revolutionary realignment," as the film industry went on a theater buying binge led by MCA (Universal), Columbia Pictures, Tri-Star, Gulf & Western (Paramount), and Warner Brothers. As such, *Variety* noted that Hollywood returned to "the origins of its pre-TV growth and prosperity in the most massive marriage of distribution and exhibition since the 1948 *Paramount* decrees forced the majors to sell off their theater circuits" (Hollinger, 1987, p. 3). The result was a $1.62 billion spending spree for the nation's premier theaters and a renewed sense of entitlement around Hollywood boardrooms; with the threat of government regulation receding, the major studios began eyeing more assets for potential takeover.

Already, many of Hollywood's majors had begun to resemble entertainment conglomerates by acquiring properties in other media-related industries. MCA/Universal, for example, was continuing to build their cable holdings with their interest in the USA network and February 1986 purchase of cable superstation WOR in New Jersey for $387 million (Harmetz, 1986). They also owned home video distribution for their own films and produced cable programming that could be shown on their networks. Paramount owned a string of television stations and theme parks, along with a very successful television production company that had such ratings successes as *Cheers* and *Family Ties*. Warner Communications, Inc. produced and distributed television programming in addition to being part owners of six stations at the time. Additionally, they owned the nation's third largest cable system, co-owned The Movie Channel, and had significant interest in MTV and Nickelodeon that were distributed by their cable system. Disney was also an owner of broadcast and cable channels, in addition to a producer of film, cable and broadcast entertainment.[14]

With the merger of Time and Warner in 1989—a combination that created the world's largest media company (with a stock market value exceeding $15 billion and annual revenues at the time over $10 billion [Bruck, 1990]) the new model for the tightly diversified global media conglomerate was established and a new era in Hollywood had officially arrived. The deal was official 1 month after Reagan left office and kicked off a watershed year for media-related mergers and a new wave of conglomerate takeovers in the industry. Thomas Schatz (1997) noted that 414 deals were struck in 1989, totaling more than $42 billion, including the Time-Warner merger and Sony's acquisition of Columbia and CBS records for $5.4 billion.

With the Time-Warner deal at the end of the decade, it became clear that the tide of deregulation in the 1980s had swept in an entirely new relationship between the entertainment industry and government regulators. With an unprecedented series of media marriages in the 1990s, the new party line was official: deregulation had evolved into nonregulation. In 1995 alone, the entertainment industry completed well over $70 billion worth of mergers, the three biggest being Disney and CapCities/ABC ($18.83 billion), Time-Warner and Turner Broadcasting ($6.88 billion), and Westinghouse and CBS ($5.04 billion [Cox, 1996]). Even *The New York Times* observed, "not since the wave of industrial takeovers that created the great oil, steel and auto companies at the beginning of the century has corporate America been reshaped by a sweep of merger activity as the one taking place today." (Wayne, 1998, p. A2). The deals in entertainment represented almost 20% of all completed mergers for the year across all sectors of the economy (McCauley, Rudd, & Iacono, 1999).

Not surprisingly, the merger activity reached a practical frenzy once fin-syn was fully repealed in 1995 by the FCC; Time Warner announced the merger with Turner almost immediately, the first major network and studio merger took place between ABC and Disney—a deal announced months before the formal end of the rules, and the number of broadcast networks owned by Hollywood studios expanded even further as both Viacom and Time Warner announced the launch of their new networks, UPN and the WB, respectively. That same year, Seagram purchased MCA/Universal, Westinghouse bought out CBS, and DreamWorks SKG was born, creating a new Hollywood studio from the vast resources of Steven Spielberg, Jeffrey Katzenberg, and David Geffen.

This trend was further stimulated by the Telecommunications Act of 1996, which liberalized regulations and allowed for the union of broadcast, cable and telephone companies into newly configured media empires.[15] The Act also eliminated the former ownership cap of 12 television stations altogether and increased the nationwide audience reach limitation from 25% to 35%. According to one legal analyst, the effect of the 1996 legislation "was to wave the green flag at the convergence race—the race to amass the largest information company—by reducing or eliminating ownership barriers." See Table 6.1 (Garon, 1999, p. 491) for a timeline of the most significant media deals, and Table 6.2 for a summary of the holdings of the major global media conglomerates.

TABLE 6.1. *A Timeline of Relevant Media Industry Deals and Developments*

1970	Financial Interest and Syndication laws enacted along with the Prime-Time Access Rule Viacom created when CBS is forced to divest its syndication operation
1976	Department of Justice files suit against NBC, CBS, and ABC, which results in consent decrees being signed with all networks by 1980. They mirror fin-syn restrictions and impose further limitations on the number of prime-time hours the networks are allowed to produce
1983	Tri-Star is formed through a partnership of HBO, CBS, and Columbia Pictures
1984	ABC buys ESPN ($227 million deal)
1985	FCC increases station ownership limits
	Rupert Murdoch buys Fox for $575 million
	Fox announces deal to buy six of Metromedia's seven television stations (Los Angeles, Chicago, Dallas-Ft. Worth, Houston, Washington, District of Columbia, and New York City) for $1.85 billion. These stations would form the core of the 4th broadcasting network.
1986	Capital Cities Communications Inc. purchases ABC for $3.5 billion
	GE buys RCA Corp., parent company of NBC for $6.4 billion
	Turner buys MGM for $1.2 billion
	News Corp. launches FOX, the fourth broadcast network
1987	Sumner Redstone, president of National Amusements Theater Chain, takes over Viacom
1988	Ted Turner launches TNT
1989	Time, Inc. acquires Warner Communications, Inc. for $14 billion to form Time-Warner
1994	Viacom Inc. buys Paramount Communications Inc. for $10 billion
1995	Launch of UPN (Viacom/Chris-Craft) and WB (Time Warner) networks
	Disney announces deal to buy Capital Cities/ABC for almost $19 billion
	Westinghouse announces deal to acquire CBS Inc. for $5.4 billion
	Official repeal of the Financial Interest and Syndication Rules
	Time-Warner announces merger with Turner, $7.6 billion deal
1996	President Clinton signs the Telecommunications Act into law
1997	News Corporation becomes the largest owner of TV stations in the US
1999	Viacom Inc. announces deal to buy CBS Corp. for $34.5 billion
2000	America Online Inc. announces merger with Time Warner Inc. in $165 billion deal
2004	NBC and Universal merge in May, owned 80% by GE and 20% by Vivendi
	Sony leads an investment group that buys MGM in a $5 billion deal
2005	Sumner Redstone, CEO of Viacom, declares, "The Age of the Conglomerate is Over"

Sources: Variety, The New York Times, The Los Angeles Times, The Wall Street Journal, Annual Reports, Fortune and Who Owns the Media? (Compaine & Gomery, 2000)

Table 6.2. *Global Entertainment Conglomerates: Select Holdings 2006*

TIME WARNER

Broadcast (TV/radio)

Half of the CW network

Cable television

- Time Warner Cable (one of the largest cable system in the United States, controlling 22 of the largest 100 markets and serving 14.4 million customers, industry leader in Video on- demand services)
- Turner Broadcasting (includes CNN, Headline News, TBS, TNT, Turner Classic Movies, The Cartoon Network)
- HBO and Cinemax pay cable channels (more than 39 million U.S. subscribers)
- Eight local news cable channels

Film and television content

- Warner Brothers Pictures
- Warner Independent Pictures
- New Line Cinema
- Warner Bros. Television
- Castle Rock Entertainment
- A library of more than 6,000 films, 25,000 television programs, and thousands of animated shows
- Warner Bros. Animation - Looney Tunes and Hanna-Barbera

Internet

- America Online (Internet provider service with more than 25 million subscribers worldwide)
- Road Runner high-speed Internet service (cable-based)
- Numerous other Internet brands and services, including Compuserve, AOL instant messenger, McAfee VirusScan, Mapquest, Netscape, PGA.com, NASCAR.com

Publishing

- Time Inc., the largest magazine publisher in the world with more than 150 magazines reaching audiences of more than 300 million people. (*Time, People, Sports Illustrated, Fortune, In Style, Entertainment Weekly*)
- DC Comics, publisher of *Superman, Batman, MAD Magazine,* and 60 other titles
- Time Warner Book Group, including Warner Books, Little Brown and Co.

Other

- Warner Bros. International Cinemas (650 screens overseas, 130 screens in United States)
- Warner Home Video
- Warner Bros. Interactive Entertainment
- The Atlanta Braves
- Retail stores, including more than 150 Warner Bros. stores
- Moviefone

DISNEY

Broadcast (TV/radio)

- ABC television and radio networks
- Ten television stations and 72 radio stations
- Radio Disney, ESPN Radio, ABC News Radio

Cable Television

- Disney Channel (United States and international), ESPN, ESPN2 and ESPNews; ABC • Family Channel, Toon Disney, SOAPNet, plus holdings in Lifetime, A & E Television Networks, and E! Networks

Content

- Walt Disney Pictures/Television, Touchstone Pictures/Television, Miramax, Dimension, and Buena Vista
- The Muppets
- Pixar

Other

- Publishing, including Disney Publishing and Hyperion Books
- Music labels, including Hollywood Records, Buena Vista Records, and Walt Disney Records
- 11 Theme parks and resorts on three continents
- Disney Cruise Line
- ESPN Zone
- Disney Toys
- Baby Einstein
- Walt Disney Internet Group, including Oscar.com, ESPN.com, ABCNews.com, movies.com
- Buena Vista Games, Buena Vista Interactive
- Disney Direct Marketing

VIACOM

Cable Television

- MTV Networks/International, MTV2, VH1, Nickelodeon, TVLand, BET, Spike TV, CMT, Comedy Central, Logo

Film and Television Content

- Paramount Pictures (including library of more than 1,100 titles and 18,000 hours of TV)
- Dreamworks SKG

Publishing

- Famous Music Publishing (more than 125,000 copyrights in catalogue including music from *Love Story*, *The Godfather*, *Titanic*, *Mission: Impossible*, *The Brady Bunch*, *Star Trek* and *Cheers*

Other

- Paramount Home Entertainment (video distributor)

CBS - 2005 REVENUE: $14.5 BILLION

Broadcast (TV/Radio)

- CBS Television network (more than 200 affiliated stations)
- Half of the new CW television network
- CBS Television Stations (39 stations–21 CBS, 11CW, 3 MyNetworkTV, 4 Independent)
- CBS Radio (179 radio stations across the US)

Cable television

- Showtime, The Movie Channel, Flix, Sundance Channel (co-owned with NBC-Universal and Robert Redford)

Film and Television Content

- CBS Paramount TV (70,000 hours of programming, original libraries of ABC, CBS, and NBC)
- CBS Entertainment, CBS Productions, CBS News, CBS Sports, CBS Digital Media
- King World Productions (domestic syndicator of shows such as *Wheel of Fortune*, *Jeopardy!*, *Dr. Phil*, and *The Oprah Winfrey Show*)

Publishing

- Simon & Schuster

Other

- CBS Outdoor (more than 1 million billboards all over the world)
- CSTV Networks, Inc. (digital sports media)

NEWS CORPORATION

Broadcast (TV/radio)

- Fox broadcasting network, Fox Sports Radio Network
- 35 television stations in United States

Cable television

- Fox News, Fox Sports, Fox Movie Channel, National Geographic Channel, FX, MyNetwork TV
- Ownership and/or major interests in satellite channels and services reaching Europe, United States, Asia, and Latin America including DirectTV, BSkyB, and STAR TV

Content

- 20th Century Fox, Fox Filmed Entertainment, Fox Searchlight, Fox Television, Blue Sky
- Library of more than 2,000 films

Publishing

- More than 175 newspapers and magazines all over the world, including the NY Post and The Times
- Harper Collins publishers

Other

- Myspace.com
- Largest shareholder in Gemstar (owner of TV Guide and TV Guide Channel)

- Music labels
- Outdoor/billboards
- News America Marketing

OTHER NOTABLE HEAVYWEIGHTS AND THEIR KEY HOLDINGS

Bertelsmann AG: Publishing, music and broadcasting operations in 50+ countries

Comcast: Largest cable system in United States after buying AT&T Broadband

NBC Universal: NBC television network, 10 NBC television stations, Telemundo, 15 Telemundo stations, Bravo, USA Network, CNBC, Sci-Fi Channel, Trio, MSNBC (with Microsoft), Universal Pictures, Focus Features, Rogue Pictures, Universal Studios Home Entertainment (4,000+ films and 40,000 episodes of television), NBC Universal Television—producers of many successful series including the most successful drama franchise in history: Law and Order, interests in five theme parks including Universal Studios and Universal Orlando

Sony: One of world's largest producers of home entertainment and consumer electronics owns Columbia TriStar, Sony Pictures Entertainment, Sony Pictures Television, Screen Gems, Sony Pictures Classics, Game Show Network, Sony Home Entertainment (maker of PlayStation), co-owners of MGM—including 8,000 title film library and co-owners of Sony BMG Music Entertainment (i.e., Arista, Columbia, Epic, Jive, and RCA Records)

THE CONGLOMERATE ERA, PT. II–HOLLYWOOD LOVES A SEQUEL

Under FCC Chairman Michael Powell (2001–2005), the FCC launched another effort to further relax media ownership limits in 2003. The plan was to allow one company to own up to three stations in major cities and for a broadcast network to own local stations reaching up to 45% of the national audience, up from 35%. However, the ensuing public uproar over this decision put a dent in those plans. A federal appeals court issued a stunning blow to Chairman Powell with an emergency order blocking the ownership rules from taking effect as scheduled, and then Congress introduced a resolution to nullify the new rules at the same time a Senate Committee voted to block their passage. President George W. Bush tried to step in on behalf of his chairman and threatened to veto efforts to roll back the FCC reforms but he soon backed off, as opposition to this was so widespread and across the board, uniting political opponents and inspiring millions of citizens to voice their outrage. It was said this might be the first and only time when the National Rifle Association and the National Organization for Women were on the same side of an argument.

Ultimately, a quiet "compromise" was handed down and the ownership cap in broadcasting would be increased to 39% . . . the exact percentage of stations, coincidentally, that Viacom owned and just above that held by News Corporation. Essentially, this decision sanctioned the status quo while giving the appearance that regulation was actually taking place. In fact, the tail has been wagging the dog for some time. Yet, Powell still resigned with much of his deregulatory agenda consigned to failure. The Bush Administration declined to intervene the last time around on behalf of Powell's final deregulatory efforts and the rules remain in limbo at present, headed back to the FCC for another rewrite after the Supreme Court also decided not to get involved.[16]

However, 1 month after this setback, the forces of consolidation rebounded forcefully in late 2003 when NBC's parent company General Electric Co. announced it would be purchasing the entertainment arm of Vivendi Universal for $3.8 billion. This deal merged the French company's Hollywood studio, cable TV networks, and amusement parks with the broadcast network, creating a media giant with about $13 billion in annual revenue. NBC beat several other potential suitors for the Universal entertainment businesses, including Viacom, Liberty Media Corp., MGM, and Comcast. The new company includes the NBC television network and its cable channels CNBC, MSNBC and Bravo and Spanish-language broadcaster Telemundo as well as Vivendi's Universal movie and TV studios (producer of NBC's hit show *Law & Order*) plus the USA, Sci-Fi, and Trio cable channels and several amusement parks.

When looking at the case of NBC and Universal, the guiding principle behind media acquisition strategies in the conglomerate era becomes crystal clear: Why rent when you can own? After all, the three *Law & Order* shows alone brought in more than $1 billion in advertising revenue in 2004 and NBC was poised to make almost one third of its prime-time profits from the franchise. (Consoli, 2005; McClintock, 2003). This became even more important when NBC went from first place to last in all key ratings measures and reported a loss of nearly $1 billion in advertising dollars that same year. (Wallenstein, 2005). Indeed, all of the entertainment conglomerates work steadily to enact synergies up and down the vertical chain. Time-Warner, for example, can incorporate its many holdings in broadcast television (WB network), cable (Time Warner Cable, TNT, HBO), publishing (*Entertainment Weekly, People*), the Internet (AOL), home video, theaters, and even video games to promote a feature film across a vast array of media platforms with privileges and access that would be impossible for unaffiliated companies to attain.

Sometimes this synergy is obvious (i.e., the exploitation of all things Time-Warner in the *Batman* franchise) and sometimes it is not, such as the appearance of the *Sex and the City* cast on the cover of *Time* magazine—which allowed the company to advertise its popular HBO program in one America's most significant news magazines, enhancing both properties with one savvy marketing ploy (although it may also affect perceptions of *Time* magazine as an objective reporter of news when it promotes a commercial relationship as the cover topic). Either way, it is this mantra of convergence and its entrenchment in the boardrooms of the "big six" that renders the announcement of the conglomerate era's demise perhaps a bit premature. Now that the model of owning and cultivating content to funnel through massive distribution networks has found itself on relatively safe, established terrain, it would not seem probable that the entertainment industry would begin dismantling itself en masse. Stocks have certainly been underperforming, and companies such as Viacom, Clear Channel, and even Time Warner have begun spinning off some assets and refocusing their holdings . . . but these events hardly signal a revolution in political economy.

Instead, the conglomerate era will likely see more of an evolution . . . a shift that takes hold as the industry's core assets assume different degrees of importance and prominence in the company's overall revenue profile. New technologies such as the Internet and videogames are growing components of an entertainment conglomerate's holdings. Television continues to transform and embrace new distribution platforms, such as videophones and computer screens. The full impact of digital technology in all facets of the business has yet to be realized yet in terms of forging new models of production, distribution and cross-promotion. And of course, this influence will ultimately be negotiated and determined by the shifting political winds and the resulting changes (or lack thereof) in antitrust and regulatory policy. In essence, it appears the industry may be writing a new chapter in a familiar Hollywood epic . . . one in which the cast of characters and story arcs are quite likely to remain the same, but there is still enough drama and plot twists to make it worth watching.

NOTES

1. Other companies looking to distance themselves from corporate scandal followed suit: WorldCom also changed its name in April 2003 to MCI (the name of its long-distance carrier) and cigarette maker Philip Morris Companies changed its name to Altria Group.
2. There are notable exceptions to this generalized approach, such as the market-specific regulatory guidelines for airlines, telecommunications (public utilities) and intellectual property.
3. Numerous regulatory mechanisms for broadcast and cable television also fell by the wayside during this era that are not discussed at length in this chapter because they relate primarily to content as opposed to the focus of my study which is structural economy. They include the Fairness Doctrine, regulations on advertising as it related to children's programming and political candidates, and the erosion of "public access" requirements for cable TV. Yet they are still examples of the incompatibility of the new order and the status quo; most of the old models for regulatory policy as they applied to the media were casualties of antitrust interpretation during the Reagan Administration, whether they related to the industry's structure of content.

4. Monopsony power relates to the power of the one buyer over many sellers; it is that power possessed by a buyer of a product who purchases such a large percentage of the producers' output or controls the market in such a way that it can control the price it will pay.
5. Even home video was still largely separate from the major studios—as the battle over the legality of videocassette recorders waged on, home video largely remained the province of independent companies, not the Hollywood studios that were suing the makers of video recording devices. However, on January 15, 1984, the Supreme Court ruled that videotape did not constitute copyright violation and the studios were handed the luckiest legal defeat in their history. That same videotape that they were trying to outlaw would soon overtake theatrical revenue and become one of the most critical elements in their balance sheets.
6. The networks also signed consent decrees between 1976-1980 stemming Justice Department lawsuits which enforced and further solidified fin-syn and also severely limited the hours of "in-house" prime time programming the networks could produce. It is quite difficult to distinguish between the effects on the industry that resulted from the lawsuits (and eventual consent decrees) against the networks and the effects that resulted from fin-syn, as the issues, resolutions and time periods were virtually identical., although the consent decrees were removed in 1993 and fin-syn was not fully repealed until 1995.
7. For an outstanding overview and analysis of the rules and their implications, see Einstein (2004).
8. Fin-syn was actually about to be rescinded in 1983 but President Reagan surprised even his hand-picked FCC commissioners and stepped in on behalf of his close friends in Hollywood (after a closed-door meeting with his ex-agent and head of MCA, Lew Wasserman) and declared the rules were to stay in place, which they did to varying degrees until 1995.
9. When broadcast licenses were first being issued, the FCC barred any individual or organization convicted of monopolistic practices from owning one. Although no formal declaration was ever issued, the film studios were still in the throes of the *Paramount* case and Warner Bros.—which had intended to construct its own network—had its application suspended in 1948 because of the "freeze." Subsequently, the major studios were helpless as the radio networks solidified their position and power in the broadcast industry over the next four years. For more on this issue, see Hilmes (1990), Gomery (1984), and Anderson (1994).
10. The "rule of sevens" or 7-7-7 rule which allowed one company to own seven television, AM and FM radio stations was changed to the "rule of twelves" and was a crowning achievement for Fowler but not the elimination of all restrictions that he wanted. The 12-12-12 rule was also bound by the provision that the twelve stations together could not reach more than 25% of the nation's homes.
11. For a detailed account of this period in broadcast history, see Auletta (1992).
12. Even though he lost this battle to takeover Warner Communications, the "greenmail" that Murdoch was paid to leave the studio alone helped him to finance his purchase of 20th Century Fox less than 1 year later.
13. DuMont, like Fox, was also known as the "fourth network" during its operation from 1946 to 1956. It had actually been a pioneer television network, but was never as powerful or successful as any of its rivals. This was largely due to the fact that it did not own a radio network which meant it did not have a stable of stars and established pro-

grams like the other networks. Further, many of its stations were consigned to the UHF spectrum, which were much less desirable stations than the VHF channels 1–13.

14. For a more lengthy discussion of the historical relationship between Hollywood and the broadcast/cable industries, see Hilmes (1990).
15. One extremely clear and thorough study of this act is found in Aufderheide (1999).
16. For an outstanding overview of the political and legal battles surrounding the 2003 debates over media ownership, see Scott (2004).

REFERENCES

A user's guide to fin-syn and its varied net results. (1993, April 2). *Daily Variety*, p. 34.

Anderson, C. (1994). *Hollywood TV.* Austin: University of Texas Press.

Aufderheide, P. (1999). *Communications policy and the public interest.* New York: Guilford Press.

Auletta, K. (1992). *Three blind mice.* New York: Vintage Books.

Auletta, K. (2001, April 23-30). The lost tycoon. *The New Yorker*, p. 154.

Bork, R. (1993). *The antitrust paradox.* New York: The Free Press.

Bruck, C. (1990, January 8). Deal of the year. *New Yorker.* p. 80

Charting the future: The prospects for a fourth network. (1986, February 10). *Broadcasting*, pp. 36+.

Compaine, B., & Gomery, D. (2000). *Who owns the media* (3rd ed.). Mahwah, NJ: Erlbaum.

Conrad, K. (1997, April). Media mergers: First step in a new shift of antitrust analysis? *Federal Communications Law Journal, 49*(675), 7.

Consoli, J. (2005, July 26). Dick Wolf castigates TV critics. Retrieved September 3, 2005, from http//www.mediaweek.com/mw/search/article_display.jsp?vnu_content-ID=1000992089

Cox, P. M. (1996, Fall). What goes up must come down: Grounding the dizzying height of vertical mergers in the entertainment industry. *Hofstra Law Review, 25*(261), 13.

Department of Justice Press Release. (1985, February 6). p. 2.

Einstein, M. (2004). *Media diversity.* Mahwah, NJ: Erlbaum.

Fox, E. M. (1986, October). The politics of law and economics in judicial decision making: Antitrust as a window. *New York University Law Review, 61*(554), 8.

Fox, E. M., & Sullivan, L. A. (1987, November). Antitrust—retrospective and prospective: Where are we coming from? Where are we going? *New York University Law Review, 62*(936), 7.

Garon, J. M. (1999). Media & monopoly in the information age: Slowing the convergence at the marketplace of ideas. *Cardozo Arts & EntertainmentLaw Journal, 17*(491).

Geisst, C. R. (2000). *Monopolies in America.* New York: Oxford University Press.

Gomery, D. (1984). Failed opportunities: The integration of the U.S. Motion Picture and Television industries. *Quarterly Review of Film Studies*, 219-227.

Guider, E. (2005, May 30-June 5). Voiceover. *Variety*, p. 17.

Harmetz, A. (1986, March 3). Hollywood seeks control of outlets. *The New York Times*, p. D1.

Hilmes, M. (1990). *Hollywood and broadcasting: From radio to cable.* Chicago: University of Illinois Press.

Hollinger, H. (1987, January 14). Merger mania facing Congress. *Variety*, p. 3.

Horowitz, R. (1989). *The irony of regulatory reform.* New York: Oxford University Press.
Hovenkamp, H. (1985, November). Antitrust policy after Chicago. *Michigan Law Review, 84*(213), 1.
Lowry, B. (2005, August 1-7). Tuning in. *Variety*, p. 12.
MacDonald, J.F. (1994). *One nation under television.* Chicago: Nelson-Hall.
McCauley, R. N., Rudd, J. S., & Iacono, F. (1999). *Dodging bullets: Changing U.S. corporate capital structure in the 1980s and 1990s.* Cambridge, MA: MIT Press.
McClintock, P. (2003, November 18). Merger mystery. *Variety*, p. A1.
Moore, T. (1987, July 6). Citizen Murdoch presses for more. *Fortune*, pp. 90+.
Moran, A. (1996). Terms for a reader. In A. Moran (Ed.), *Film policy: International, national and regional perspectives* (pp. 1-22). New York: Routledge.
Parsons, P. R., & Frieden, R. M. (1998). *The cable and satellite television industries.* Boston: Allyn & Bacon.
Prince, S. (2000). *A new pot of gold: Hollywood under the electronic rainbow, 1980-1989.* Berkeley: University of California Press.
Schatz, T. (1997). The return of the Hollywood studio system. In E. Barnouw (Ed.), *Conglomerates and the media* (pp. 73-106). New York: The New Press.
Scott, B. (2004). The politics and policy of media ownership. *American University Law Review, 53*(645).
Sharp, K. (2004, January 25). A plot fit for Hollywood: The bloodbath of the moguls. *San Francisco Chronicle*, p. D1.
Sutel, S. (2005, July 8). Redstone: Age of the media conglomerate is over. *Associated Press.* Retrieved September 3, 2005, from http://www.forbes.com/associatedpress/feeds/ap/2005/07/08/ap2130170.html
Usborne, D., & Wilson, S. (2000, January 1). Media giants in world's biggest merger. *The Independent*, p. 1.
Vogel, H. (1998). *Entertainment industry economics* (4th ed.). New York: Cambridge University Press.
Wallenstein, A. (2005, August 23). *The fall season: NBC-Fall TV preview guide.* Hollywood reporter.com
Wayne, L. (1998, January 19). Wave of mergers is recasting face of business in U.S. *The New York Times*, p. A1.

7

POLITICS OF LEGISLATING MEDIA OWNERSHIP

Congresswoman Diane E. Watson
33rd Congressional District of California

Shawn H. Chang

BACKGROUND: THE 2003 FCC PROPOSED MEDIA OWNERSHIP RULES

In June 2005, the public battle over the future of our nation's media ownership landscape came to a temporary halt when the U.S. Supreme Court decided not to review the *Prometheus Radio Project v. Federal Communications Commission* case, in which a lower court had struck down the Federal Communications Commission's [FCC] June 2, 2003 order on media ownership rules. Under the new rules, one company would have been allowed to own a major newspaper, eight radio stations, three television stations, and the cable company in a single market. In essence, one company could control all programming and news content in a single media market made up of millions of individuals. It represented a drastic departure from prior media ownership limitations, where a single company was prevented from owning a daily newspaper and television or radio station in the same market, among other cross-ownership restrictions. Not only would the free flow of information be adversely impacted by the FCC rule changes, but many local, independent, and minority-owned media companies could be forced out of the market.

Indeed, one of the main reasons that drew me to the fight over the 2003 media ownership rules was the lack of diverse voices in today's media. Specifically, the conglomeration of information outlets has led to a precipitous decline of minority ownership and representation in media over the last 20 years. For example, media consolidation has already adversely impacted the radio industry. Since the passage of the 1996 Telecommunications Act, the number of radio station owners has dropped 34%, whereas in almost every market, five companies control more than 70% of the market share. The picture is especially bleak for minority business owners. In 2000, 175 minority broadcasters owned 426 stations, or about 4% of the nation's 10,577 commercial AM and FM radio stations. At the end of the last decade, only 23 or (1.8%) of 1,288 full-power commercial television stations were owned by minorities (The Minority Telecommunications Development Program, 2000).

Several weeks before the FCC issued its 2003 media ownership rules, I drafted a letter, co-signed by 28 members of the Congressional Black Caucus, to FCC Chairman Michael Powell voicing our opposition to the rule-making process and requesting a delay of the vote until more hearings were held. Subsequently, close to 150 members of the Democratic Caucus joined in and sent a letter also asking the FCC to delay its decision. In addition, my office received more than 1,100 individual letters and calls from constituents opposing the FCC's decision. However, all of these requests were ignored. Nevertheless, the overwhelming public outrage generated expansive legislative, legal, and popular challenges to the rule, culminating in the Prometheus Radio Project v. Federal Communications Commission case, and the Third Circuit and Supreme Court's decisions.

Although the court's decision was a great triumph for localism, competition, and diversity in America's media, what is less known is that included in the package of media ownership rules proposed by the FCC, one rule was already modified by Congress before the courts' involvement and is now firmly in place. In fact, Sec. 629 of the FY2004 Consolidated Appropriations Act instructs the FCC to modify one of the rules—the National Television Ownership rule—by setting a cap in which no single company can own stations reaching more than 39% of our nation's home, raising the cap from 35%. But how did this 39% cap come about?

THE 39% MEDIA OWNERSHIP CAP

In all honesty, I suspect very few members and staff in Congress really knew why 39% was the chosen limitation level. When the House passed H.R. 2799, its version of the Commerce, Justice, and State Appropriations bill in July 2003, it included a provision, adopted by the Appropriations Committee on a bipartisan vote of 40 to 25, to suspend changes to the national TV ownership cap, effectively restoring the pre-existing cap of 35% established since 1996. However, when the House and Senate could not agree on several appropriations bills and all of them were rolled into the omnibus appropriations bill, H.R. 2673, the language of the legislation was secretly changed through back-room deals and negotiations. Instead of the 35% that the Appropriations Committee adopted and the full House approved, a "39% compromise" on the ownership cap was reached within the conference committee, thereby ignoring the will of the public and the vast majority of members of Congress.

So, why 39% again? Well, many of us not part of the negotiation suspected that it had something to do with the fact that Viacom and News Corp were both already over the 35% limit, unlawfully standing at 38.8% and 37.8%, respectively. Raising the cap to 39% legalized the status quo, accommodating News Corp and Viacom's violation of the law.

Nevertheless, the 39% limitation on national TV ownership was the only portion of the FCC's June 2 decision overturned by an act of Congress. It was also the least critical element of the FCC's media ownership rule changes, since it did not concern any of the rules dealing with cross-media ownership. So what happened? After all, 3 million people had contacted the FCC voicing their opposition to the rules, and my office alone received thousands of letters in opposition to the rule. I can imagine that other Congressional offices, regardless of their political orientation, received a similar outcry from their constituents. So, why didn't Congress do anything?

MEDIA OWNERSHIP AND THE LEGISLATIVE PROCESS

Immediately before the June 2 FCC vote, I joined Congressman Bernie Sanders and Congressman Maurice Hinchey in introducing H. Res. 218, a resolution expressing the sense of Congress that the FCC should have delayed its adoption and implementation of the new media-ownership rules. The resolution received no Committee and floor consideration despite having 135 co-sponsors. In the weeks following the vote, we introduced a Resolution of Disapproval to overturn the FCC's new rules. A Resolution of Disapproval is a device used by Congress to overturn any rules promulgated by a federal agency. It has to be adopted by both chambers of Congress. Once again, that resolution received no hearings and no votes in the Committee, despite having had more than 100 co-sponsors. Even when it was later adopted in the Senate by an overwhelming bi-partisan margin of 55 to 40, the House speaker refused to bring the Resolution up for a floor vote. At that point, we decided to circulate a letter asking Speaker Hastert to bring up the resolution. That letter attracted 205 bipartisan signatories, but short of the 218 required for a discharge petition that forces a bill out of Committee onto the floor for debate and an eventual vote.

When that effort failed, we decided to offer an amendment, known as the Hinchey-Price Amendment, to the Commerce, Justice, and State (CJS) Appropriations bill that would prohibit any federal dollars from going into the enforcement of the rules by the FCC for 1 year. This is the point in which we, a coalition of House media reformers, ran into a surprising opposition—the Democratic leadership. Indeed, Democrats were urged not to support our amendment because the president had indicated that he would veto the entire CJS appropriations legislation if the prohibition of the media ownership rules became part of the final bill. To me, such a veto threat was extremely suspicious. After all, the president has never exercised his veto authority, and the CJS bill eventually passed the House by a veto-proof margin. Nevertheless, because of the opposition from the Democratic leadership, the Hinchey-Price amendment was defeated with 60 Democrats voting against the bill and 34 Republicans voting for it.

The House members concerned about the issue of media ownership and consolidation exhausted all of their options in trying to bring a vote to the floor. We were blocked at every turn, not just by the Republican leadership, but also by the Democrats. Simply put, the voice of the public cannot compete against the voice of the telecommunication lobbyists. Indeed, a report by the Center for Public Integrity last year documented that communications companies spent more than $169 million in 2003 alone on lobbying, the year in which the FCC attempted to further deregulate media ownership rules (see Dunbar, this volume).

INDECENCY AND MEDIA OWNERSHIP

Jumping ahead to 2004, Congress was confronted once again with public outcry over the state of American media, this time regarding the issue of broadcast indecency caused by Janet Jackson's so-called wardrobe malfunction. The Republican leadership's solution to the issue was to increase indecency fines on both the broadcast licensees and the performers. So the House voted, by a margin of 391 to 22, to increase penalties to $500,000 for each indecency violation. Although the Senate never took up the bill last Congress, the House voted again this year to increase the fines, this time with 38 members voting against H.R. 310, the Broadcast Decency Enforcement Act, and 389 members voting for it. Once again, the Senate had not acted on similar legislation.

I was one of the few members who voted for the bill in 2004 but voted against the bill this year. Indeed, during that period of time in between the two votes, I became acutely aware of the level of self-censorship created by the political and policy uncertainties surrounding "indecent programming" and the potential for huge fines. I was particularly disturbed by the decision of 66 ABC-affiliates to reject the movie Saving Private Ryan on Veteran's Day last year out of fear of being fined by the FCC because of some crude language.

Despite voting against the bill, several of us once again attempted to attach media ownership related amendments to the indecency legislation, because it was the only telecommunications-related legislation moving through the House. My argument was simple: A consolidated media market controlled by profit-driven conglomerates is bound to produce indecent, shock-value programming for the sake of viewership. When big media grows bigger, and when the race for audiences turns to the lowest denominator in trash programming to appeal to the broadest possible audience; those conglomerates move further away from quality programming and the principles of diversity, localism, and competition crucial for serving the public interest.

During debate on the legislation, I offered an amendment with Representative Hinchey to the Rules Committee that would effectively put in statute existing ownership rules such as restrictions on newspaper and broadcast television cross-ownership and limitations on the number of broadcast television ownership in a single market. I also offered an amendment with Congresswoman Louise Slaughter of New York to restore the fairness doctrine, which was an FCC policy established in 1949 to impose a set of obligations on broadcasters to afford reasonable opportunity for discussion of contrasting points of view on controversial issues of public importance. The doctrine was abandoned in the 1980s by the FCC, and a Congressional effort to restore it was vetoed by President Ronald Reagan in 1987. Unfortunately, both of the amendments that we offered were rejected by the Rules Committee members along a party-line vote. Table 7.1 summarizes the various legislation related to this issue of media ownership caps.

Table 7.1. *Comparison of Media Cross-Ownership Law and Legislative Actions*

RULE	OLD	NEW	H.R. 2052 (BURR/DINGELL)/ OBEY AMENDMENT ADOPTED BY HOUSE CJS APPROPRIATIONS BILL	H.R. 2462 (SANDERS/HINCHEY/ BROWN/WATSON ET AL.)	HINCHEY-PRICE AMENDMENT REJECTED BY HOUSE 7/22/03
National TV ownership cap	No company may own TV stations reaching more than 35% of the national audience	Cap relaxed so a broad-cast network can own a group of TV stations reaching 45% of the national audience	Maintains the current 35% national audience cap and asks broadcast stations in excess of the limitation to be brought into compliance within 1 year	Invalidates all rules adopted on June 2 by FCC and reinstates previous rules, prohibits the use of biennial review	Not applicable
Local TV ownership limit	One company may own up to two TV stations in the same market if only one ranks in market's top four and if there are eight remaining separately owned stations after the merger	One company can now own two TV stations in same market if only one ranks in market's top four and the market has at least 5 stations. One company can own three TV stations in the same market if only one ranks in the market's top four and at least 18 stations are in the market	No provision	Invalidates all rules adopted on June 2 by FCC and reinstates previous rules, prohibits the use of biennial review	Prevents use of funding to the FCC through CJS appropriation to carry out the Local TV Ownership Limit Rule

Local broadcast-newspaper cross-ownership limits and Local TV-radio cross-ownership limits	One company may not own a - daily newspaper and TV or radio station in the same market Depending on the number of "voices" in the market, including TV, radio, newspaper, and cable outlets, one company may own a TV station plus one, four, or seven radio stations	Ban on cross-ownership is lifted in markets with nine or more TV stations In markets with four to eight TV stations, a company can own one of the three following combos: 1. A newspaper, a TV station, and up to half of the maximum number of radio stations allowed in a market 2. A newspaper and the maximum number of radio stations allowed in a market 3. Two TV stations and the maximum number of radio stations allowed in a market No cross-ownership allowed for markets with three or fewer TV station[s], but a company can seek a waiver if it can show that the TV station does not serve the area covered by the radio station or the newspaper	No provision	Invalidates all rules adopted on June 2 by FCC and reinstates previous rules, prohibits the use of biennial review	Prevents use of funding to the FCC through C-J-S appropriation to carry out the cross-ownership rules

Table 7.1. *Comparison of Media Cross-Ownership Law and Legislative Actions* (continued)

RULE	OLD	NEW	H.R. 2052 (BURR/DINGELL)/ OBEY AMENDMENT ADOPTED BY HOUSE CJS APPROPRIATIONS BILL	H.R. 2462 (SANDERS/HINCHEY/ BROWN/WATSON ET AL.)	HINCHEY-PRICE AMENDMENT REJECTED BY HOUSE 7/22/03
Local radio ownership limit	One company may own 8 stations in markets with 45 total; 7 stations in markets with 30–44 total; 6 stations with 15–29; 5 stations in markets with 14 or fewer	Maintain the caps. However, FCC redrew the maps defining radio markets through an adjusted Arbitron system to prevent the sort of intense concentration that followed the lifting of national radio ownership limits in 1996	No provision	Invalidates all rules on June 2 by FCC adopted and reinstates previous rules, prohibits the use of biennial review	
Dual-Network Ban	Mergers prohibited among top four networks —ABC, CBS, Fox, and NBC	No change	No provision		No provision
FCC Review Process	Biennial review as man-mandated by Congress: Sec. 202 (h) of the Tele-Communications Act of 1996 mandates FCC to reveal and modify regulation deemed no longer necessary as a result of competition every 2 years			Prohibits the use of biennial review	No provision

LESSONS LEARNED

So what are the lessons from all of these battles? First, it is clear that the House leadership will shut down debate on media cross-ownership issues at about any cost. They do not want a vote, they do not want any hearing, and they certainly do not want any debate to draw attention to the topic for the simple reason that they are on the losing side of the argument. Second, the public can no longer rely on write-in or call-in campaigns to push the issue, because one or two members of Congress can control the entire agenda of both the House and the Senate. Third, we need to identify and form strong alliances with entities across the political spectrum. One could argue that the 2003 FCC vote drew so much attention partly because of the vociferous opposition from the conservative Parents Television Council and the National Rifle Association, who opposed the rules out of fear that issues of importance to them will be ignored by the "liberal" media. Another good example is the future prospect of passing the indecency bill this Congress. The current debate on indecency regulation has dragged the National Cable and Telecommunication Association (NCTA) into the picture because of discussions on extending indecency fines to cable and satellite programming. Because of NCTA's strong presence on the Hill, any broadcast indecency legislation that includes such a provision would likely be defeated. So for media reformers who do not want to see increased indecency fines on television programming, paradoxically making alliance with powerful lobbying forces and conservative voices may be the best strategy in this climate.

What are our next steps? I believe several critical regulatory and legislative actions on telecommunication issues will redefine our nation's media policy for generations to come. Several pieces are already moving through the FCC and Capitol Hill. Anyone concerned about our current state of media cross-ownership should follow these developments closely. First, the indecency legislation is awaiting Senate action, and the discussions about extending indecency regulation to cable and satellite television continue. Because of the underlying goals of providing family-friendly programming, several members are also discussing an "a-la-carte" tiering system for cable programming that would allow parents to pick and choose packages of TV shows that they think are suitable for children (see Dunbar, this volume).

Second, Congress is looking to rewrite the 1996 Telecommunications Act because of the rapid advancement in telecommunication technology. Additionally, Congress is looking for ways to set deadlines for the time when broadcasters will have to return their analog spectrum and move over to a digital television airwave. Since the mid-1980s, Congress and the FCC have played critical roles in the development of digital television, which provides sharper pictures, a wider screen, CD-quality sound, better color rendition. In 1997, the FCC decided to grant existing, full-power television stations additional spectrum to facilitate the transition from analog to digital television. The same year, Congress set a deadline of December 31, 2006, for the completion of this transition. With digital television, the number of channels could potentially be expanded by the hundreds. As Congress considers setting a new deadline to reclaim the analog spectrum and force the digital television transition, the increased spectrum capacity will create an opportunity for far more public interest programming from the broadcasters, such as civil, educational, and children's programming. Additionally, the potential increase in the number of digital television channels transmitted by broadcasters will also raise the question of whether cable companies will be mandated to carry all of broadcasters' additional channels. The debate over broadcasters' claim for "digital must-carry" will also be a point of contention on the media ownership debate.

Finally, the Supreme Court's recent decision on NCTA v. Brand X could greatly transform the face of the Internet as we know it. The 6–3 ruling reaffirmed the FCC's earlier contested decision exempting cable companies from common carriage laws that would have regulated cable companies' broadband services as regular phone services. As a result, cable companies may now close their wires to competitive services and even be allowed to control access to Internet content. If that becomes the case, I fear the open marketplace of ideas we take for granted in our understanding and use of the Internet will soon fragment into multiple walled gardens guarded by big cable companies. As the FCC begins to consider extending the same monopolistic ownership and control to broadband services provided by incumbent telephone companies; Congress must begin considering network neutrality rules that keep the wires open, encourage free-flowing information, and expand access to all Americans. The battle for media ownership will certainly continue from traditional forms of media, such as newspapers and broadcast stations, to new forms of media such as the Internet and Internet protocol-enabled technologies.

CONCLUSION

Learning from these past lessons, the public interest and consumer rights communities, as well as members of Congress, are engaged and ready to fight for the future of an open, equitable, and transparent media system. In March 2005, the National Conference on Media Reform organized by the Free Press drew more than 100 media activist groups and 2,500 participants from across the nation. In May, more than 110 groups gathered at the National Press Club in Washington DC to announce the formation of the Media and Democracy Coalition. At the same time, a new Future of American Media Caucus was formed in Congress by Congressman Hinchey to ensure that media reform advocates have a place at the table during the upcoming telecommunication debate on Capitol Hill. I am also a founding member of that Caucus.

One thing is clear: any upcoming debate on telecommunication and media policies in this country will be closely scrutinized by millions of Americans who are tired of our current media policies. It is time for us to ensure there are no more "39%" secret negotiations.

REFERENCES

The Minority Telecommunications Development Program. (2000). *Changes, challenges, and charting new courses: Minority commercial broadcast ownership in the United States* (pp. 34-37). Washington, DC: National Telecommunications and Information Administration, U.S. Department of Commerce.

III

OWNERSHIP AND INFLUENCE ISSUES

8

ARE THE AMERICAN MEDIA BECOMING MORE CONCENTRATED?

Eli M. Noam

A LOST GOLDEN AGE?

Media concentration is not new. It has been part and parcel of historic discussion of media. The first media merger in North America took place in 1741, when the *Boston Gazette* consolidated with the *New England Weekly* (Michael & Emery, 1984). In Philadelphia, Benjamin Franklin became the earliest American media mogul. As Walter Isaacson—journalist, biographer and later head of CNN—wrote, "Franklin's print shop had by [1748] grown into a successful, vertically integrated media conglomerate. He had a printing press, publishing house, newspaper, an almanac series, and partial control of the postal system. He also had built a network of profitable partnerships and franchises from Newport and New York to Charleston and Antigua. Money flowed in" (Isaacson, 2003, p. 126). Given Franklin's pioneering role in electricity, he undoubtedly would have moved into electronic media, too, had he lived to see them.

In the 19th century, American media were much more concentrated than nostalgia suggests. In 1897, the nation's leading periodical, *Munsey's Magazine*, had a circulation of 700,000, almost 40% of U.S. magazine circulation. Frank Munsey, the proprietor, was eulogized by a Kansas newspaper publisher: "Frank Munsey contributed to the journalism of his day the great talent of a meat packer, the morals of a money-changer, and the manners of an undertaker. He and his kind have about succeeded in transforming a once noble profession into an 8 percent security. May he rest in trust" (White, quoted in Herrick, 2003, p. 26). Press historian Gerald Baldasty wrote that "By the late 19th century, the formula of the newspaper-as-business-first-and-foremost was firmly established" (Underwood, 1993, pp. 42-43).

Even theater, a medium one would expect to be heavily local, became concentrated in the 19th century due to changing transportation technology (i.e., railroads). A system of traveling companies emerged after 1870, replacing the traditional system of local repertory companies and entrepreneurs. Some 250 companies toured the country, but by the 1890s two firms dominated this traveling theater business and the booking, owning and leasing theaters in important cities (Poggi, 1966).

The telegraph, first among electronic transmission media, quickly became monopolized by the Western Union Co. Railroad financier Jay Gould bought it in 1881, giving him a telegraph monopoly and control over the *Associated Press. The New York Times* opined anxiously that Gould would control "what has been the free press of America" (Klein, 1986, p. 394).

In wireline telephony, one firm, AT&T, became a national near-monopolist, and soon the world's largest private company. In radio-telegraphy, similarly, the Marconi Company held a patent monopoly for maritime communications. Marconi for a time refused to allow the operators to communicate with those stations that were using other firm's equipment (Headrick, 1991).

As the 20th century dawned, new technologies established new media and distribution. The film industry witnessed almost immediately an attempt to organize a wall-to-wall monopoly. The Motion Picture Patent Company, organized by J.P. Morgan, brought together the interests of Edison, Pathé, Eastman Kodak, Vitograph, and Biograph. The cartel, known as the "Edison Trust," pooled its patents, transacted only with distributors that paid high royalties, and dealt with theaters that agreed to charge high prices and follow censorship rules. But it was unable to suppress independent producers. These firms, which soon became "Hollywood," still dominate the film medium as an oligopoly.

When broadcast radio emerged in the 1920s, its technology, content, and distribution was considerably dominated by a single company, RCA. Later, when television emerged, it replicated the radio's network system, and three companies, all located near each other in Manhattan, dominated an enormously influential medium.

Also in the late 1940s, a new type of information technology began to emerge—computers—that would soon enable new forms of communication. This industry quickly became dominated by a single firm, IBM, leading to several federal antitrust attempts to break up the firm. In the 1980s, IBM introduced its personal microcomputer, but with its microprocessor and operating software supplied by Intel and Microsoft, and thereby launched these two firms on the road to their own dominance.

The purpose of this brief sketch of American media and communications history is to provide a context to today's controversies. Media concentration is not a new phenomenon. On the contrary, it has a long and contentious history, commented on and fought over at each stage in the country's past.

TODAY'S DEBATE

Of course, just because a problem is old does not make it less important. The opposite may be true. But the historic perspective may moderate some of the rhetoric of falling sky and lost golden age that pervades today's debates.

Recent years have witnessed the expansion of large media firms in the United States. This development led to fear that American communications media are increasingly controlled by an ever-shrinking number of firms, and that those firms are capable of affecting public opinion, democracy itself, the national agenda, and global culture.

Concentration of the means of information has become an issue around the world, the information-age version of the industrial-age struggle over control of the means of production. That issue had been a central element of Marxist theory and politics. It led in some countries to revolution; in others, to the nationalization of key industries, such as the United Kingdom's heavy industry sector after World War II. The notion was that ownership of coal and steel—the "commanding heights" of business—could be leveraged into control of the economy as a whole, and with it, of the political system. With hindsight, such fears of dominance of heavy industry companies—and the policies they engendered—seem overblown. This is not to suggest that today's fears over ownership of the means of information are baseless. But it points to the need for a careful analysis of facts and dynamics.

What then are these facts about American media concentration?

To many observers, with the size of major media firms increasing and numerous mergers taking place, the obvious conclusion is one of major concentration. But to other observers, with media channels increasing enormously as exemplified by the Internet, the obvious conclusion is quite the opposite.

This leads us to recent debates. In 2003, Michael Powell, chairman of the Federal Communications Commission (FCC), narrowly pushed through his commission a set of rules that relax broadcast television ownership restrictions. The result was a political firestorm in which the political right and left joined forces against free market advocates and major media companies. It led to a lopsided congressional vote of 400 to 21 opposing parts of the FCC's new rules. The White

House, after some brave talk of a presidential veto, retreated. At the same time, federal courts overturned the FCC's rules (*Prometheus Radio Project vs FCC,* 2004).

The forces of opposition, while desperate, were united by their dislike of media companies. The Economist editorialized: "Behind almost every argument about why the FCC endangers democracy lurks a grudge about content: it is too conservative; it is too liberal; it is too violent; it under-represents feminists, or the Catholic Church. Merely cataloging these conflicting grievances shows the impossibility of ever resolving them" ("Media madness," 2003).

One would think the differences among these perspectives to be irreconcilable. But their proponents also have common needs: as political participants and activists they are major users of media, and media access and diversity is no abstraction to them.

Part of the vehemence of the debate over media concentration lies in the self-image of its advocates on both sides. Opponents of the FCC's new rules view themselves as engaged in a digital Thermopylae, a last line of defense against homogenized news controlled by five giant profit-driven media conglomerates. They fear a situation like that of Italy, where Silvio Berlusconi used his media empire to achieve power and office, or Russia, where Putin acquired control over the media to sustain his power and office. In contrast, defenders of the FCC see themselves as removing the shackles of the state. They argue that we are in the midst of a historic blossoming of information technology. Both sides are deeply convinced that "the media" are tilted against them. Both sides project themselves as defenders of free speech, either protecting media from the heavy hand of government, or, alternatively, protecting diversity from being choked off by communications empires.

Who is right? Is media becoming controlled by the few or open to the many? It is necessary to step back, look at the facts carefully and without prejudgment, and ask questions such as the following: Are American media more concentrated today, nationally and locally, than they were in the past? Where are the bottlenecks? What are the causes and drivers?

Media concentration has become an ink blot into which people project their hopes and fears about the role of media. The media system is held partly responsible—either as a contributing cause or as an absent remedy—to a host of problems including violence, gender and ethnic stereotyping, materialism, hedonism, escapism, low political participation, lack of political debate, inequality, obesity, bulimia, drug dependency, and so on. In consequence, many people tend to expect too much from a de-concentration of media. Because so many of society's ills are attributed in part to media, it follows that media reform, especially de-concentration, is seen as social reform. Yet these expectations are not realistic. It is valid to argue that unconcentrated media provide greater source diversity, which is important in a democracy. The argument could stop here and be persuasive. But many people earnestly believe much more, that other ills of media, and therefore of society generally, would be remedied by media de-concentration. Inevitably, they will be disappointed. De-concentration should not be confused with higher quality, greater content diversity, or greater independence.

AN EMPIRICAL INVESTIGATION OF MEDIA CONCENTRATION

To get a clearer picture we will now engage in an empirical investigation of media concentration trends. It is part of my much larger study (Noam, 2007).

First, we must operationalize concentration. A number of indices for concentration exist, such as the Gini-coefficient (a measure of variation from equal distribution) and the C4 index (the percent of a market controlled by the top four companies). The Antitrust Division of the U.S. Department of Justice (DOJ) has been using a different measure, the "Herfindahl-Hirschman Index" (HHI; often known as "Herfindahl"). This widely used HHI concentration measure is equal to the sum of the squares of the market shares of all market participants:

$$\text{HHIj} = \sum_{i=1}^{f} S_{ij}^{2}$$

where f = number of firms participating in an industry,

S_i = each firm's i market share in the industry j

The HHI can range between zero—where the share of each firm is infinitely small—to 10,000, when a single firm accounts for 100% of the market. The DOJ's antitrust enforcement guidelines classify market concentration levels as follows:

HHI <1,000	Unconcentrated market
1,000 <HHI <1,800	Moderately concentrated market
1,800 <HHI	Highly concentrated market

When we use the measure of the DOJ ranges for "unconcentrated," "highly concentrated," and "moderately concentrated," it should be read as a shorthand, and not as a commentary on the appropriateness of these particular thresholds.

An HHI below 1,000 would have at least 10 participants, and an HHI of 1,800 would have at least 6 participants. A typical highly concentrated industry with a market leader with 60%, a runner-up with 30%, and two niche players with 5% each, would have an HHI of 4,550.

The concentration trends of individual industries are important by themselves. However, each industry has its own specific developments. Thus, the concentration in newspapers and academic journals might rise, while that of magazines declines and that of books remains stable. To get the big picture then, it is useful to look at trends for larger media categories. In this case we use "Print Publishing" as the broader category. Accordingly, we combine the individual industries along the dimensions of broader sectoral categories. Such measures average the various related industries, weighing each industry based on size, in terms of revenues. The weighted average HHI is defined as:

$$\text{W.A.H.H.I.} = \sum_{j=1}^{n} \frac{m_j}{\sum m_j} \sum_{i=1}^{f} S_{ij}^2$$

where j = an industry

m_j = total revenue of an industry

S_i = each firm's market share of an industry

n = number of industries in a subset of the information sector

f = number of firms in an industry.

The second step of the analysis is the actual data collection. We must determine, for various media and information industries, the market shares of its major firms over recent years of the US market. We cover the past two decades 1984–2004. We also cover 100 industries from the mass media, telecom, Internet, and Information Technology (IT) sectors. We go beyond mass media and also observe the parallel trends in telecommunications, the Internet, and IT. This broader approach does not eliminate the narrower one. We still obtain results specific to mass media in general and to specific media industries such as radio stations.

Covering 100 information industries for a 20-year period has been an ambitious undertaking, far broader than anything done previously for the United States or any other country.

THE FINDINGS

We report first the findings for the entire information sector, comprised of all 100 sub-industries. Figure 8.1 provides the weighted average trends of concentration for this sector.

FIGURE. 8.1. Concentration (HHI) trends for the total information sector, 1984-2004.

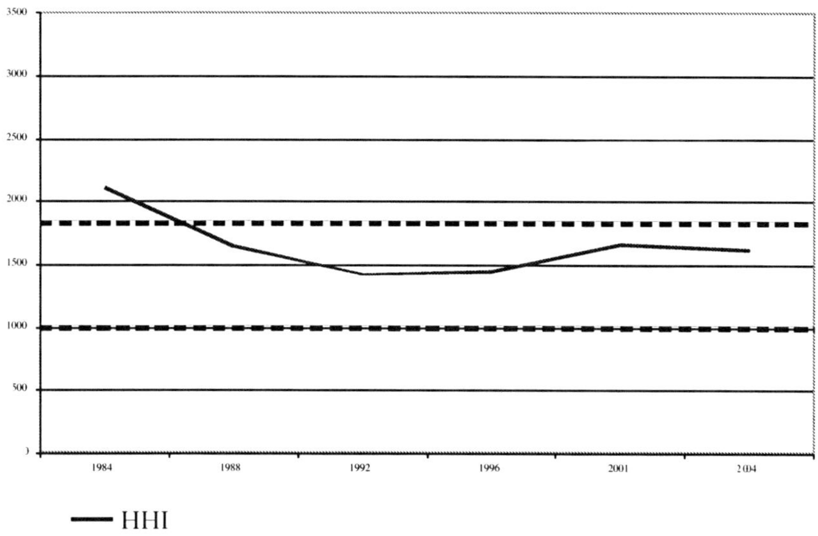

What we find is that, in a way, both sides in the debate over concentration are right when it comes to the entire information sector. Whether one finds more or less concentration today depends on the choice of the base year: Is it 1984, or 1996? Compared with 1984, the information sector is less concentrated today. But compared with 1996, it is more concentrated. And compared with 2001, it is less concentrated again.

Moving beyond the aggregation for the entire information sector, let us look at the trends for four major subsectors: mass media, telecom, Internet (cybermedia), and IT. As Fig. 8.2 displays, the findings show the following:

- The average concentration in the mass media subsector has grown steadily after 1988.
- However, this concentration is the lowest of the four subsectors.
- The highest concentration exist in the telecom and the IT subsectors; both are above the DOJ's threshold levels of "highly concentrated" industries.
- If retailing industries such as video rental stores are excluded from the mass media measurements, concentration trends are a bit lower and rising less sharply, but the difference is small.
- This comparison should not be interpreted to imply that mass media concentration is low, because a case can be made that concentration thresholds should be lower for mass media than for other industries given its role in democracy and the importance of access to media.

FIGURE. 8.2. Concentration (HHI) trends for the four subsectors of total information industry, 1984-2004

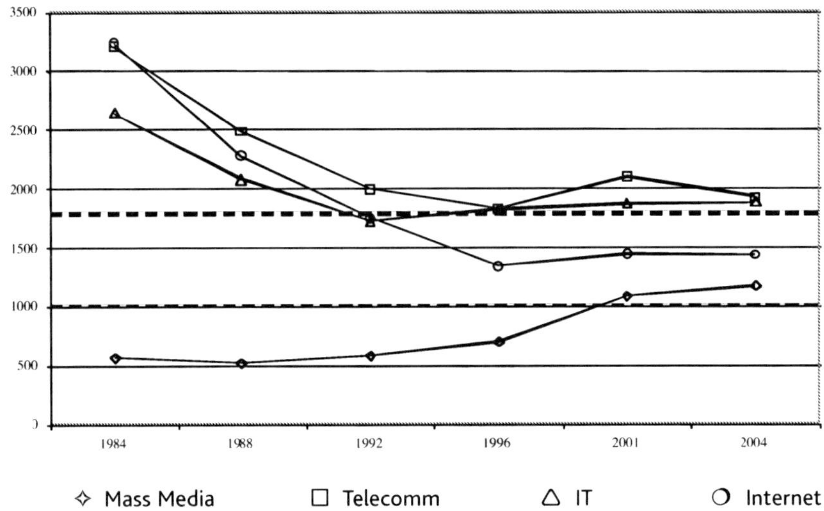

Next, we look at the concentration trends among those mass media that provide news, excluding entertainment industries such as music or film, or non-news industries, for example, educational book publishing.

Figure 8.3 shows declining concentrations for TV networks, increasing concentration for radio stations, and fairly flat or only mildly rising trends for almost all other media—newspapers, trade and paperback books, radio networks, cable TV channels, magazines, and TV stations. The weighted average for these eight media is the bolded line, which rose slightly from 1988 to 2001, and was flat thereafter, at the low level of HHI = 500.

The weighting is done on the basis of revenues. Other weighting could be done by some measures of "attention" or "news content." This would be methodologically more subjective, however. And in terms of results, it would almost certainly show a lower average concentration, since the high-concentration media tend to be those with relatively low-news content but high revenues, on average, such as radio and TV networks, and cable channels. Media revenues are more concentrated than media news content.

It should be noted that these concentrations are based on national markets. This is a reasonable analysis for books, magazines, and TV networks, but less so for newspapers, which are mostly local in nature. A San Francisco newspaper does not compete with a Houston one. We therefore also analyze local media concentration. We investigated 30 local markets of different sizes. Figure 8.4 summarizes the findings.

FIGURE. 8.3. Concentration (HHI) trends of news media, 1984-2004

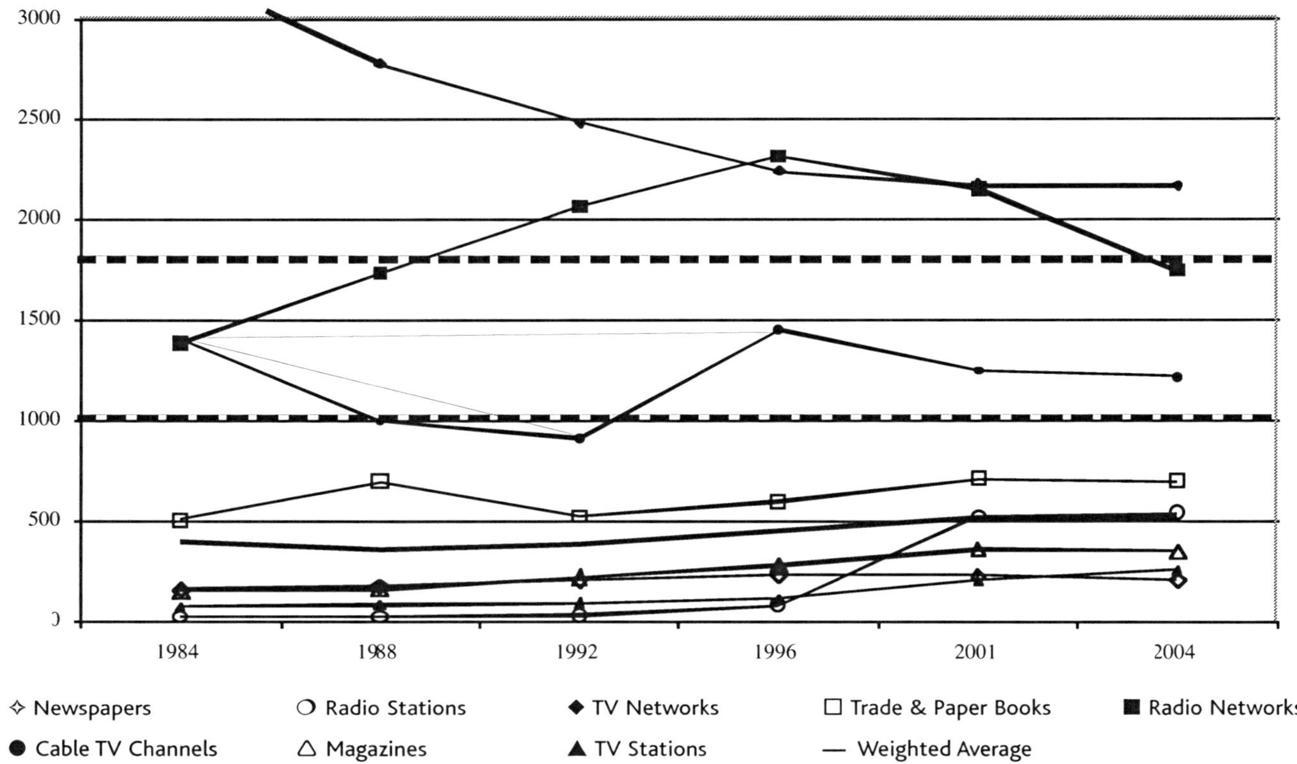

FIGURE. 8.4. Concentration (HHI) trends trends for local mass media, 1984-2004.

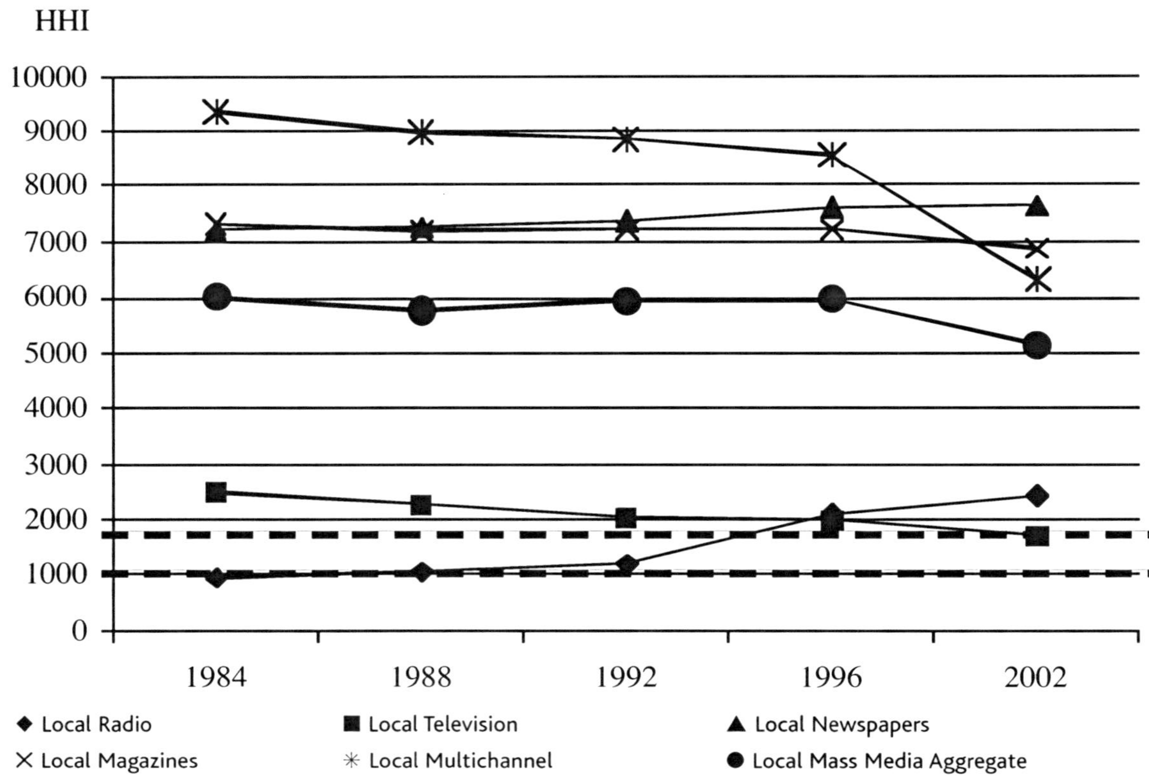

- Local media concentration are very high, particularly for local newspapers, magazines, and multichannel (cable/DBS; for cable alone, concentration would be higher).
- Local radio concentration has risen considerably, although it is much lower than that of newspapers.
- Local TV stations concentration has declined (even as national TV station concentration has risen.)
- Overall local media concentration has declined somewhat, and would have been flat if we exclude DBS as a competitor to cable distribution.

VERTICAL CONCENTRATION

A purely horizontal concentration measure cannot take account of the growth of firms that are active across a host of industries, as is the case for firms such as Time Warner, Viacom, Sony, and Microsoft.

Figure 8.5 shows a steady and clear upward trend for the top 5 firms after 1988. After 1992, with liberalization policy in place, concentration rose pro-

FIGURE. 8.5. **Share of top 5, 10, 25 mass media companies of overall mass media sector.**

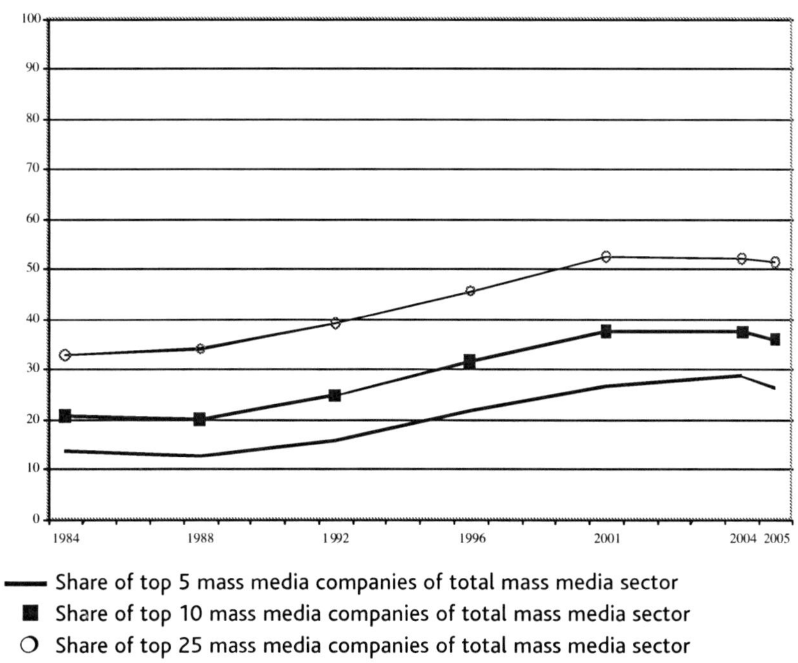

nouncedly. In 1984, the top 5 firms accounted for 13%. By 2004 their share had more than doubled to 29%. For the top 10 mass media firms, the combined share rose from 20% to 38%. And for the top 25 firms, it rose from 32% to 51%. The increase was largest for the top 5 firms. The share of firms 6–10, and 11–25, however, declined slightly. The real growth was in the top 5 firms, which gained, on average, 3.2 % over the period. The next 20 firms gained, on average, only 0.15 %. Thus, the top 5 clearly were the gainers. Even so, their aggregate share was 29 % of the overall mass media sector, much lower than in the telecom and IT sectors where the top 5 companies, in 2004, held an aggregate share of 55.7% and 36.9%, respectively. The top 5 firms' share, on average, was 5.8% of overall mass media. In most industries this would be a low number, even if it has doubled since 1988.

CONCLUSION

Those media free-marketeers who deny the trend of top-5 concentration will need to re-examine their claims. One can argue that mass media are still fairly unconcentrated by DOJ measures but one cannot argue that a substantial rise has not occurred.

Media reformers, on their part, need to deal with the fact that overall concentration, although clearly rising, is much lower than some of the alarmist rhetoric suggested. Criticism has been shrill even when media concentration was quite low. Ben Bagdikian (1983) wrote about *The Media Monopoly*, not as a prediction but as an observation, and in a period when media concentration was actually declining.

The important questions are: is the rise in mass media concentration merely part of a cycle of concentration and deconcentration, a general ebb and flow of empire building and their eventual demise, and hence, will the problem, by and large, take care of itself? Or, to the contrary, are there more fundamental forces at work that lead information and media industries towards further concentration?

In favor of the former view are the failures of powerful firms, rarely mentioned by media critics, leading them to be absorbed or much reduced in power—such as the rise and fall of RCA; Vivendi; Maxwell; Western Union; Lucent; AT&T; DEC; AT&T; CBS; ABC; Reader's Digest; Gulf & Western; MCI Worldcom; or the spin-offs by companies such as Cablevision, Ziff-Davis, Thompson; AT&T; Viacom; Time Warner; and Clear Channel. Companies and industries go through life cycles, and to focus on their vigorous growth phases alone gives a distorted view.

Many of the merger deals of the 1990s were fuelled by the stock market, which both rewarded these deals and financed them through appreciated shares. With stock prices back to earth, deals became scarcer, thereby slowing additional acquisitions, and requiring some sell-offs. Of course, for the same reasons some companies failed, others became cheaper to buy, and their departures raised concentration in their industries.

Thus, it could be argued that what we have witnessed was merely a temporary upward trend in concentration, that corrective forces are already at work and that they are leading to a phase of deconcentration, as part of an eternal ebb and flow.

But a second interpretation is also possible. That interpretation is that the concentration trend in mass media is part of their move to a market structure that is more similar to that of the other parts of the information sector; that they are increasingly sharing the same dynamics as the rest of the information sector; that the result will be an ongoing concentration to the still higher levels prevailing in the rest of the information sector; and that the trend in those other sectors is towards greater concentration despite (or perhaps because of) their innovation and entrepreneurialism. Thus, the dynamics of digital convergence lead the mass media industries to assume the market structure characteristics of the rest of the information sector. This would mean that concentration trends will continue, even if in fits and starts. And because concentration, by traditional antitrust measures, is not especially high in many cases, antitrust by itself is not likely to block this trend.

REFERENCES

Bagdikian, B. (1983). *The media monopoly.* Boston, MA: Beacon Press.

Headrick, D. (1991). *The invisible weapon: Telecommunications and international politics, 1851-1945.* New York: Oxford University Press.

Herrick, D. F. (2003). *Media management in the age of giants.* Ames: Iowa State Press.

Isaacson, W. (2003). *Benjamin Franklin: An American life.* New York: Simon & Schuster..

Klein, M. (1986). *The life and legend of Jay Gould.* Baltimore,: John Hopkins University Press.

Media madness. (September 11, 2003). *The Economist.* http://www.economist.com/displayStory.cfm?Story_ID=2052042

Michael, E., & Emery, E. (1984). *The press and America: An interpretative history of the mass media.* Englewood Cliffs, NJ: Prentice Hall.

Noam, E. (2007). *Media ownership and concentration in America.* New York: Oxford University Press.

Poggi, J. (1966). *Theater in America: The impact of economic forces, 1870-1967.* Ithaca, NY Cornell University Press.

Prometheus Radio Project v. FCC (4th Circuit Court of Appeals, June 24, 2004). http://www.ca3.uscourts.gov/opinarch/033388p.pdf

Underwood, D. (1993). *When MBAs rule the newsroom: How the marketers and managers are reshaping today's media.* New York: Columbia University Press.

9

COMPETITION IN TELEVISION AND RADIO

Trends and Policy Implications

Benjamin Compaine

Do individuals have access to more sources of news, information, and entertainment from their homes, cars, or workplaces now than 20 years ago? Is there a crisis impending due to the perception of increased consolidation of the media industry? Are the proposed rules of the Federal Communications Commission (FCC) for lessening broadcast station ownership limits a reasonable response to court rulings and the changing media landscape or a threat to diversity of viewpoints and to democracy in the United States?

This chapter addresses these issues. It seeks to assess whether media consumers in the United States are better able to find the variety of news, entertainment, and information that they want and need, from more sources, with reasonable cost, than a generation or more ago, when presumably the media were less concentrated and, by inference, "better."[1]

The subject of media ownership trends is as timely in 2006 as it was in 1978, when the Federal Trade Commission (FTC) held a two-day public symposium as part of its inquiry into the consequences of the mergers in the print media—growing newspaper chains, book publisher combinations, and magazine acquisitions. (FTC, 1978). Throughout the 1980s and 1990s, various federal agencies were called on to pass judgment on dozens of proposed mergers and acquisitions (Compaine & Gomery, 2000). U.S. Congress (1996) weighed in with specific orders to the FCC to eliminate many ownership restrictions on radio and television broadcasters. And in 2003, the FCC issued new rules on radio and broadcast television ownership limits. Implementation of those rules was stayed by various appellate courts (FCC, 2004). Thus, the subjects remain current, contentious, and dynamic.

An examination of media economics, media metrics, and media competition for traditional television and radio programming compared with the new distribution channels of the Internet (via satellite, cable modem service, digital subscriber lines, fiber optics, etc.) shows that the distinctions between "old" and "new" media are already blurring and are providing increasing opportunities for choices by consumers. The relevant issue then becomes how one reconciles the notion of less competition with the reality of great consumer choice and options.

WHAT IS KNOWN ABOUT MEDIA COMPETITION?

In 1982, it was already clear that the blurring of the boundaries among the traditional media would create unsettled boundaries among previously well-defined technology-defined media products and services:

> Computers and connected terminals in homes and offices increasingly allow users to *select* the information they wish to receive, at precisely the time they wish to use it. Computers have made it economically feasible to mail identical, "personalized" messages to millions of recipients using the postal system that at one time was reserved for point-to-point communication. The telephone can give countless users almost simultaneous access to the same computer data base. The telephone and computer are also being combined to provide "electronic mail," perhaps doing for mail what the Xerox machine did for memos. Video and audio cassette recording devices allow individuals to record broadcast programs for replay at a time of their own choosing. (Compaine, Sterling, Guback, & Noble, 1982, p. 3)

Less than 20 years later, the online "Drudge Report" received a greater readership than the weekly circulation of *Time* magazine (Drudge, 1998). In 2003, the three dominant commercial broadcast television networks—ABC, CBS, and NBC—claimed only two thirds of the prime time ratings they held in their heyday of the 1960s and 1970s (Wolzien & Mackenzie, 2004). Meanwhile, by June 2003, nearly 85% of households got their television signals from a cable or satellite service instead of from terrestrial broadcast signal ("In the matter," 2004). Radio stations, limited by licenses allocated by the FCC, are sharing their audiences with unregulated Internet-only "radio" stations as well as national satellite-distributed radio services. And the Internet itself, originally accessed via pokey conventional telephone lines, is increasingly accessed by much faster connections not only from the telephone companies but by cable and satellite providers as well.

Between the mid-1980s and the end of the following decade the media industry in aggregate grew faster than the U.S. economy, rising from below 10% of the gross domestic product in 1987 to nearly 13% by 1996 (Huang, 1999).

WHAT IS SAID ABOUT MEDIA COMPETITION?

In this context, however, a debate rages whether the media industry has become more concentrated (see, e.g., www.opendemocracy.net). Actually, there is less a debate than an explicit assumption that ownership in the media industry has become more concentrated despite the new technologies and industries that have joined the older print, and broadcast and motion picture sectors.[2] It is regularly contended that "five large corporations" control 80 percent of the prime time television audience,"[3] and thus control how and what information people have access to. Media critic Robert McChesney sums up a widely held perception: "Over the past two decades, as a result of neoliberal deregulation and new communication technologies, the media systems across the world have undergone a startling transformation. There are now fewer and larger companies controlling more and more, and the largest of them are media conglomerates, with vast empires that cover numerous media industries" (Policing the thinkable, 2004).

A similar viewpoint is voiced by Stanford law professor and FCC critic Lawrence Lessig (2003) who decries an era "when fewer and fewer control access to media." Senator Byron L. Dorgan (D-North Dakota) said rules proposed by the FCC would lead to "galloping concentration" in the media industry (Ahrens, 2003). Five conglomerates, we are told, "control most prime time TV programming and one company . . . dominates radio" (Hatch, 2003, p. 845). Those in the "creative community" of packagers, directors, writers for television, etc. complain that "these conglomerates strongly favor their own in-house production when ordering new shows, whether their shows are better or not" (Rintels, 2003, p. 4).

The critics of current trends in media structure have two related arguments. One, typified by Dorgan and Lessig, focuses on consolidation as purportedly leading to a reduction in diversity of content. Bagdikian (1992) seems to have best articulated this concern in writing that "our view of the social-political world is deficient" if there is a regular omission or insufficient inclusion of certain elements of reality. And that is happening, he believes, when "the most important institutions in the production of our view of the real social world," the mass media, are becoming "the property of the most persistent beneficiaries" of the mass media's biases. Another stream of concern, typified by McChesney and Pat Mitchell, chief executive of the Public Broadcasting Service (PBS) among many others, is that it is the quality and/or ideology of the content of the media companies that is most disturbing. Less often cited is the argument presented in antitrust terms; that is, that media mergers in general could be driving up the cost for advertisers and/or consumers.

"Yet, in the paradox of today's media landscape," observes a *Congressional Quarterly* research report, is that "consumers have more choices than ever" (Hatch, 2003, p. 845). The rise of new communications technology, coupled with new producers of information, may have fragmented the overall mass audience once captured by the old mass media. The number of owners of television stations has declined from 540 to 360 since the 1970s, and the number of companies owning daily newspapers has declined from 860 to 300 (Hatch, 2003). Yet, even these reduced numbers would seem not to support the assertion of being "close to a monopoly or oligopoly situation" proclaimed by an FCC commissioner (Hatch, 2003). And what is the disconnect between the concern of Pat Mitchell that "The American public seems less informed than ever" (p. 852) with the reality of the change from 1980 with 30 minutes per night of prime-time news from each of three national networks to 24–7 news shows from three additional national networks in 2004? How then does one reconcile, if at all, the apparent paradox of "few and fewer" resulting in "more choices?"

APPROACH OF THIS STUDY

The debate on media competition constantly keeps returning to television, and to a lesser extent, radio. Newspapers, motions pictures, book and magazine publishing, the multidimensional Internet—all also are part of the media competition landscape. The largest circulation newspaper, *USA Today*, sells 2.3 million copies on a typical weekday, with readership of perhaps 5.6 million (Moon, 2004). The broadcast network with the lowest evening news show rating, CBS, has 6.8 million viewers ("Evening News," 2004). A single episode of a moderately successful prime-time TV series is seen that night by more people than the cumulative sales of a book on the fiction bestseller list ("Evening News," 2004).[4] Paradoxically, most people live in places that have only one or two local daily newspapers, but three or more locally based and separately owned television stations and a dozen or more separate owners of local radio stations. Thus, this study looks particularly at television and radio. Furthermore, it considers any mode of transmission—traditional terrestrial broadcast, satellite, or cable—in its approach to these media. Finally, the study incorporates the Internet to the extent that it too is a relevant conduit for connecting users with video and audio content.

Much of the data presented in this study includes raw secondary data that has been newly aggregated and analyzed in the context of the implications of media ownership structure. The primary question that this chapter seeks to address is as follows: "Do consumers across the United States have greater or fewer choices for sources of video and audio content than two decades previously?" Twenty years prior to 2006 presents a time of three commercial television networks that characterized the national television landscape since the advent of television and just about the time the major cities were wired for cable. The ensuing discussion addresses why this matter of choice—rather than actual viewership or listenership—is the relevant operational question for judging the policy implications of media structure and regulation.

Old Metrics and New

The metrics typically used to address media competition have included traditional antitrust measures such as revenue of the largest four or eight firms in a sector and the Herfindahl-Hirschmann Index (HHI), a somewhat more fine-tuned measure also based on revenue. The percentage of circulation of newspapers, the number of publications owned, the advertising share of radio stations, or the number of journals in a discipline, are other common measures. They have been used not only to address economic power—or lack thereof—but by extension, the range of potential editorial and cultural diversity that many competing owners are presumed to encourage.

The notion of diversity of political opinion lay behind the press freedom clause of the First Amendment. The fear at the time of the Founding Fathers was that only government might have the power to limit that diversity.[5] But today, there is concern in some quarters that the range of opinions to which the public has access is being limited by large media conglomerates.

"Diversity" has been the overall goal of much broadcast regulation. "In setting its licensing policies, the [Federal Communications] Commission has acted on the theory that diversification of mass media ownership serves the public interest by promoting diversity of programming and service viewpoints" (Hamburg & Brotman, 1996; see also *FCC v. National Citizens*, 1978). Preventing undue concentration of economic power has also been served by the FCC's ownership policies. But over the years, the contentious policy issues have been most concerned with "diversity."

Diversity and how it can be objectively measured has never been well established. Competition and localism have been thought to foster diversity. But neither policy has been proven to guarantee diversity of content. Often, diversity has inferred a "higher quality" than the prevailing standard, particularly in television. Take this slightly edited comment by a media critic:

> You will see [on television] a procession of game shows, violence, audience participation shows, formula comedies . . . blood and thunder, mayhem, violence, sadism, murder, western bad men, western good men, private eyes,

> gangsters . . . and cartoons . . . And most of all, boredom. True, you will see a few things you will enjoy. But they will be very, very few. (Minow, 1961)

This critic's list of program genres suggests that television programming was quite varied: game shows, comedies, westerns, audience involvement shows, detective stories, cartoons. This despite the time frame, 1961. The critic, Newton Minow, then chairman of the FCC, however, let on that most of it bored him. The mass audience at the time may have disagreed.

The time period of the diversity reference is important. In 2003, an FCC commissioner held that the FCC's media ownership rules are "about the ability of a very small number of companies to control [the content of] our media. To dictate what the entertainment is going to be, to dictate what the civic dialog is going to be, to dictate that we'll no longer have as much . . . diversity as we had *before*" (Hatch, 2003, p. 847, italics added). The reference point of "before" could refer to the period of Minow's era of quite varied if boring programming, or some other "before" that needs to be identified in order to compare with "after."

Perhaps the FCC's more ambitious regulations to foster diversity and localism ended with at best mixed results in the mid-1990s, having had a 25-year run without achieving their goal. These were the frequently paired policies codified as the Prime Time Access Rule (PTAR) and the Financial Interest and Syndication Rule (fin-syn). Combined, these regulations were intended to lessen the role of the three networks by limiting the hours of programming they could supply individual stations during the prime-time viewing period and prohibiting them from having an ownership stake in most prime-time entertainment programming (Hamburg & Brotman, 1996). The expectation—or hope—was that local stations would fill their non-network time slots with local programming and that independent producers would be providing higher quality programs than the networks had been broadcasting.

These regulations are credited with helping to foster the growth of powerful financially strong production companies and syndicators (King World Productions, Viacom, and Lorimar among them; Walker & Ferguson, 1998). They may have helped along independent television stations in the 1970s by making a steady supply of old network re-runs available. They also aided the major Hollywood movie studios to become prominent players in television production (Walker & Ferguson, 1998). But none of these outcomes substantially improved the quality or diversity of content on television.

By the 1980s, it was already evident to many that such enforced attempts at diversity were ineffective or even counterproductive. Based on a staff study issued in 1980, the FCC in 1983 had signaled its intention to repeal the rules (*Schurz Communications, Inc. v. FCC*, 1992). Congressional pressure squashed such a policy change. Eventually, the U.S. Court of Appeals, in *Schurz Communications, Inc., v. FCC* told the FCC that the fin-syn rules could no longer be justified and they soon were repealed, along with PTAR.

There are empirical studies that make a strong case that simply having more sources of programming does not create diversity in programming. There is strong empirical evidence that it was the waiver, and ultimate elimination, of these rules that permitted vertical integration in production and distribution, which created the economic model that brought about the fourth and subsequent broadcast networks that television's harshest critics had long sought. An academic study by Einstein (2004b) measured diversity as the number of different genres of content (situation comedies, musicals, variety shows, news, documentaries, reality, etc.). Figure 9.1 shows that diversity (higher values) decreased in the years after the initiation of the fin-syn and PTAR rules and increased substantially within several years of the end of the rules. Figure 9.2 further shows that the three networks tended to have the same programming types–less diversity–at any given time slot during the years that the policies were in effect than before or since.

Finally, "Although many media critics have decried the consolidation of the media industry because of its negative effects on diversity, this research just does not support that theory" (Einstein, 2004b). The study did find that the sources of programming for all the current broadcast networks have greatly consolidated since the Fin-Syn rules ended. However, this consolidation has been taking place in the context of the trends seen in Figures 9.1 and 9.2—an increase in diversity from the levels when there were far more players supplying programming. On the other hand, diversity, as seen in Figure 9.1, has not reached the level of the 1960s. The new equilibrium—for traditional broadcast networks–has been less diversity. Einstein's study confirmed what others had found: despite more outlets, diversity of genres has actually declined (De Jong & Bates, 1991; Einstein, 2004a)—and

FIGURE. 9.1. Network program diversity pre/post fin-syn (from Einstein, 2004b).

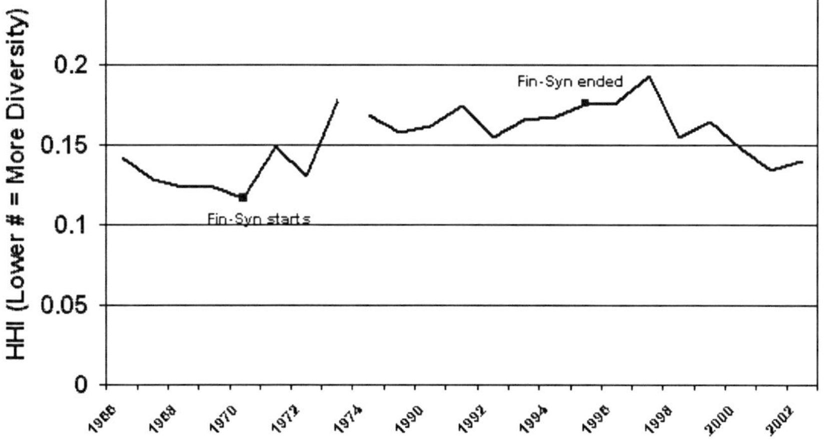

FIGURE. 9.2. Network program diversity across time slots pre/post fin-syn (from Einstein, 2004b).

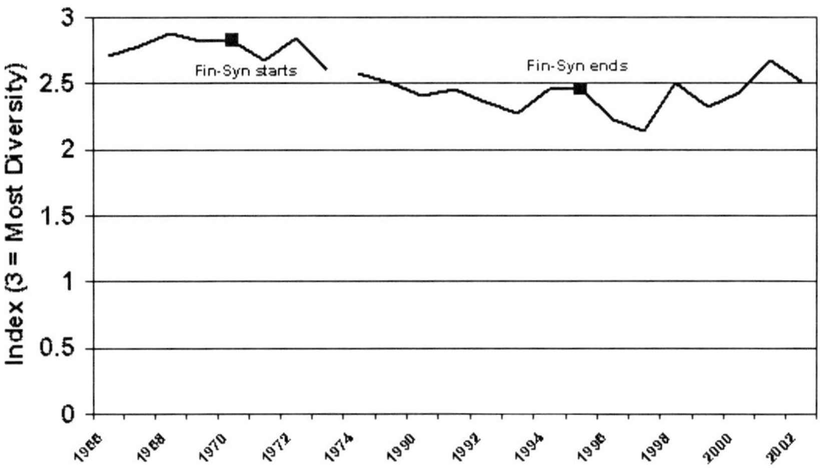

diversity does not increase as the number of sources creating programming either (Napoli, 1999). It is the economic model of selling TV advertising that pushes programmers to genres that attract large audiences that matters.

There also remains an inherent vagueness about measuring diversity. Einstein admitted that measuring diversity as it has often been done—by the variety of genres—does not capture much of what is often imputed by the term. Einstein (2004a) concluded that the predominant measures do little to inform us of whether there is truly a multiplicity of viewpoints (e.g., both *All in the Family* and *Suddenly Susan* are both situation comedies but present different political perspectives; and much of Spanish-language cable programming is similar to English-language cable programming). Indeed, by this measure of diversity, the two leading all-news networks, CNN and Fox News Channel, are both in the news and information genre and would not be seen as creating any diversity, but are widely perceived as having distinct voices.

DOES OWNERSHIP MATTER?

The academic research that has been reported does not support the contention that media ownership by chains or conglomerates leads to any consistent pattern of lowered standards, content, or performance when compared with media owned by families or small companies. For example, a review of 17 studies concluded that there were few differences in the editorial page slant of newspapers owned by groups and those that were independent (Compaine & Gomery, 2000). On the other hand, editors of chain-owned newspapers were found to have greater editorial latitude in determining editorial policy than those at family-owned newspapers. Whereas publicly owned newspaper chains may be more focused on profitability than those that are family controlled, they are less likely to have an ideological agenda they want to promote. Overall, studies show that large group, corporate, and public ownership have had a neutral to positive effect on the editorial side of newspapers (Compaine & Gomery, 2000).

The findings in video programming are along the same lines. To what extent do commonly owned newspapers and television stations in a community speak with a single voice about important political matters? One study looks at coverage of the 2000 presidential campaign in 10 communities where a local television station owned a local newspaper. The finding was, for the markets studied, "common ownership of a newspaper and a television station in a community does not result in a predictable pattern of news coverage and commentary about important political events in the commonly owned outlets" (Pritchard, 2002).

Ownership Does Matter

One of the most thorough studies of the role of ownership was a 5-year study by the Project for Excellence in Journalism. It looked specifically at television news at the local station level. As most broadcast television programming at local stations comes from the networks–ABC, CBS, NBC, Fox, CN, and Pax—the one area for decisions about original programming at the local station level comes in the form of the local news shows.

The Project for Excellence in Journalism (2003) study confirms that ownership does matter, but not always in ways that support conventional hypotheses. The study attempted to create objective measures of "quality" and then used a content analysis to apply these to 23,000 stories in 172 news programs over 5 years. Among the study's findings were the following:

- Stations with cross-ownership—in which the parent company also owns a newspaper in the same market—tended to produce higher quality newscasts.
- Ownership type made no measurable difference in terms of the diversity of people depicted in the news and little difference in the range of topics a station covered. In general, there is striking uniformity across the country in what local television stations define as news.[6]
- Stations owned by the largest groups produced higher quality early evening newscasts than those owned by the smaller groups. Smaller station groups tended to produce higher quality late evening newscasts than stations owned by larger companies (Project for Excellence, 2003).
- Network-affiliated stations tended to produce higher quality newscasts than network-owned and -operated stations.

- Local ownership offered little protection against newscasts being very poor and did not produce superior quality.
- Stations of privately owned companies and publicly owned companies did not perform significantly differently from each other (Project for Excellence, 2003).

Even within these overall conclusions there is something for everyone. For example, the notion of "quality" is highly dependent on the criteria used and how they are weighed. Although small company stations were found overall to have higher quality than those owned by the largest companies, those owned by this latter group rated highest on the criteria of "offering communities a variety of viewpoints in their newscasts." And medium-sized owners were better than smaller owners when it came to enterprise reporting and the greatest localism (Project for Excellence, 2003).

For the most part, "television" is defined and described by the programming provided by networks. In any market, most programming most of the time is the same, whether the station is owned by a large multimedia conglomerate or an independent local owner, whether the station is owned and operated by a network or is simply an affiliate of a network.

THE VIDEO LANDSCAPE IN 2006

In the 1960s—the time frame of Minow's "vast wasteland"—there were three "powerhouses" in the mass medium of television: the ABC, CBS, and NBC networks. These three controlled 90% of the prime-time viewership (households actually watching television; Coreth, 2004). That is, 90% of those watching TV were tuned in to one of three networks. By 2004, those three networks accounted, on average, for a share of less than 40% of the prime-time audience (Coreth, 2004). Over the extended period 1960 to 1980, these commercial broadcast networks dominated the television landscape. As illustrated in Figure 9.3, they had a combined prime-time rating of 56% of all households with television sets being tuned to one of those networks on an average weekday night (Wolzien & Mackenzie, 2004).

Today, Ratings for Networks are Way Down

But with the development of cable, competition from new networks, video rentals, and other distractions (e.g., gaming, Internet use), the audience share of those big three networks has declined steadily. For the 1970-1971 ratings period, the top-rated TV series was ABC's *Marcus Welby, M.D.*, which had a Nielsen rating

FIGURE. 9.3. Network Ratings, 1970 and 2003

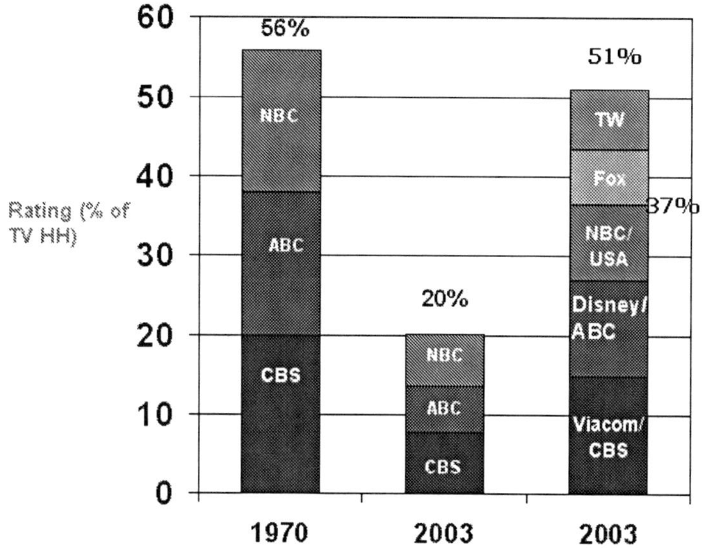

Source: See endnote 5

of 29.6 ("TV ratings," 2004). The average rating for the top 20 shows was 19.0, spread over *three* networks ("TV ratings, 2004; my calculation). For the 2003-04 season, the average rating for the top 20 shows was 11.6, spread over *five* networks (Paszaz Entertainment, 2004), a 39% decrease.

Figure 9.3 shows that by 2003, the three old broadcast networks were averaging only a cumulative 20% rating (Paszaz Entertainment, 2004). Starting in the 1980s and continuing to the present, those three networks either invested in the start-up of programming services distributed by cable operators, acquired such services, or were themselves merged into other entities that were in the video content business. CBS, which had divested itself of Viacom after a 1970 FCC ruling prohibiting television networks from owning a financial interest in the shows they broadcast (Sterling, 1981), was acquired by its spin-off in 1999 (Compaine & Gomery, 2000). By that time Viacom had become a major provider of cable programming, as well as motion pictures and books. Ironically, CBS had over the years divested itself of its books and magazine publishing holdings, so it was very much a stripped down version of the conglomerate it had become in the 1980s.[7] ABC, which had previously merged with Capital Cities to become Capital Cities/ABC, was acquired by the Walt Disney Company in 1996 (Compaine & Gomery, 2000). NBC, acquired by General Electric in 1986, had remained largely to itself until its acquisition of the cable networks and movie studio holdings of Vivendi Universal in 2004 (Ahrens, 2004).

Table 9.1 identifies the networks owned by the entities that, in 2003, included the three traditional broadcast networks. Viacom included CBS, Nickelodeon, MTV, TNN, BET, and TV Land, among others. Besides ABC, Disney owned ESPN and the Disney Channel, plus minority interests in others. The table also incorporates the networks NBC acquired in May 2004. The inventory encompasses both advertiser supported networks and those that are subscriber supported, such as Viacom's Showtime and The Movie Channel.

Table 9.1, therefore, shows that even adding in the audience share of all the new cable/satellite-distributed networks under the same ownership as the older broadcast networks—and subtracting the loss of viewership affecting the same networks—the surviving entities had an audience share in 2003 of about 37%, one third lower in household viewership than in the pre-cable era.

Table 9.2 provides a similar itemization for the networks of News Corp. and Time-Warner. The Fox Network, which did not exist until 1986, is a traditional broadcast entity, whereas Fox News and FX are cable/satellite distributed networks. Time-Warner started the over-the-air WB network in 1995 to add to its cable networks, CNN, TNT, HBO, and the like. Time-Warner is unlike the other four network operators in that it is the only one that owns no local television stations. The WB network was created exclusively from affiliates. News Corp. and Time-Warner together add 14.3% of households to the share of the now five major television program distributors. These five entities combined had a prime-time audience rating in December 2003 of 51.2%, or nearly 5% less than the combined rating of the three networks of the "vast wasteland" era..

That the make-up and competitiveness of television has changed dramatically since the 1980s can be further seen in by examining the 10 programs with the highest ratings since Nielsen began measuring audiences (see Table 9.3). These

TABLE 9.1. ***Share of All TV Households of Owners of the Three Traditional Broadcast Networks, 1970 and 2003 (% of Prime-Time Audience, TV HH = 108.4 million)***

	1970 Rating	2003 Rating
Viacom total		**15.0%**
CBS	20%	7.8
MTV		0.7
Nickelodeon/Nick at Nite		1.5
TV Land		0.6
BET		0.6
VH1		0.4
UPN		1.7
Spike TV		0.7

Comedy Central		0.1
Country Music TV		0.3
Noggin		>0.05
Showtime		0.3
Movie Channel		0.1
Disney total		**12.4%**
ABC	19%	6.0
ESPN, ESPN2		2.3
Lifetime*		1.6
Soapnet*		0.2
Toon Disney		0.2
ABC Family		1.0
Disney Channel		1.3
NBC/Universal total		**9.5%**
NBC	17%	6.4
MSNBC		0.2
CNBC		0.1
Bravo		0.4
USA Network		1.5
SciFi		0.9
Total, three traditional netsworks	**56%**	**36.9%**

Source. Nielsen Media, *Television Activity Report* for December 2003. See Compaine and Gomery (2000). Totals may not add due to rounding.

Table 9.2. *Share of All TV Households of Owners of the New Broadcast Networks December 2003 (% of Prime-Time Audience, TV HH = 108.4 million)*

Time-Warner Total	**8.7%**
WB	1.8%
CNN	0.6%
CNN-Headline News	0.2%
TBS	1.5%
TNT	1.5%
Cartoon Network	1.1%
HBO	1.4%
Multimax	0.5%
News Corp. Total	**5.6%**
FOX	4.2%
Fox News	0.9%
FX	0.5%

Source. Nielsen Media, *Television Activity Report* for December 2003. See See Compaine and Gomery (2000).

TABLE 9.3. *Top 10 Network Telecasts of All Time (By Household Rating)*

	Program	Date	Network	Rating %
1	*M*A*S*H* Special	Feb. 28, 1983	CBS	60.2
2	*Dallas*	Nov. 21, 1980	CBS	53.3
3	*Roots* Pt. VIII	Jan. 30, 1977	ABC	51.1
4	Super Bowl XVI Game	Jan. 24, 1982	CBS	49.1
5	Super Bowl XVII Game	Jan 30, 1983	NBC	48.6
6	XVII Winter Olympics	Feb. 23, 1994	CBS	48.5
7	Super Bowl XX Game	Jan. 26, 1986	NBC	48.3
8	*Gone With the Wind*-Pt.1	Nov. 7, 1976	NBC	47.7
9	*Gone With the Wind*-Pt.2	Nov. 8, 1976	NBC	47.4
10	Super Bowl XII Game	Jan. 15, 1978	CBS	47.2

Source. Nielsen Media Research, Inc. (2004).

include four Super Bowls, the last one being XX in 1986, the *M*A*S*H* special in 1983, and a *Dallas* episode in 1980. Only 1 of the 10 postdated the start of the fourth network in 1986 and that was one evening of the Winter Olympics from early 1994. Four of the 10 were from the 1970s (Nielsen Media Research, 2004).

THE RADIO LANDSCAPE

Industry Background

For most of the second half of the 20th century, radio was a rather sleepy industry. In the early 1950s it had been eclipsed by television as the medium with the greatest national interest to audiences and to advertisers. From its peak in 1945, when 15% of all advertising revenue was devoted to radio, to its trough just 10 years later at just 6% of advertising expenditures, radio seemed like a moribund medium (Sterling & Haight, 1978). Slowly, over the next five decades, the radio industry first stabilized and modestly increased its competitive position, so that in 2003 almost 8% of advertising revenue went to radio broadcasters (Newspaper Association of America, 2004).

Over the same time period, the number of radio stations proliferated. In 1945, there were 973 stations in the United States, 94% of which operated on the AM band, whereas less than 1% were noncommercial (Sterling & Haight, 1978).

By 2003, there were just under 14,000 stations, 20% of which were noncommercial. Slightly more than 36% of the stations were AM and the remainder used FM spectrum (Arbitron, 2004).

It was the radio industry that created the network model later adopted by the television industry. Today, radio is largely a local medium serving primarily local advertisers and local audiences. But at its pre-television peak, radio was dominated by four national networks: NBC, CBS and a second tier of ABC and Mutual (Sterling, 1982). Although no entity owned more than seven AM stations, by 1947, 94% of all stations—and all of the largest—were part of one or more networks. As early as 1937, the networks, and their handful of 10 Clear Channel owned and operated stations, accounted for fully 50% of industry revenues (Sterling, 1982). Even in 1947, when the number of stations had increased by two thirds, these networks still accounted for 25% of industry revenue. Stations owned by newspapers accounted for another 25% of industry revenue and stations owned by radio equipment manufacturers had another 13%. Even more germane for understanding policy today, the nature of the long-term contracts that the networks had with their affiliates gave these four networks "final say on [most of] what local stations programmed" (p. 303).

As with local television station ownership, FCC regulations kept a tight cap on the number of radio stations a single owner could control until the 1990s. Until 1985, a single owner was limited to seven AM and seven FM stations, and not more than one of each service could be in the same market. In cases where the AM and FM were in the same market, they could not be used to simulcast the same programming (Sophos, 1990). Over the next decade, the FCC expanded the limits, in stages, to 20 of each service. In the Telecommunications Act of 1996, Congress directed the FCC to lift the number of stations that could be owned by any entity nationally, although limits were imposed on ownership within local markets (Hamburg & Brotman, 1996).

Table 9.4 tracks the growth in the number of radio stations with the ownership limits provided by FCC regulation and by Congress. In the early days of radio, when newspaper publishers were encouraged to start radio stations as a way to promote growth of the fledgling medium, there were no limits on the number of stations that could be owned. It was not until 1940 that the FCC created the first such regulation and that was to limit the authorized, but nonexistent, FM band owners to five stations. In 1944, the FCC refused to allow CBS to purchase an eighth AM station, although a seven-station limit was not codified and made final until 1954, along with a similar limit on FM and TV stations (five of which could be VHF (Sterling, 1982).

For decades, the limits on radio ownership remained constant even as the number of stations expanded. In 1944, when the *de facto* limit was seven AM and six FM stations, there were 910 stations (Sterling & Haight, 1978). By the time the limit was raised to 12 AM and 12 FM in 1985, there were 8,600 commercial stations alone (U.S. Census, 1995). Until the further loosening and final elimination of the national ownership caps, radio station groups were held by regulation to

TABLE 9.4. *Radio Stations and National Ownership Limits, 1937-2003*

	Total Stations	Station Ownership Limit	Max National Ownership %
1937	646	No Limit (NL)	N/A
1947	1,212	AM=NL/FM=6	N/A
1960	4,306	AM=7/FM=7	.33
1970	6,889	AM=7/FM=7	.20
1980	8,748	AM=7/FM=7	.16
1990	10,631	AM=12/FM=12	.23
2003	13,898	No limit	8.8[a]

[a]Maximum national ownership limit for 2003 is actual % of stations owned by largest station group.
Sources. Stations 1937-1970, Sterling and Haight (1978), Table 170-A; 1970-2003, Arbitron, Inc..

levels of a fraction of 1% of stations as shown in Table 9.4. Of course, their audience reach and share of advertising revenue could be much higher, as enterprises with such strict limits would tend to own stations in larger markets and program them to reach the maximum audience.

Radio Advertising Revenue

Consider the financial fortunes of the radio industry by looking at advertising revenue of the industry in the framework of the growing number of stations that divide up the total. The impact of new competition from television for the mass audience eroded radio as the primary electronic medium for national network advertising. At the same time, radio's share of the total pool of advertising expenditure fell from 15% in 1945 to 6% in 1955. The number of new local stations that were starting up meant that the total revenue was split into more slices, leaving less, on average, for each. "The very growth of radio was one of its worst problems, since more stations meant more operators scrambling for available advertising dollars" (Sterling & Kittross, 1978).

Advertising revenue for the industry increased slightly from $605 million in 1950 to $692 million in 1960, or about 1.4% annually (Table 9.5). However, when adjusted for inflation to current dollars, industry revenue actually decreased 11% over that period. At the same time, the growth in stations meant that average revenue per station was declining through the 1950s and into the 1970s. Not until the 1990s—the period when ownership restrictions started to ease—did average per-station ad revenue reach the real dollar level of 1950. For most of this period, radio did not have the characteristics of a thriving industry.

TABLE 9.5. *Commercial Radio Stations and Advertising Revenue, 1950–2000*

Year	Number Commercial Stations	Advertising Revenue (million)	2000 Current $ (million)	Average Revenue per Station, Current $ (million)
1950	2,819	$605	$3,700	$1.3
1960	4,144	692	3,295	0.795
1970	6,476	1,308	4,839	0.747
1980	7,871	3,702	6,972	0.886
1990	9,379	8,839	10,879	1.16
2000	10,577	19,295	19,295	1.82

Sources. Stations, see Table 5; Advertising revenue: Universal McCann (formerly McCann-Erickson); Conversion to current dollars from GDP Price Deflator: http://www.jsc.nasa.gov/bu2/inflateGDP.html Budget of the U.S. Government, Fiscal Year 2005, Table 10.1—Gross Domestic Product and Deflators Used in the Historical Tables- 1940–2009.

Out From the Shadows: Noncommercial Radio Gets Stronger

Noncommercial radio has become a major media force (see Table 9.6). Primarily on the FM band, they have grown faster than the number of radio stations overall, accounting for 10% of all stations in 1980 and tripling in number and doubling in proportion to more than 20% of stations by 2000. Noncommercial radio stations are often affiliated with higher education, municipalities, or public television stations.

Noncommercial station growth has been accompanied by the emergence of National Public Radio (NPR) as a programming network that rivals the reach of the largest commercial owners. NPR's most well-known programming includes "Morning Edition" and "All Things Considered." The role of NPR is important in the context of radio competition.

- NPR provides content for much of the programming day on a national basis for the 780 public radio stations it serves. (About NPR, 2004).
- NPR's reach is so pervasive it claims "Just about anywhere you find yourself, you'll find NPR." (NPR, 2002, p. 2).
- NPR reaches 26 million listeners in a typical week, double the number of 10 years previously. In the 1980s, NPR claimed an audience of about 2 million weekly .(NPR's growth, 2005).
- Public radio listeners outnumber the combined circulation of the top 35 U.S. daily newspapers. ("Public radio," 2004).

TABLE 9.6 *Noncommercial Radio Stations, 1937–2003*

	Total Stations	Noncommercial Stations	As % of Total Stations
1937	646	0	0
1947	1,212	10	0.8
1960	4,306	162	3.7
1970	6,889	413	6.0
1980	8,748	877	10.0
1990	10,631	1,252	11.8
2000	13,307	2,730	20.5

Sources. Stations 1937-1970, Sterling & Haight, (1978); 1980-2000, calculated from U.S. Census Bureau, ***Statistical Abstract of the United States***: 2003, Table 1126: Utilization of Selected Media: 1970 to 2001.

For comparison, Infinity Broadcasting, the second largest owner of radio as measured by the number of stations owned, has 25% the number of stations as those that carry NPR and attracts a weekly audience of 76 million listeners (Viacom, 2004). On the other hand, the nationally syndicated talk show of political commentator Rush Limbaugh has a weekly audience between 15.5 and 20 million listeners on about 600 stations (Carney, 2003; Mayer, 2004).

NPR programming is available to roughly the same audiences as Infinity Broadcasting in the top 25 markets (Infinity, 2003; NPR, n.d.). In practical terms, the NPR stations are providing a national blanket that offers uniform programming accessible, as NPR promotes, virtually anywhere in the United States.

The Local Station Context

From the standpoint of individual users of radio, the context becomes what choices do they have and from whom when they turn on their radio in the car, the kitchen, or wherever they want to have access. In the Telecommunications Act of 1996, Congress was very specific in its instructions to the FCC for creating new caps on local ownership:

1. In a radio market with 45 or more commercial radio stations, a party may own, operate, or control up to 8 commercial radio stations, not more than 5 of which are in the same service (AM or FM).
2. In a radio market with between 30 and 44 (inclusive) commercial radio stations, a party may own, operate, or control up to 7 commercial radio stations, not more than 4 of which are in the same service (AM or FM).

3. In a radio market with between 15 and 29 (inclusive) commercial radio stations, a party may own, operate, or control up to 6 commercial radio stations, not more than 4 of which are in the same service (AM or FM).
4. In a radio market with 14 or fewer commercial radio stations, a party may own, operate, or control up to 5 commercial radio stations, not more than 3 of which are in the same service (AM or FM), except that a party may not own, operate, or control more than 50% of the stations in such market.

Since implementation of this mandate, radio station operators have indeed acquired additional licenses in many markets. In 2004, the largest operator of radio stations was Clear Channel Communications, owner of more than 1,200 stations (8.8% of the almost 14,000 radio stations nationwide; 12% excluding noncommercial stations), as well as outdoor billboards and concert promotion. Detractors complain of the homogenized play lists used on many Clear Channel stations around the country and their lack of local content (Hatch, 2003). Senator Russell Feingold asserts, "The homogenization of radio is a prime predictor of the ravages of deregulation" (cited in Hatch, 2003, p. 852). However, the practical impact on content and diversity of content has been inconclusive.

Whether radio was less homogeneous in the past may be addressed by looking at its history. With four networks providing the bulk of radio programming, radio in the 1930s looked much like television did prior to the 1980s: hundreds of local stations that relied on national networks for their content. In 1937, for example, during typical weekday evening prime-time hours, the Milwaukee AM radio station WTMJ (owned by the *Milwaukee Journal* newspaper) broadcast all NBC-Red network programming. The bulk of its afternoon schedule was network as well (Sterling & Kittross, 1978).

After television gained audience share in the 1950s, radio localism began to take the form of "top 40" formats. Packagers such as Todd Storz and Gordon McLendon helped stations across the United States implement their formula. From 20 such stations in 1955 there were hundreds by 1960 (Sterling & Kittross, 1978). Although the names of the disc jockeys may have been different—Hy Lit in Philadelphia, Alan Freed in New York—the sound was the same. They subscribed to national "tip sheets" to identify the emerging hits and top 40 play list services. "Time and time again an MOR-format [middle-of-the-road] station would take the plunge and achieve dramatic increases in listeners and income" (p. 339).

Empirical research data does not support the widely held assumption that changes in local ownership regulation has *prima facie* resulted in less diversity in radio. Several studies using data from the first round of local consolidation of radio stations found that consolidation had a positive effect on the diversity of formats available to listeners at the local level (Drushel, 2003; Waldfogel & Berry, 2001). These studies confirm the classic hypothesis proposed by Peter Steiner (1952): that a monopoly radio operator in a community would be more likely to provide a variety of formats than would several competitors all looking to adopt the format that would gain the largest audience. That is, in markets where often 20 or 30 owners were competing, a smaller portion of the audience would be served because more of the owners would be trying to program to the handful of formats that attract the largest audiences. Some amount of consolidation should—and apparently has—allowed owners to increase their aggregate audience by programming for smaller and unserved audiences as they acquired more stations in a market. A major academic study of radio format diversity in local markets between 1995 and 2000 found that slightly more than 50% of the population had the same or more choice in radio formats available at the end of the period—and less than 50% had less choice (Wirth, 2002). Yet another study concluded that in comparing the actual music play lists of similar-format stations in low and moderately concentrated local radio markets to those considered highly concentrated, in only one of the three measured criteria did the more highly concentrated markets appear to be slightly less diverse than the lesser concentrated markets (Chambers, 2003).

Consider station ownership in eight large, medium, and very small markets (see footnote 8 for a description of the selection process). There are 213 commercial radio stations in these eight markets, ranging from 46 in the 75th ranked Monterey-Salinas-Santa Cruz, California metropolitan area to 11 in 250th ranked State College, Pennsylvania (home to Penn State University) and 275th ranked Cookeville, Tennessee. The stations are owned by 62 different organizations. Clear Channel Communications owns 37 of the stations, or about 17%. The next largest total is CBS subsidiary, Infinity, with 9% of the stations. None of these enterprises owns stations in all eight markets. Clear Channel has no presence in Kansas City, the 29th largest market with 25 measured commercial stations, or in State College.

On the other hand, the largest presence in Yakima, Washington, the 198th market, is New Northwest Broadcasting, with 7 of the 18 stations. New Northwest does not have any properties in the other seven markets in the table, but does own 29 other stations in 11 other Washington, Oregon, and Alaska cities ("R&R directory," 2004). Clear Channel owns six stations in Yakima. Four other companies own the other five commercial stations in Yakima.

Boston, the largest market by population of this group, had 34 measured stations. Three groups, Clear Channel, Infinity, and Greater Media each owned 5. The latter owned no other stations in these eight markets and owned 19 stations spread over six markets. Overall, Boston had 16 separate "voices" besides the public and university-owned stations available to listeners.

A similar analysis of each of these markets—and by implication any other market—finds that there are a substantial number of voices and even more variety in program formats. In a small market such as Yakima listeners can choose among 12 different formats, from adult contemporary to regional Mexican. It has three news/talk format stations provided by three different owners. Even tinier

Cookeville, TN with a population of 83,000 (12+ years old), had 11 commercial stations, run by seven different owners, with 10 formats serving a spectrum of interests.

The consolidation of radio station ownership at the local level looks substantial when compared to the pre-1996 baseline of an industry artificially fragmented as a result of past regulatory limits. However, for an observer untainted by the way it "used to be," the implication of the data is that there would appear to be meaningful numbers of owners in communities serving the vast bulk of the population.

One popular perception is that radio has become more homogenized or less likely to provide local news (Hatch, 2003). However, this again needs to be compared with a baseline of what radio programming was in the decades between the ascendancy of television and the Telecommunications Act of 1996. Regardless of who owns the stations, radio is used primarily as a source of music, often as background to another activity. Only 10% of Americans consider radio as their primary source for news, compared with 57% who say the same for television and 23% for newspapers (Nielsen Media Research, 2002).

New Competition and Alternatives to Broadcast Radio

Another player in the mix of radio competition is the growth of satellite radio. The two national services, Sirius and XM Radio, had more than 6.2 million subscribers in mid-2005, more than doubling the total of mid-2004 (Kingsbury, 2004; "SIRIUS satellite radio," 2005; "XM satellite radio," 2005). The reality of the potential of satellite radio as a competitor to terrestrial broadcast radio was highlighted by the announcement in October 2004 that the controversial syndicated radio show of Howard Stern was leaving Infinity Broadcasting for a slot at Sirius (Bary, 2004). Besides an attractive financial package, at Sirius Stern is not subject to the FCC's limits imposed on his content, because Sirius does not use the "public" frequencies that regular radio does.

Various portable devices, successors to the older portable cassette and CD players, are used for downloading and storing personal archives of music and spoken-word content. An estimated 47 million portable digital music players were sold worldwide in 2004 ("MP3 player," 2005), up from 17 million in 2003 (Salkever, 2004). This is formidable competition for the non-talk radio segment of the radio industry (and increasingly so for the talk radio segment, which is beginning to provide downloadable voice programming through podcasting). Finally, there is the increasingly robust phenomenon of Internet-transmitted radio, which has already gone beyond the restriction of having to sit by a computer to hear it.

THE INTERNET AND WORLD WIDE WEB

The Internet is the popular shorthand term for not a single medium but a collection of services and service providers. These include, but are not limited to, backbone providers, Internet Service Providers (ISPs), utilities (e.g., search engines), merchants (e.g., Amazon.com), portals (e.g., Yahoo! or MSN), traditional information providers offering advertising supported content (e.g., *The New York Times* online) or subscription supported content (e.g., *The Wall Street Journal* online), customer service and convenience applications (e.g., view bank transactions online); and information providers who did not and could not exist without the Internet (e.g., Internet radio stations). The Internet itself is the sum of the processes through which bits of data are sent from computer to computer via sundry networks using the common TCP/IP format. The World Wide Web (WWW) is one component of the Internet that enables the use of hyperlinks to locate documents and data files on remote servers connected to the Internet.

In the United States, about 75% of adults had access to the Internet in 2004 (NetRatings, 2004). An estimated 140 million were active Internet users. (Greenspan, 2004). At the same time, more than 50% of households that subscribe to an Internet service did so with a broadband connection. This represented a 47% increase from the same period in 2003 (Nielsen NetRatings, 2003). This has several ramifications: Households using broadband have an always-on connection and thus spend more time online. They also can use the Internet for what is termed *rich media*, including higher quality audio and video streams and downloads.

Web Sites With Highest Traffic

Table 9.7 gives some sense of the most popular destination sites of Internet users from home (only slightly different from those sites accessed from the workplace; Nielsen NetRatings, 2005). They use it to access sites that are provided by traditional media companies, by other large although not traditional media companies, and a host of new players, from travel agencies to U.S. government agencies. Microsoft-affiliated sites receive visits from 6.6 million more users than second-ranked Time-Warner. Microsoft's sites would include their MSN service as well as visits to the corporate site for updates and downloads on Microsoft products. Time-Warner's audience of 85 million monthly visits from home users would include visits to the AOL and CNN Web sites, as well as to the sites of the Time, Inc. publications. InterActiveCorp is the parent of many e-commerce services, such as the travel agencies Expedia and Hotwired, financial service provider LendingTree, and event planning service Evite. The substantial use of sites run by various U.S. government agencies is a positive indicator of the use of online access for making it easier to obtain information and services from government.

TABLE 9.7 *Home Audience Reach of Top 10 Web Sites for July 2005, Aggregated by Parent Companies*

Parent Name	Unique Audience	Reach % (000)	Time Per Person (hours)
Microsoft	91,823	65.61	01:26:20
Time Warner	85289	60.94	03:46:50
Yahoo!	82770	59.14	02:00:36
Google	63158	45.13	00:18:31
eBay	41140	29.40	01:25:08
U.S. Government	32422	23.17	00:15:53
InterActiveCorp	28555	20.62	00:17:56
Amazon	27412	19,59	00:22:52
RealNetworks	26253	18.76	00:42:14
Viacom	24823	17.74	00:15:13

Source. Nielsen NetRatings from http://www.nielsen-netratings.com/reports.jsp?section=pub_reports& report=parent&period=monthly&panel_type=2

Internet Radio and Video

Both early research and continuing trends in the data support the contention that "by delivering diverse programming to a significant portion of the market, Internet radio broadcasters complement traditional radio and provide more overall diversity to audiences" (Compaine & Smith, 2001). Aggregate tuning hours for Internet radio were up by 118% in 2004, with the top 10 sites averaging 137.5 million hours during the first 2 months of the year, compared with 63 million hours in 2003 (Greenspan, 2004b). Delivery of video has followed closely the adoption of broadband to the home. Internet users with broadband accessed an average of 15.4 video clips per month in 2004, up 43% from a year earlier. Nearly 80% of the video streams provided during the first half of 2004 were to broadband users ("Accustream report," 2004). Video streams served during January and February 2004 by the top 10 sites averaged 523 million, compared with 292 million in 2003, which itself was up 104% from 2002. Music video captured the largest share of the streaming audience in 2003, followed closely by news ("Accustream report," 2004).

Internet users with broadband connections have been found to spend almost as much time per day online (2 hours) as they spend with television (2:45 hours) and radio (2:23 hours). On average, they spend 11 minutes with newspapers (Arbitron/Edison, 2003).

Both the absolute numbers and the steep positive trends on adoption, penetration, and broadband validates the use of the Internet as a serious avenue for distributing and accessing what have collectively been called the mass media: news, entertainment, information, and commerce—as text, audio, and video—bypassing many of the geographic, regulatory, and capital barriers associated with older processes of distributing the same type of content.

Internet radio is particularly relevant, given the high profile that radio mergers have had in the policy debates since 1996 (e.g., Clear Channel Communications). Internet radio encompasses traditional, licensed radio stations in the United States and elsewhere that provide real-time access via the Internet, as well as radio-like stations that operate only via the Internet, popularly called "webcasting." These include professional ventures that aim for substantial audiences and revenue as well as "garage" ventures that are operated as hobbies, small businesses, or to promote other agendas.

Cutting the PC Cord for Streaming Media

So long as streaming media, whether radio or video, is restricted to reception on a personal computer (PC), its role as a potential competitor with broadcast or cable is limited. The widespread implementation of the 802.11 wireless protocol (best known as Wi-Fi), as well as other wireless technologies for wider range broadband, such as Wi-Max (Worldwide Interoperability for Microwave Access; O'Hara, 2004), have begun the process of cutting the cord and making truly portable Internet radio—to be used around the house, at the beach, or in the car. The first such devices are on the market (Howard, 2004). Equipment manufacturers have introduced portable "boom boxes" for the satellite-distributed XM and Sirius radio services as well (at http://www.live365.com/info/index.html).

Internet as New Video Competition for Cable and DBS

Even as the old-line television networks adjust to the impact of multichannel distribution alternatives such as cable and satellite, these new players are looking over their shoulder at the Internet and the telephone companies as potential competitors. As cable companies encroach on the old regional exchange carrier's turf by offering consumers local and long distance telephone service using their wires to the home, the telephone companies are looking at their own wires for a response. The 500 Kbps to 1.5 megabits-per-second (Mbps) downstream speed of today's cable and DSL broadband is fast enough to offer decent quality video in a 5-inch square window on a computer screen. It is expected that it will take capabilities of 5 Mbps to stream over the Internet a DVD quality signal that would compete with cable—and even faster for a high-definition picture ("New service," 2004). "The fiber platform for broadband will pervade our world and how we live," is the flat-out assessment of the president of retail markets for Verizon ("Video services," 2004). To that end, Verizon initiated a build out of fiber optic cable that passed more than 3 million homes by the end of 2005 at a cost of $2.5 billion ("New services," 2004). Verizon began offering video services in 2005 over its new fiber lines. Verizon says it will not develop its own programming, providing an opportunity for current and potential new program producers ("VerizonFiOs TV," 2005; "Video services," 2004).

In addition to this fourth very high-speed channel to the home (beyond terrestrial broadcasting, cable TV and two major national DBS providers), there are several ventures looking at ways to provide consumers with high-quality video using the existing Internet infrastructure. TiVo, the maker of a popular digital video recorder (DVR), has initiated a service that further blurs the line between programming delivered over traditional cable and satellite channels and content from the Internet. The service connects the current DSL, cable modem, or DBS Internet input to a personal DVR, storing downloaded movies, music, or other programming that users can then watch or listen to at their convenience. Thus, the download speed does not have to provide real-time streaming, only a speed fast enough to allow downloading in minutes or a few hours, perhaps during the night ("New service," 2004).

A similar service was offered by Akimbo in 2005 (see http://www.akimbo.com/order.html). The package includes a $200 box that, like the TiVo system, plugs into a TV set and, through a home network, to a high-speed cable or DSL Internet connection. Akimbo charged $10 monthly (or $200 for a lifetime subscription), for which users can select movies, sports clips, and other programs to download to their set-top boxes. From the viewpoint of consumer options, it is significant that the Akimbo business plan expects that content creators will view this as a new form of distribution, as much of the programming will be content that is not repurposed from other video outlets ("Start-up Akimbo," 2004).

The clear implication of these trends around the Internet is that current characterizations about the nature of markets for audio, video, and even text content are likely to continue shifting.

CONCLUSION

The data assembled for this chapter is far from exhaustive of all the research and data that address the topic of media competition and ownership. It does present strong evidence that the choices for consumers have increased for obtaining their news, information, and entertainment, not only in absolute numbers of television channels but in the number of different providers of that programming.

In brief, by many useful metrics, the empirical reality does not support a simple assertion that in the United States, in 2004, consumers of content via the media had fewer choices of sources or fewer choices for any type of content that has been available in the past. Even in the radio industry, which of the old media has seen the greatest change in ownership structure in the past decade, the contrast is largely a function of the extremely fragmented state the industry had been kept in by federal ownership caps that had not changed in decades despite the prodigious expansion of outlets. With the cap loosened, the industry has, like a bottle of seltzer that had been jostled before opened, spurted to find the equilibrium it could have been moving to more gradually—and less visibly—if the cap had been loosened slowly over time. By standards of American industry—and even much of the other media—radio in most places is still highly competitive and diverse.

As Tables 9.1 and 9.2 show, on a typical weekday evening, about 51% of adults in the United States. are watching television programs that originated from the five largest television programmers. The rest are watching PBS, The Weather Channel, or any of the hundreds of other viewing choices on cable or DBS, or reading the newspaper (only one of which, the *New York Post*, is published by any of these five enterprises).

The data does confirm that a relatively small number of firms account for a large share of the television audience at any given time—although a smaller share that even fewer firms divided up in the not-so-distant past. Thus, there is more competition sharing a smaller mass audience than as recently as the 1980s. Indeed, individuals have ready and inexpensive access to so much information that they need search engines and on-screen directories to help find and manage it.

The term *mass media* implies that there must be a "mass audience" somewhere. There are a handful of forms of entertainment that cut across income, education, race and other cultural differences among Americans: The Super Bowl, the Olympics, some comedies, dramas, and news events that can strike a common chord. Those are logically tapped by a relatively small number of content providers who have the skills and the distribution mechanism to serve a mass audience. If 60 firms all tried to serve the mass audience there would not be enough of a mass to sustain them. They would all be providing similar, common denominator books or magazines or video productions or music. Diversity is the antithesis of mass. So it would seem that in fact society is always going to have a relatively small number of firms that will serve large markets, while there are and should continue to be thousands of entities serving niches.

Those niches, meanwhile, can aggregate into significant numbers. For example, the old television networks and their local affiliates continue to be profitable even though their audience is way down from 20 years ago. At any given moment, more than half the audience for television is watching something other than broadcast network programming. But that half is cut into 40 slices: a percent for CNN, half a percent for Bravo, 0.1% for the Sundance Channel. The mass audience is smaller these days, as the traditional broadcast networks retain more of an audience than other individual networks can provide at one time for an advertiser.

Washington Post columnist Robert Samuelson proposed that those who keep trying to throw roadblocks into regulatory change "confuse size with power. It's true that some gigantic media companies are getting even bigger at the expense of other media companies. But it's not true that their power is increasing at the public's expense" (Samuelson, 2003).

The findings of this chapter are consistent with the policies that have been advanced by the FCC in recent years, although at odds with the position of many in Congress. The latter has the ultimate jurisdiction over media ownership and structural matters, within the limitations imposed by the First Amendment. However, prior to any legislation in this arena Congress might ponder the advice of the majority decision of the Supreme Court in the *Brand X* case, "The questions the Commission resolved in the order under review involve a 'subject matter [that] is technical, complex, and dynamic.' . . . The Commission is in a far better position to address these questions than we are. Nothing in the Communications Act or the Administrative Procedure Act makes unlawful the Commission's use of its expert policy judgment to resolve these difficult questions" (*National Cable & Telecommunications Association et al. v. Brand X*, 2005).

NOTES

1. Twenty-five years is a convenient period. Although admittedly somewhat arbitrary, it does takes us back to the time just before cable reached most large cities, the effects of satellites were widespread other than for telecommunications, and just on the cusp of the personal computer. Certainly, if we looked back 100 years we would not need much research to state flatly that there was less access to information.
2. For example: "Today, a remarkable number of leading Americans of all political and ideological stripes are joining that call, expressing grave concerns about the harmful effects of media concentration on our nation's democracy and culture" (Rintels, 2003), or "The news industry in America is already far down the road to media concentration—too far, in our view" ("The troubling arc," 2004).
3. I have seen the 5 companies equals 80% content figure bandied about. In this case the source is Hatch (2003). However, Hatch cited back to Fisher (2003). Fisher's article, however, says that five companies "boast 75 percent of primetime TV viewers," which is at odds with Hatch's 80% figure. Either way, this figure represents the share of the TV audience at that moment. It ignores the 40% of households that may be using other media (reading books, listening to CDs, etc.) or using no media. And it measures households, not people.
4. For the week of May 24-30, 2004, the 10th rated show, *Still Standing*, on CBS had 12.1 million viewers. The biggest selling fiction book of 2003 was *Harry Potter and the Order of the Phoenix*. It sold 11 million copies in the United States from its publication in June through November ("Harry Potter sales," 2003). "Harry Potter sales reach 250 million" (2003, December 3). *Book Publishing Report*. Accessed (by subscription) July 16, 2004 at http://www.highbeam.com/library/doc3.asp?docid=1P1:89464398.
5. "The way to prevent these irregular interpositions of the people is to give them full information of their affairs thro' the channel of the public papers, and to contrive that

those papers should penetrate the whole mass of the people. The basis of our governments being the opinion of the people, the very first object should be to keep that right; and were it left to me to decide whether we should have a government without newspapers, or newspapers without a government, I should not hesitate a moment to prefer the latter." Thomas Jefferson in a letter to Edward Carrington, January 16, 1787. *The Founders' Constitution*, Vol. 5, Amendment I (Speech and Press), Document 8 Accessed August 2, 2004 at http://press-pubs.uchicago.edu/founders/documents/amendI_speechs8.html. The University of Chicago Press.

6. This is consistent—and likely for the same reasons—with the findings in Herbert Gans' (1980) classic study. Gans' empirical study of reporters and editors at national television news networks and news magazine concluded that their story selection and editorial judgment were remarkably similar because they all were similarly acculturated to the news process and faced similar demands—such as content limitations of air time or column inches.
7. In 2005, Viacom was split again into two companies: CBS as the broadcaster and Viacom with Cable interests.
8. The eight were selected from the Arbitron database of measured stations available at their Radio and Records.com Web site (www.radioandrecords.com). The selection process was random, as the term is popularly used, although not a statistical random sample of the universe, which in August 2004 included 287 geographic markets down to the size of 57,000 population of 12 years and older. My intention was to look at a selection of stations in large, medium, and small markets. The Web site allows users to enter a market size (e.g., 31, which turned out to be Salt Lake City). The resulting table is intended to be a snapshot of these markets. There were no markets that I initially selected that were subsequently not included in the table for any reason. I entered eight sets of numbers and used those eight markets. Thus, this is not a statistically valid sample of any sort, but nor is it created to select markets with any preconceived structure, other than population size. The stations listed in each market are those rated by Arbitron. Excluded are educational and not-for-profit stations and some commercial stations, mostly small ones, that Arbitron does not include in its measurements.

REFERENCES

About NPR. Accessed August 20, 2004, from http://www.npr.org/about/.

AccuStream report: User consumption of broadband video streams up 42% in first half of 2004. Accessed August 24, 2004, from http://www.accustreamresearch.com/news/aug17-04.html.

Ahrens, F. (2003, September 17). Senate votes to block FCC media rules. *The Washington Post*, p. A1.

Ahrens, F. (2004, May 13). GE, Vivendi give rise to a giant; New NBC Universal a $43 billion concern. *The Washington Post.* Accessed July 5, 2004, from http://www.highbeam.com/library/doc3.asp?DOCID=1P1:94510773&num=7&ctrlInfo=Round5a%3AProd%3ASR%3AResult&ao=.

Arbitron, Inc. (2004). Station trends. *Radio Today 2004.* Accessed July 30, 2004, from http://www.arbitron.com/downloads/radiotoday04.pdf.

Arbitron/Edison Media Research (2003, September 23). *Internet and multimedia 11: New media enters the mainstream*, p. 9. Accessed from http://www.arbitron.com/downloads/Internet_ Multimedia_11.pdf.

Bagdikian, B.H. (1992). *The media monopoly* (4th ed.). Boston: Beacon Press.

Bary, A. (October 11, 2004). Shock stock: Just like Howard Stern, shares of Sirius Satellite Radio are mighty pricey. *Barron's Online*. Accessed October 20, 2004, from http://online.barrons.com/article/SB109728355122340948-search.html?collection=barrons% 2F30day&vql_string=stern+sirius%3Cin%3E%28article%2Dbody%29.

Carney, S. (August 6, 2003). In Rush Limbaugh's world, he's always right. *Talk Radio/Radio Talk*. Accessed August 19, 2004, from http://www.radiotalk.org/news/alwaysright.html.

Chambers, T. (2003). Radio diversity in the era of consolidation. *Journal of Radio Studies, 10*(1), 33-45.

Compaine, B. M., Sterling, C. H., Guback, T., & Nobel, J. K., Jr. (1982). *Who owns the media? Concentration of ownership in the mass communications industry* (2nd ed., p. 3). White Plains, NY: Knowledge Industry Publications.

Compaine, B. M., & Gomery, D. (2000). *Who owns the media? Competition and concentration in the mass media industry* (3rd ed.). Mahwah, NJ: Erlbaum.

Compaine, B., & Smith, E. (2001). *Internet radio: A new engine for content diversity?* International Telecommunications Society Conference, Dublin, September. Accessed from http://itc.mit.edu/itel/docs/2001/compaine_smith_radio.pdf.

Cotreth, D. (June 7, 2004). *Impact of technology on media*. Panel at Deutsche Bank Securities 12th Annual Media Conference 2004—Final. Accessed July 30, 2004, from http://www.highbeam.com/library/doc3.asp?DOCID=1P1:95731319&num=12&ctrlInfo=Round5a%3AProd%3ASR%3AResult&ao=.

De Jong, A. S., & Bates, B. J. (1991). Channel diversity in cable television. *Journal of Broadcasting & Electronic Media, 35*, 159–166.

Drudge, M. (June 2, 1998). *Anyone with a modem can report to the world*. Speech presented to The National Press Club, Washington, DC. Accessed May 5, 1999, from http://www.frontpagemag.com/archives/drudge/drudge.htm.

Huang, D. (1999). Size and growth trends in the information industries, 1987-1996. In B. Compaine & W. Read (Eds.), *The information resources policy handbook* (pp. 347-361). Cambridge, MA: MIT Press.

Drushel, B. E. (2003). The Telecommunications Act of 1996 radio market structure. *The Journal of Media Economics, 11*(3), 3-20.

Einstein, M. (2004a). Broadcast network television, 1955–2003: The pursuit of advertising and the decline of diversity. *The Journal of Media Economics, 17*(2), 145–155.

Einstein, M. (2004b). The financial interest and syndication rules and changes in program diversity. *The Journal of Media Economics, 17*(1), 1–18.

Evening news ratings race heating up. *CNN.com*. June 3, 2004. Accessed July 16, 2004, from http://www.cnn.com/2004/SHOWBIZ/TV/06/03/nielsens.ap/.

FCC v. National Citizens Committee for Broadcasting, 436 U.S. 775, 780, 98 S.Ct. 2096, 56 L.Ed.2d 697 (1978).

Federal Trade Commission, Bureau of Competition. (1978). *Proceedings of the symposium on media concentration* (2 vols.). Washington, DC: U.S. Government Printing Office.

Fisher, M. (2003, May 18). Sounds familiar for a reason. *The Washington Post*, Outlook Section. Accessed July 16, 2004, from http://www.highbeam.com/library/doc3.asp?DOCID=1P1:73985312&num=1&ctrlInfo=Round6%3AProd%3ASR%3AResult&ao=.

Gans, H. (1980). *Deciding what's news.* New York: Vintage Books.

Greenspan, R. (2004a, June). *Active Internet users by country. Nielsen//NetRatings.* Accessed August 19, 2004, from http://www.clickz.com/stats/big_picture/geographics/article.php/3386051#table.

Greenspan, R. (2004b, April 15). *Streaming more mainstream.* ClickZ Network. Accessed from http://www.clickz.com/stats/big_picture/applications/article.php/3341031#table. As measured by AccuStream iMedia Research.

Hamburg, M., & Brotman, S. (1996). *Communications law and practice.* New York: Law Journals Seminars Press.

Hatch, D. (2003, October 10). Media ownership. *CQ Research*, p. 845.

Howard, B. (2004, August 3). Time for the Internet boom box. *PC Magazine.* Accessed August 25, 2004, from http://www.pcmag.com/article2/0,1759,1619716,00.asp.

Infinity: Viacom Form 10-K for Year Ending Dec. 31, 2003, pp. I-19—I-22. Accessed from http://www.viacom.com/pdf/form10KMar2004.pdf.

In the matter of annual assessment of the status of competition in the market for the delivery of video programming. MB Docket No. 03-172, *Tenth annual report*, pp. 4-5, FCC 04-5. Released January 28, 2004.

Kingsbury, K. (August 4, 2004). *Satellite radio captures ears of millions.* CNN.com. Accessed August 8, 2004, from http://www.cnn.com/2004/TECH/science/07/30/satellite.radio.

Lessig, L. (2003, October 20). Lessig blog. Accessed July 13, 2004, from http://www.lessig.org/blog/archives/cat_mediacon.shtml.

Mayer, W. (June 22, 2004). Why talk radio is conservative. *Public Interest.* Accessed August 20, 2004, from http://www.highbeam.com/library/doc0.asp?docid=1G1:119025431&refid=hbr_flinks1.

Minow, N.N. (May 9, 1961). *Television and the public interest* (Speech). Washington, DC. Accessed August 2, 2004, from http://www.americanrhetoric.com/speeches/newtonmiow. htm.

Moon, C. (2004, June 22). *Mid-year review presentation.* Accessed July 16, 2004, from http://www .gannett.com/go/press/mymr04/mymrcm04.htm.

MP3 player shipments surging, says iSuppli. (2005, March 15). Accessed August 15, 2005, from http://www.eet.com/news/latest/showArticle.jhtml?articleID=159900293.

Napoli, P. (1999). Deconstructing the diversity principle. *Journal of Communication, 49*(4), 7–34.

National Cable & Telecommunications Association et al. v. Brand X Internet Services et al. (2005, June 27). Accessed August 15, 2005, from http://laws.findlaw.com/us/000/04-277. html.

National Public Radio. (2002). *Annual Report 2002*, p. 2.

NetRatings, Inc. (2004, March 18). *Three out of four Americans have access to the Internet, according to Nielsen//Netratings.* Accessed from http://www.nielsen-netratings.com/pr/pr_040318.pdf.

Newspaper Association of America. (2004). *Facts about newspapers 2004.* Accessed July 30, 2004, from http://www.naa.org/info/facts04/expenditures-allmedia.html.

New service by TiVo will build bridges from Internet to the TV. (2004, June 9). *The New York Times.* Accessed June 9, 2004, from http://www.nytimes.com/2004/06/09/technology/09net.html?ei=5007&en=7bd25f7836498ba2&ex=1402113600&partner=USERLAND&pagewanted=all&position=.

Nielsen Media Research. (2002, September 22). *Consumer survey on media usage.* Federal Communications Commission, p. 22. Accessed January 18, 2004, from http://hraunfoss. fcc.gov/edocs_public/attachmatch/DOC-226838A17.pdf.

Nielsen Media Research. (2004). *Top 10 network telecasts of all time (ranked by household rating).* Accessed August 5, 2004, from http://www.nielsenmedia.com/ratings/topnetworktelecasts.ht.

Nielsen NetRatings, Inc. (2003, August 18). *U.S. broadband connections reach critical mass, crossing 50 percent mark for web surfers, according to Nielsen//Netratings.* Accessed from http://www.nielsen-netratings.com/pr/pr_040818.pdf.

Nielsen NetRatings, Inc. (2005, July). *United States: Top 10 parent companies, month of July 2005, work panel.* Accessed August 9, 2005. (Note that this URL shows the latest monthly figures. It is not archived for July 2005.)

NPR. *NPR Stations.* Accessed from http://www.npr.org/stations/pdf/nprstations.pdf.

NPR's growth during the last 30 years. Accessed August 2, 2005, from http://www.npr.org/about/growth.html.

O'Hara, B. (2004, January 12). Will wi-max make wi-fi obsolete? *Network World Fusion.* Accessed August 20, 2004, from http://www.nwfusion.com/columnists/2004/0112wizards.html.

Paszaz Entertainment Network, compiled by Nielsen Media Research. Accessed August 5, 2004, from http://www.pazsaz.com/tvseasn.html.

Policing the thinkable. www.OpenDemocracy.net, Accessed July 21, 2004, from http://www.opendemocracy.net/debates/article-8-24-56.jsp.

Pritchard, D. (Sept. 2002). *Viewpoint diversity in cross-owned newspapers and television stations: A study of news coverage of the 2000 presidential campaign.* Washington, DC: Federal Communications Commission. Accessed October 15, 2002, from http://hraunfoss.fcc.gov /edocs_public/attachmatch/DOC-226838A7.pdf.

Project for Excellence in Journalism. (April 29, 2003). *Does ownership matter in local television news: A five-year study of ownership and quality.* Washington, DC: Author.

Public radio listening grows amidst downward trend on the commercial Dial. The WBEZ Alliance. Accessed August 16, 2004, from http://www.thirdcoastfestival.org/pages/extras/press/03/radiofacts.html.

R&R directory. RadioandRecords.Com. Accessed August 3, 2004, from http://www.radioandrecords.com/RRDirectory/Directory_Main.aspx.

Rintels, J. (2003). Reply comments of the Center for the Creative Community Regarding Competition and Media Concentration in the Television Industry, In the matter of 2002 Biennial Regulatory Review—Review of the Commission's Broadcast Ownership Rules and Other Rules Adopted Pursuant to Section 202 of the Telecommunications Act of 1996, MB Docket No. 02-277, Feb 3, 2003.

Salkever, A. (2004, May 27). iPod: Leader, but not ruler. *Business Week Online.* Accessed August 25, 2004, from http://www.businessweek.com/technology/content/may2004/tc20040527_8900_tc056.htm.

Samuelson, R.J. (2003, August 6). The myth of "big media." *The Washington Post,* p. A17. Accessed September 10, 2003, from http://www.washingtonpost.com/ac2/up-dyn?pagename=article&node=&contentId=A21980-2003Aug5.

Schurz Communications, Inc. v. FCC, 982 F.2d 1043 (7th Cir. 1992). In T. G. Krattenmaker (1998). *Telecommunications law and policy* (2nd ed.) Durham, NC: Carolina Academic Press.

SIRIUS satellite radio reports record revenue for second quarter 2005, August 2, 2005. Accessed August 15, 2005, from http://www.shareholder.com/sirius/ReleaseDetail.cfm? ReleaseID=170152&cat=Earnings&newsroom=.

Sophos, M. (1990). The public interest, convenience, or necessity: A dead standard in the era of broadcast deregulation? *Pace Law Review, 10*, 661, Section III. Accessed August 19, 2004, from http://www.wdfh.org/art_3.htm.

Start-up Akimbo systems gets funding for Internet-TV system. (2004, July12). *The Wall Street Journal*, p. B5. Accessed July 12, 2004, from http//online.wsj.com/article/0,,SB108958511185860813,00.html.

Steiner, P. (1952). Program patterns and preferences, and the workability of competition in radio broadcasting. *The Quarterly Journal of Economics, 66*(2), 194-223.

Sterling, C.H. (1981). Cable and pay TV. In B. M. Compaine, C. H. Sterling, T. Guback, & J. K. Nobel, Jr., *Who owns the media? Concentration of ownership in the mass media industry* (rev. ed.). White Plains, NY: Knowledge Industry Publications.

Sterling, C. H. (1982). Television and radio broadcasting. In B. M. Compaine, C. H. Sterling, T. Guback, & J. K. Nobel, Jr., *Who owns the media? Concentration of ownership in the mass communications industry* (2nd ed., p. 303). White Plains, NY: Knowledge Industry Publications.

Sterling. C. H., & Haight, T. (1978). *The mass media: Aspen institute guide to communication industry trends.* New York: Praeger.

Sterling, C. H., & Kittross, J. M. (1978). *Stay tuned: A concise history of American broadcasting.* Belmont, CA: Wadsworth.

The troubling arc of media concentration. (2004, March 31). *The Seattle Times.* Accessed and filed April 1, 2004.

TV ratings, 1970-1971. Classictvhits.com. Accessed March 20, 2004, from http://www.classictvhits.com/tvratings/1970.htm.

U.S. Census Bureau. *Statistical abstract of the United States: 1995*, Table 897.

U.S. Congress. (1996). Telecommunications Act of 1996, sec 202. Available from http://www.fcc.gov/Reports/tcom1996.pdf.

Verizon FiOS TV will offer a new customer experience, Seidenberg says. (2005, April 18). Accessed August 15, 2005, from http://newscenter.verizon.com/proactive/newsroom/release.vtml?id=90601.

Viacom, Inc. (April 15, 2004). Letter of stockholders, p. 3. Accessed August 6, 2004, from http://www.viacom.com/pdf/stockholderLetterApr2004.pdf.

Video services are big lure. (2004, August 19). *The Wall Street Journal*, p. B1. Accessed August 19, 2004, from http://online.wsj.com/article/0,,SB109286816672195246,00.html?mod=article-outset-box.

Waldfogel, J., & Berry, S. (2001). Do mergers increase product variety? Evidence from radio broadcasting. *The Quarterly Journal of Economics, 116*(3), 1009-1024

Walker, J., & Ferguson, D. (1998). *The broadcast television industry.* Boston: Allyn & Bacon.

Wirth, T. (2002). Direct format competition on the radio dial and the Telecommunications Act of 1996: A five-year trend study. *Journal of Radio Studies, 9*(1), 44.

Wolzien, T., & Mackenzie, M. (January 28, 2004). TV ratings study: Viacom, Disney up across ad-based networks; Surprising warning signs for TWX's HBO. *Bernstein Research Call*, p. 2.

XM satellite radio holdings inc. announces second quarter 2005 results and increases year-end subscriber guidance to 6 million. July 28, 2005. Accessed August 15, 2005, from http://www.xmradio.com/newsroom/screen/pr_2005_07_28.html.

10

CURRENT MEDIA OWNERSHIP TRENDS

Amy S. Mitchell

Gauri Malhotra

BACKGROUND

This chapter is a summary of the full 2005 Annual Report on the State of the News Media (www.stateofthenewsmedia.org), the work of the Project for Excellence in Journalism (www.journalism.org) and is the second in our annual effort to put in one place as much original and aggregated data as possible about each of the major American journalism sectors. The Project's intention is to inform, not to persuade. The Project is a part of the Pew Research Center and funded by The Pew Charitable Trusts. This full report involved a wide range of collaborators (see http://www.stateofthemedia.org/2005/authors.asp), and outside sources (footnoted throughout). This chapter summarizes and integrates the state of and changes in media ownership within the nine sectors of the news industry: network television, local television, cable television, newspapers, radio, magazines, online, ethnic, and alternative media.

INTRODUCTION

The year 2004 was one of unusual stability for media owners. After a decade of consolidation, the level of acquisition activity was the lowest in years. History will probably recall then-Federal Communications Commission (FCC) Chairman Michael Powell as an ironic force in stalling consolidation. In June 2003, Powell released a new set of rules that would have dramatically loosened limits on station ownership and allowed even more consolidation. The proposed rules focused on three areas. The first was the ownership cap, which Powell proposed raising from 35% of the country's households to 45%. (Two companies—Viacom and Fox—already had waivers allowing them to exceed the cap; see Watson & Chang, chapter 7, this volume.) The second area was ownership of multiple TV stations in a single market, known as *duopoly* ownership. Although this was already permitted in the largest markets, the Powell proposals would have allowed duopolies in all but the smallest markets. The third area of regulation, cross-ownership, would have relaxed rules forbidding companies from owning a TV station and a newspaper in the same market.

Many members of Congress balked at the FCC's proposals, and a vocal minority suggested they would exercise their authority to prevent the FCC from putting the regulations into effect. The proposed changes were halted by Congress in 2003 and blocked by the courts in 2004. Powell resigned in 2005, and under the new chairman, Kevin Martin, the FCC's approach is still uncertain.

But the forces behind conglomeration are powerful. For one, buying other companies is the fastest and easiest way to make a company grow, something that Wall Street tends to enjoy, at least initially. For another, there continues to be a sense that as audiences fragment, owners can still sell a mass audience to advertisers by aggregating it under a corporate roof and offering package advertising deals. The fact that many deals prove more difficult in practice than in planning, as AOL and Time-Warner discovered, seems less powerful than the money to be made up front by almost everyone involved.

NETWORK TELEVISION

The most salient trend in network news ownership is that journalism is an increasingly small part of what the corporations that own the networks do. Some observers have used corporate annual reports as a proxy, albeit an uncertain one, to gauge the importance of a division to a corporation's overall identity. By that measure, news is indeed a relatively small part of these entities. The Walt Disney Company's 2003 95-page annual report contains 22 sentences about news, 14 of them about the network, 2 about local news, 3 about broadband, 1 about wireless, and 2 about radio. General Electric's (GE) annual report that year says the least. In its 117 pages, the narrative does not mention NBC news even once by name, although there is a reference to "advertising reductions because of coverage of the Iraq war ($0.1 billion)." (Networks can lose advertising revenue during major news events by pre-empting regular programming.)

A more quantitative assessment of value is to break out how much of the revenue of the owners of the Big Three networks comes from broadcasting—that is, all of the revenue they receive from whatever TV stations and TV networks they may own aside from cable. Only part of that revenue, of course, is from news. Looking at 2004 data from *Advertising Age's* annual "100 Leading Media Companies" (2004), broadcast revenue as a percentage of the overall revenue of parent companies has either dropped very slightly or held steady since 2002.

Viacom's broadcast revenue, $7.8 billion for 2003, is 29% of the company's total revenue of $26.6 billion. Broadcast provided $4.8 billion, or 18%, of Disney's total revenue of $27.1 billion. Broadcast provides only a small portion, 5%, of GE's overall revenue: Only 4.6%, or $6.2 billion, of GE's total $134.2 billion in revenue comes from broadcasting, and just a fraction of that comes from news ("100 Leading Media," 2004). This may help explain the lack of any reference to NBC News in the company's annual report.

When cable is added to the mix, the contributions to overall revenue change. Although the networks complain about cable competition, some of that competition comes from their own corporate siblings.

Viacom's television revenue (broadcast and cable combined) of $13.4 billion constituted a full 50% of the company's overall revenue in 2003, a 1% decrease from 2002. But the additional revenue is derived almost entirely from entertainment. Viacom, which does not own a cable news channel, does own, among others, such popular cable stations as MTV and Nickelodeon, both of which have run very modest news operations ("100 Leading Media," 2004).

In 2003, 38% of Disney's revenue came from television. That was a 3% gain from 2002 and totaled $10.3 billion. But it, too, owns no cable news channel. Disney's cable portfolio, among other channels, includes ESPN, which does a significant amount of sports journalism, as well as A&E and, of course, the Disney Channel ("100 Leading Media," 2004). Its digital start-up ABC News Now was in its infancy.

GE's cable revenues, as with broadcast, are substantially lower than those of either Disney or Viacom. The company, whose cable holdings include two channels offering news (CNBC and MSNBC) earns just 6% of its overall revenue from television. The numbers take into account the acquisition by GE of Vivendi Universal. The $8.2 billion earned in 2003, although an increase from 2002, did not change the percentage of television's overall contribution to GE's revenue.

More recent numbers are available that isolate the revenues of the three broadcast TV networks alone. In 2004, according to *Broadcasting and Cable*, NBC TV by itself had revenues of $5 billion, CBS $4.5 billion, and ABC $3.2 billion (Higgings, 2004).

NBC

NBC got much bigger in 2004 with its $14 billion acquisition of the movie studio and entertainment assets of Vivendi Universal Entertainment, the ailing French entertainment giant. The result is a company consisting of TV networks, studios, cable networks, theme parks, owned and operated local stations, and a network news division. The corporation's goal was to fashion an operating division better protected from the volatility of the TV distribution business (which relies on advertising and audience ratings) by acquiring control of TV production as well (a large studio and its entertainment library of 5,000 films and 40,000 hours of TV programming). As chief executive Bob Wright explained, "A lot of the savings comes from marketing and promotion, and cross-promoting various platforms . . ." such as showcasing NBC television programs in company theme parks (Burt, 2004).

NBC Entertainment's chief, Jeff Zucker, was reported by *The New York Times* to have acknowledged that NBC had been criticized for using time on the network's news programs to promote entertainment programming like the finale of *Friends* (Carter, 2004). But he went on, the paper said, to explain, "Clearly we're not going to have Tom Brokaw do the Nightly News from Universal Riverwalk, but we might have 'Access Hollywood' do a show from there" (Carter, 2004, p. C1). But, in fact, it was as part of the *Today Show's* coverage of the 2005 Golden Globe Awards that the weathercaster Al Roker showed up on the Riverwalk demonstrating *Fear Factor* stunts.

With the Brokaw–Williams handoff now completed, NBC executives appear to have adopted a wait-and-see attitude about changes to the *Nightly News*. Williams, in public statements and interviews, appears anxious to position himself as a protector of traditional news values and presentation, at the same time acknowledging the changing media landscape: he seems as likely to make these appearances in an Internet chatroom or on Comedy Central's *The Daily Show* as before a conference of journalists or academics.

ABC

The Walt Disney Co. had a rocky 2004. There was continued pressure from dissident shareholders. There was the spectacle of chief executive Michael Eisner on the stand during a lengthy trial concerning the costly dismissal of Michael Ovitz. By year's end there was a looming publication date of journalist James B. Stewart's book, *Disney War*. The book chronicled the behind the scenes corporate workings of Eisner's control of Disney from 1984 to 2004.

As *The Financial Times* put it in April, "The network's weakness was highlighted at Disney's annual meeting in 2004, when 43 percent of shareholders withheld their support from Michael Eisner, chief executive" (Burt & Yeager, 2004, p. 24). By March 2005 the Disney board, intent on moving the company quickly forward, elected Robert A. Iger (then president and COO of The Walt Disney Co.) to succeed the embattled Eisner. Although it had been reported the Eisner would be retiring from Disney in 2006, Iger assumed the position of president and CEO on October 1, 2005.

During the 2004 annual meeting, network officials highlighted the strong performance of ABC's news division, as well as sports and daytime programming. They called the continuing problems in prime time "our biggest challenge and priority" (Burt & Yeager, 2004, p. 24). By fall, ABC appeared to be beginning to meet some of that challenge with the success of such shows as *Desperate Housewives* and *Lost.*

CBS

As with Disney and NBC Universal, CBS's parent company, Viacom, is now a media company rather than principally a broadcaster. It owns the Paramount movie studios and TV syndication business; the CBS and UPN networks; 20 CBS affiliates, 18 UPN local TV stations and an independent; MTV, Nickelodeon, Black Entertainment Television, VH1, Comedy Central, and the Movie Channel; more than 185 Infinity Broadcasting radio stations; Viacom Outdoor Advertising; Paramount theme parks; movie theaters, and Simon & Schuster, the publishing house. It also owns CBS News.

In 2004, the network's entertainment division prospered. Not only did it recapture the ratings crown from NBC but, in doing so, it attracted more of the younger viewers most prized by advertisers. CBS News was also profitable despite the fact that only two of its programs—the Sunday edition of *60 Minutes* and Charles Osgood's *Sunday Morning*—were ratings leaders. The last-place standing of the *Evening News* worsened. Morning news numbers increased steadily for the first time in years, but were still in third place. The cloning of *60 Minutes* to a second night appeared to be working until the disaster of its Texas Air National Guard segment. The division's other prime-time magazine show, *48 Hours*, held its own. *Face the Nation* was second among the Sunday morning talk shows, while *Sunday Morning* remained a favorite of critics and audiences alike.

In a November 25, 2004 article examining the network's improving health, *The New York Times* media writers Bill Carter and Jacques Steinberg wrote that "CBS is doing so well right now with its entertainment programming that the developments in its news division—led by Dan Rather's sudden announcement that he will step down sooner than expected from the anchor position after a 24-year run—are being taken entirely in stride" (p. C1). The writers quoted CBS's chairman, Leslie Moonves, as saying, "Our news division is not hurting us" (p. C1). "That may not represent a ringing endorsement," the *Times* writers observed, "but it also does not signal the kind of management displeasure that in the past led to major layoffs, big budget cuts or other batterings of CBS News" (p. C1). Although Moonves has been consistent in expressing his support for the news division, he also signaled that change was in the air. "There's no question that CBS News is at a transition point," Moonves said. "There has to be a re-examination of what network news will be in the future" (Carter, 2004a, p. C1).

Three significant events in 2005 suggest change is certainly in the air, because perhaps not in the format of the evening news. At the end of 2005, Viacom announced that it would divide its vast cross-media holdings into CBS Corporation and what was being called the new Viacom. CBS teamed with the infinity Radio Network, UPN and Viacom's outdoor advertising, promotion and book publishing properties. Leslie Moonves became chief executive. The new Viacom included MTV and Paramount's holdings and is led by Thomas E. Freston. Second, CBS, still the only network without a cable presence, chose to bypass the older technology and leapt right into an online broadband channel with a relaunch of CBSNews.com. In designing the site, CBS created an ombudsman-like online environment—"Public Eye"—that uses blog-style postings to encourage dialogue by the public on topics ranging from reaction to CBS news programming and explanations about how decisions are made to the news business in general. Finally, in late October Moonves announced that the long-time president of CBS Sports, Sean McManus, would be replacing Andrew Heyward as the president of CBS News.

LOCAL TELEVISION

The trend toward consolidation of local TV station ownership effectively stalled in 2003 and 2004. In 2003, Michael Powell, then chairman of the FCC, proposed rule changes that would have opened up the marketplace tremendously. But with many organizations strongly objecting to the proposed changes and the FCC put on hold by court battles, owners were hesitant to add to their properties.

In 2003, the total value of TV stations that changed hands amounted to $500 million; the figure for 2004 was roughly $700 million.[1] This was a significant reversal of the buying trend begun with the Telecommunications Act of 1996, a Clinton-era law that removed a number of limits on station ownership, allowing companies to buy more stations in more parts of the country than ever before. For example, it raised the cap on the reach of a company from stations covering 25% of U.S. households to stations covering 35% of the country. It also relaxed rules on common ownership of radio and television stations ("U.S. Policy," 1996). Between 1995 and 2003, ten of the largest TV station owners went from owning 104 stations with $5.9 billion in revenue to owning 299 stations with $11.8 billion in revenue.

In anticipation of further deregulation, companies including Tribune, Gannett, and Media General had acquired newspapers in markets where they also owned TV stations. In early 2005, they asked the Supreme Court to revisit the Philadelphia court ruling, arguing in part that "spectrum scarcity" was no longer a valid argument for restricting stations ownership in an era of cable television and the Internet. The largest transaction of 2004, announced 1 month after the election, reflected an expectation that the rules would be relaxed: Viacom purchased the Sacramento station KOVR from Sinclair for $285 million, thereby creating yet another duopoly (Freedom du Lac, 2004). In June 2005, however, the Supreme Court rejected the appeal.

With a new head, Republican Kevin Martin (appointed in March 2005), the FCC is expected to move forward with rewriting its ownership rules to comply with the Philadelphia appeals court's ruling. The next round of proposed changes may be more limited than Powell's first proposals, but the general trend toward deregulation is likely to continue.

In 2005, however, changes in ownership continued to be in a lull. As a report from BIAfn (2005) on the revenue for the top ten station groups from 2000 to 2004 noted, "In order to grow revenues, many existing television groups have had to continue to expand into new markets or add stations in existing markets." In other words, new revenue streams have yet to appear; groups can only make more money by cutting costs. Notable company-level developments include Emmis, which owns 15 TV stations, apparently deciding to raise cash by selling off some of its TV stations and plowing the money into its radio operations ("Emmis may sell," 2005.) and Viacom deciding to split in two and to put its owned and operated stations in a company for those looking for steady dividends rather than "growth."

For most of 2005, the FCC commissioners were split with Martin and Kathleen Abernathy on the Republican side facing the Democrats Michael Copps and John Adelstein. The vacant Republican spot was not filled until November 2005, when Deborah Tate was nominated by the President. But Abernathy's term expired the same month, perpetuating the split. It was only in February 2006, when attorney Robert McDowell filled the GOP seat left vacant by Abernathy, that the FCC had all 5 commissioners in place. In 2006, the FCC also committed to reviewing ownership rules by 2007.

CABLE TELEVISION

There were no changes in the ownership picture of cable news in 2004. No new companies entered the field, and none of the three major cable news operations changed hands.

In 2004, cable news management came under attack on a variety of fronts, including allegations (notably in Robert Greenwald's movie Outfoxed) that Rupert Murdoch had turned Fox News into a covert political operation, and Ted Turner's (2004) castigating Time-Warner for managing CNN with too much attention to short-term results and the bottom line. Nevertheless the ownership structure of the Big Three remained solid and the only talk of potential competitors entering the field was from the British Broadcasting Corporation (BBC).

The three cable news channels are all owned by media conglomerates, and all have corporate siblings heavily involved in other media sectors, from magazines (Time-Warner), newspapers (News Corp.), local TV stations (News Corp. and NBC), and movie studios (all three).

Despite those similarities, the news channels play different roles within their respective companies that are useful to note in trying to imagine how they might evolve in the future. Fox News is one of the stars of News Corp.'s TV portfolio and Roger Ailes, its chairman, is a key decision maker in the parent company as well. CNN and Headline News (sold as a package) is a solid performer for Time-Warner, but not an enormous factor. MSNBC, by contrast, contributes little financially to either of its parents, Microsoft (the MS), or GE (the NBC of MSNBC). On the one hand, that might suggest MSNBC is in a precarious position, which has led to speculation about its owners' commitment to the channel in the long term. On the other, having multiple platforms to deliver news and other content may be central to both companies' strategies.

In 2004, each company started a number of activities that attempt to create a "seamless" experience by tying cable news to its online counterparts. In most cases, the cable channels' attempts to boost their ties to their online counterparts revolved around reporters and staff people discussing what their lives are like behind the scenes—postings about the techniques that got them a particular story, the day-to-day business of putting on a television show, or their personal lives. News is not the star—the "cast" is the star.

It is noteworthy, however, that the standing of the three cable networks in the TV ratings does not correspond to their online popularity. Among all news sites, in December 2006 MSNBC.com was the most popular, with 25.6 million visitors; CNN was in second place, with 24.3 million visitors, and FoxNews.com trailed in 18th place, with 6.9 million visitors (Project for Excellence in Journalism, 2007). The next section looks at the economics of the three news channels one at a time.

CNN

Despite the trends—declining viewership and flat revenues in 2004—CNN is still a tremendously profitable organization. In 2004, its roughly $858 million in revenue represented just below 3% of Time-Warner's total revenue for the year ("100 Leading Media," 2004). Publicly, Time-Warner management expressed satisfaction with CNN's performance, and even lashed out at its rivals—in particular Fox News, which Time-Warner's CEO, Richard Parsons, called a network for "crazy people exchanging views" (Eggerton, 2004). On the other hand, there are apparent signs of strain. In late October, CNN announced that it would be shutting down its business news channel, CNNfn. Even with the backing of Time-Warner cable, the channel was never widely distributed. CNN executives said they would be consolidating operations and focusing their efforts on CNN and Headline News. 2005 audience ratings from Nielsen Media Research indicate that their efforts have paid off. From May 2005 to September 2005, Headline News has surpassed MSNBC to be come the third most watched cable news channel.

In addition to synergy across networks, CNN is closely integrated with CNN.com, which has been one of the most popular Web sites since it was first established (even before Time-Warner's merger with AOL). Its high popularity has made it easy for CNN television's sales staff to arrange cross-platform deals with CNN.com (Angwin, 2004).

Another important component of CNN's business model is CNN NewsSource. NewsSource works on a subscription model: Local TV stations around the country pay a fee to have access to the system, and they can both contribute their own video to the NewsSource system and download clips and packages from other stations in order to put together stories on news events from around the country. NewsSource handles the technical aspects of gathering, labeling, and distributing the video over satellite feed systems; it has some 700 clients, the vast majority of them television stations. Indeed, in most markets multiple stations subscribe to the service.[2] Thanks to the NewsSource system, CNN itself has access to video from throughout the country. Indeed, in some cases when a particularly vivid news story is breaking (such as a fire or car chase) CNN has been able to broadcast video from Fox-owned and operated stations that is unavailable to Fox News, because station contracts with NewsSource say CNN takes precedence.

Fox News

In 2004, News Corp., the parent company of Fox News, moved its incorporation from Australia to the United States In some ways, this could be seen as the final step in a process that began in the 1980s when Rupert Murdoch, News Corp.'s main shareholder, became an American citizen to comply with FCC regulations forbidding foreigners from owning TV stations. One reported motivation behind the decision to leave Australia was to take advantage of looser corporate regulation in the United States (Frew, 2004). Another was to make it easier for News Corp. to raise money on Wall Street (Chunovic, 2004). Investors have been unfazed by high-level changes at News Corp. after Murdoch's eldest son, Lachlan quit his executive posts with the global media giant in late July, ("Murdoch's News Corp," 2005). Roger E. Ailes, who had been chairman of Fox News Channel since its inception, was given the additional title of chairman of Fox Televisions Stations—signifying that management is very pleased with his success at the news channel (Steinberg, 2005).

Within News Corp, Fox News appears to be a more crucial element of the corporate organization than either CNN or MSNBC are of theirs. Besides becoming the most-watched cable news channel, it is also the most successful of News Corp.'s American cable channels.[3] A Smith Barney report in June 2004 estimated that Fox News accounted for a third of Fox's cable income.[4] (News Corp.'s other networks are all based on entertainment, including F/X, the National Geographic channel, various regional Fox Sports channels, and others).

Online, however, Fox News' presence is far behind both of its competitors. FoxNews.com has never broken into the top-10 most-visited news sites, whereas CNN and MSNBC are habitually in the top three with more than 20 million unique visitors per month. FoxNews.com usually hovers between 10th and 15th place (Nielsen/NetRatings, 2004). In the hope, it seems, of parlaying its TV popularity into online success, Fox News anchors and reporters are being featured more prominently on the web than they were in earlier years. The anchors of the morning show *Fox and Friends*, for instance, have a group blog where they discuss everything from their kids' elementary school graduation to how an interview went (Romano, 2004).

As of mid-2005, News Corp is launching a new unit grouping together the Internet properties owned by its Fox entertainment, news, and sports businesses, to be called Fox Interactive Media. The division will make it easier for browsers to move from one Fox-owned site to another, personalize what they see and get access to a greater variety of content such as video. News Corp has been considering its Internet strategy over the past several months with the aim of capitalizing on the quickly growing number of advertising dollars being spent on the World Wide Web (Sutel, 2005). That, however, may strike some more as marketing rather than real brand extension. And it is not yet clear that the strategy will prove successful. As the network was gaining more viewers toward the end of the year, FoxNews.com made impressive strides, going from 4.9 million visitors in July 2004 to a peak of 7.9 million visitors in November. But in December 2004 it had dropped back to 5 million viewers, less than 25% of CNN.com's visitors.

Meanwhile, Fox News has also developed a unique new model of synergy via radio. It has long had connections to radio, both in personality and in personnel. The TV hosts Sean Hannity, Bill O'Reilly, and Tony Snow all have their own talk-radio shows. In 2005, the linkage was made explicit when Clear Channel announced a deal for Fox News to provide its stations with news programming, including both hourly news updates and a nightly news program ("Clear Channel," 2004) In a separate deal in Seattle, one radio station owned by Fisher Communications announced that it would start branding itself as "Fox News" radio and featuring talk shows from Hannity, Snow, and Alan Colmes (KVI, 2004.) In 2005, Fox News Radio is estimated to have 500 affiliate stations by summer's end and expanding resources for newsgathering.

Also in 2004, News Corp. completed its purchase of the satellite-TV provider DirecTV. News Corp. has been established in the satellite business overseas for years, through its subsidiaries BSkyB in Great Britain and Star in China. Now, for the first time, it has its own distribution system in the United States and is competing with cable systems for subscribers. With its American incorporation making it easier for News Corp. to borrow money, the company may be looking forward to a period of further expansion. Fox News is expected to launch a new business channel by the end of 2007 and is getting more aggressive about its distribution to other countries, branding itself as the only "American" news network with worldwide distribution (opposing itself to CNN International, which bases a significant portion of its foreign news operations in the regions where they are aired).

MSNBC

MSNBC is in an unusual position, thanks to its co-ownership by Microsoft and NBC. Its inability to break out of third place among the cable channels has led to questions about whether it is a worthwhile investment. But some advantages seem to balance out MSNBC's prognosis.

For example, MSNBC is a useful outlet for spreading out NBC expenses. That has been particularly true for news operations. During the national political conventions, NBC was able to set up a facility to churn out material not only for NBC network news and MSNBC, but owned and operated NBC and Telemundo stations as well. That has also happened in other instances: MSNBC served as an extra channel for coverage of the Athens Olympics, airing some 130 hours of events, or 7.5 hours a day.[5] MSNBC has also been a way for Microsoft to publicize Internet services such as MSN (Microsoft Network).

In terms of content, TV/online synergy has not worked as smoothly as was optimistically expected when the channel was first created.[6] But it may be too soon to discount that potential. MSNBC.com has been one of the most popular spots on the web since its inception, thanks in part to its ties to MSN. Although the site has always featured stories from the NBC News staff and *Newsweek* magazine, in 2004 it became a showcase for MSNBC cable's talk shows and personalities to a greater degree than it had been before. As the Democratic convention began, the Web site introduced "Hardblogger," a blog featuring commentary from *Hardball*'s Chris Matthews, Joe Scarborough of *Scarborough Country*, Joe Trippi, and other MSNBC commentators, and stories from the MSNBC staff about what happens behind the scenes.

Cyberjournalist.net spoke to the importance of web news sites being connected to television: 8 of the top 20 sites have a TV channel to promote them. Nine more are connected to brands with significant print reputations that would be hard to match—*The New York Times*, *The Washington Post*, *USA Today* and *Tribune Company*. The other three are connected with portals—Yahoo, Google, and AOL. Why is MSNBC.com doing so well if its parent perennially trails other TV news channels? "They have a much savvier Web news staff than their competitors have," said Vin Crosbie, an online news consultant. It seems that MSNBC is concentrating on its web presence more than television (Mindlin, 2005, p. 3).

The BBC Factor

In 2004, there was a good deal of talk about the other English-speaking news operation that already produces a cable network, BBC, wanting in on the

American cable news market. The BBC already gained attention with a 30-minute American broadcast on more than 200 PBS affiliates both in the morning and evening. On average, roughly 930,000 U.S. households watched the version of BBC World produced for American audiences in the summer of 2004—twice the 466,000 households of November 2001 (Melvin, 2004). In 2005, the BBC secured a 3-hour block in place of the 30-minute bulletins, marking the first time BBC World News had broadcast for 3 hours straight in the United States. In 2006, BBC Worldwide signed a deal with Cablevision to distribute a 24-hour news channel called BBC World on its digital stream in the New York area, and the channel went on the air in June 2006.

The prospect of a BBC channel oriented for the United States has interesting possibilities. None of the American channels could compete with it in sheer newsgathering muscle. The BBC has 41 bureaus outside the British Isles (including 6 in the United States), whereas CNN has 28 non-U.S. bureaus; Fox and MSNBC have fewer than 10 foreign bureaus apiece (Van Klaveren, 2004). The British news style is not focused around show hosts and anchors, who are called news readers.[7] A BBC channel might well be designed to enter the cable market in much the same way *The Economist* has made huge progress in the American magazine market (which now makes up half its overall circulation) by peeling off some of the most sought-after magazine demographics. BBC, in other words, could become a high-end demographic cable alternative. That could prove particularly difficult for CNN, which has heretofore tried to convince advertisers that it was the elite cable news brand.

The main obstacle in BBC's way is getting clearances. Even in a so-called "500-channel" universe, there is fierce competition for spots on cable system lineups. Even though it has signed a deal with Cablevision and has the benefit of an existing partnership with Discovery Networks, the BBC faces particular resistance from Time-Warner Cable and DirecTV in its efforts. Discovery, in turn, is partially owned by Cox, the fourth-largest American cable provider (Gibson, 2004). With Cox's backing, it might be easier for a BBC news channel to get distribution because Time-Warner and News Corp. both have an interest in making sure their channels get distribution to Cox cable households. Given the wide reach of the three major U.S. cable companies, new cable news competitors will likely find it difficult to enter the market without the backing of another substantial corporate owner with its own distribution system.

NEWSPAPERS

For a fourth consecutive year, ownership changes in 2004 were infrequent and mostly involved smaller dailies or clusters of weeklies. This slow down in conglomeration marked a break with two huge trends of the final third of the last century that transformed the look of newspaper ownership. Starting with *Dow Jones* in 1967 and in gradual succession through the mid-1990s, more than a dozen newspaper companies went from private to public ownership. In turn, they used much of their Wall Street proceeds to finance a tidal wave of acquisitions and mergers. They did not, it should be noted, use the bulk of that money for innovating their product, reaching out to new audiences, or trying to increase readership.

In late 2003, there were a few ownership shifts worth noting. Journal Communications, which owns the *Milwaukee Journal-Sentinel* along with television stations and weeklies, went public and small papers in Merced and Stockton, California were acquired by McClatchy and Dow Jones, respectively. Freedom Communications and its flagship *Orange County Register* drew a lot of interest and a high bid from Gannett and MediaNews, but ultimately members of the Hays family, with the backing of private bankers, were allowed to buy the shares of other, dissatisfied family members who wanted out.

In early 2005, Lee Enterprises announced it had acquired Pulitzer Inc.'s 14 daily papers, which include, most notably, *The St. Louis Post-Dispatch* and the *Arizona Daily Star*. The acquisition was priced at $1.46 billion. Also, at the end of 2004, the E.T. Rollins family sold the *Durham Herald-Sun* to Paxton Media, which dismissed nearly 25% of the staff the day it took possession (Ranii & Biesecker, 2005).

Although there are hundreds of newspaper owners, in 2003 the largest 22 groups continued to dominate both in terms of the numbers of newspapers they owned (39% of dailies) as well as their percentage of the total daily (69%) and Sunday circulations (75%). Although there was very little change in the concentration of the number of daily newspapers and daily circulation among the largest newspaper groups, Sunday circulation among the largest newspaper groups declined 2% in 2003. Circulation concentration was even greater for the 10 companies with the highest total circulation. These chains owned roughly half (51%) of all daily circulation and a slightly higher percentage (57%) of Sunday circulation. These numbers are virtually the same as a year earlier.

A Return to Media Acquisition?

There has been speculation that an eventual easing of FCC regulations would prompt a wave of acquisitions or swaps of both newspapers and local TV stations. But the declining circulation and revenues and lack of direction for how to exist in the new age of electronic news consumption may leave traditional newspapers unappealing as assets.

In addition, it is hard to escape the feeling that the action of the 1970s, 1980s and 1990s has left the field of prospects for acquisitions at least comparatively picked over. Gannett, the leading consolidator of the period, may be a bellwether. Its CEO, Doug McCorkindale, has frequently complained in conversations with analysts about the "funny money" multiples now being asked for newspapers and

TV stations. Gannett has not bought anything big since Central Newspapers in 2000. Instead it has made a series of purchases to assemble Newsquest, a Gannett-like chain of regional newspapers in England, and dramatically improved their financial performance. In 2003, Gannett bought Captivate, a small company that places advertising in elevators and other public places. In 2005, it bought Home Town Communications Network, a small chain of 63 non-dailies in Ohio, Michigan, and Kentucky and announced a swap of ownership with Knight Ridder of the two Detroit dailies and a renegotiation of the joint operating agreement (or JOA, which means that two daily newspapers published in the same city or geographic area can operate certain business aspects together) which gives them a greater cut of profits ("Gannett to acquire," 2004).

At independent newspapers, smaller private chains, and even the smaller public companies protected by a special family-controlled class of stock, both financial and generational factors bear on the decision about whether to sell or continue. Younger family members losing interest in newspapering or thinking management is underperforming, can prompt a sale. But by now many of the surviving independents could be characterized as rugged, determined, and even stubborn. Eased inheritance laws also make a transfer of control from one generation to the next less punishing than it used to be.

Another possibility is that extremely wealthy individuals could become owners of individual papers, just as many own sports franchises. Wendy McCaw used some of the proceeds of her divorce from the telecommunications billionaire Craig McCaw to buy the *Santa Barbara News-Press* in 2000. Forbes 400 regular Phillip Anschutz acquired the battered *San Francisco Examiner* franchise in 2004, later bought several suburban weeklies outside Washington DC, and announced plans to publish free papers under the Examiner name in other cities including Baltimore, Maryland.

Private Companies

Revenues, earnings, and profit margins are generally not available for private newspaper companies. But they are among the biggest chain owners in the industry—by circulation, 4 of the top 12, and 9 of the top 20. The largest (with circulation rank) in 2004 were Advance (4th), Hearst (7th), MediaNews Group (8th), Cox Newspapers (11th), Freedom Communications (13th), Community Newspaper Holdings Inc. (14th), and Morris Communications (19th) (Ginocchio, 2003, for rankings). Advance operates 26 dailies, and has a reputation for relatively generous news spending and markedly improving most of its properties over the last 20 years. Hearst has 12 dailies, and drove its competitors under in Houston, San Francisco, and San Antonio and is hanging on in Seattle. Hearst is also a successful operator of magazines. The MediaNews Group developed a large empire built of 45 papers over a period of 30 years, running many distressed franchises that no one else wanted, with bare-bones staffs. Cox Newspapers owns 17 papers, and has a reputation for above-average news investment and quality. Freedom Communications owns 28 papers. Community Newspaper Holdings Inc. owns 96 small newspapers. *The Daily News*, New York (18) is a one-newspaper company but a big one. Morris Communications runs a chain of 27 mid-sized and small papers, mostly in the south, known for tight operations and high profit margins.

RADIO

For radio ownership, 2004 was a year of stability on the surface. But there was some movement in the second and third tier of companies, as the ethnic media and specialized companies continued to thrive.

With 1,194 stations, Clear Channel owned nearly 900 more than the next ownership group, Cumulus Broadcasting (305), followed by Citadel Broadcasting Corp (220), Infinity Broadcasting (183), American Family Association, Inc. (109), Entercom (103), Saga Communications Inc. (86), and Cox Broadcasting (78). In fact Clear Channel, which reaches 188 markets (followed by Cumulus (60), Citadel Communication (49), Infinity Broadcasting (42), and Entercom (21) (BIA Research, Inc.), dominates the industry, as does no other single company in any other medium.

Although 2005 brought little change to the top 20 ownership groups, the realities of massive consolidation showed some signs of setting in. First, the fiscal realities of radio ownership were less bright than they were during the 1980s and 1990s. In early 2005, Viacom and Clear Channel Communications announced dramatic write-downs for their radio holdings. Both claimed that this was a result of new reporting regulations, not a signal of fiscal difficulties, but Viacom has since announced its intention to split into two companies. Infinity radio network was to be grouped with UPB, CBS, Viacom's outdoor advertising, promotion, and book publishing units. In late 2006, Clear Channel announced it was being acquired by two private equity firms and by the year's end, the company was discussing selling off 448 of its smaller-market radio stations. Also in 2006, the Disney company finalized its decision to sell ABC Radio, consisting of 22 stations, to Citadel Broadcasting.

Second, newer audio offerings were leading analysts to scale back 2005 predictions. This was not surprising as devices like iPods, portable satellite radio receivers, and media-rich phones pushed audio technology forward at a pace unimagined a few years ago. According to the RBR Newsletter (2005), "Universal McCann's Bob Coen started the year with a prediction that radio revenues would rise around 5% this year. He cut that aggressive view back to 3.5% last month for local radio, but held onto the 5% for network and national spot combined." About 1 week prior to that report, Marcia Ryvicker, an analyst for Wachovia Securities, cut her 2005 growth estimate to 1.5%

Finally, bigger does not always prove better. Based on financial information from 2003, Emmis Communications, for example, did not make the top 20 in number of stations owned, but sat at number 10 in revenue. Some of the largest companies in number of stations owned (including three of the top five—Clear Channel, Viacom, which owns Infinity, and Citadel) saw the greatest decline of overall station ratings, whereas some smaller groups saw the greatest increase. Two smaller companies, Entravision (53 stations) and Univision Communications (72 stations) made news in an August 2004 *Barron's* article for some of the industry's greatest ratings gains, demonstrating that although huge station numbers might be impressive, success in radio still depends on finding the right format to attract an audience.

Spanish-language programming, along with religious and urban formats, continues to be a potential area for audience growth. Cultural and lifestyle trends in the greater population make these types of programming profitable and attractive. For example, it has been estimated that Hispanic buying power will total some $992 billion by 2009 (Humphreys, 2004). The 2004 presidential campaign made it very clear that urban minorities and people of faith are key audiences that wield increasing power in the country. In September 2004, Clear Channel announced plans to convert as many as 25 of its stations to Spanish-language programming. In October 2004, Infinity entered what was referred to as a "joint strategic alliance" with Spanish Broadcasting System to assist with and consult on Infinity's movement into Spanish-language programming.

In an industry where stations often change formats overnight, big company movement into Spanish-language programming has been slower than originally projected. According to 2005 data from BIAfn, only 16 of the 25 named Clear Channel stations had been converted to Spanish as the primary format. Infinity has just three Spanish-language stations listed.

In the end, a boom in Spanish-language programming might not do much to improve the state of radio news stations. Of Clear Channel's Spanish-language holdings, only one operates with a news format, as does one of the Infinity stations. For ownership groups, this is most likely a simple matter of business and marketing. Arbitron's 2005 Radio Today report lists Hispanic listenership to news/talk/information public radio stations at less than 6%.

MAGAZINES

The top 10 magazine companies in 2003 were the same as in 2002, although there was some shuffling among the ranks. But in the end, every company in the magazine game is playing on a completely different field from the number one company. Time-Warner is the largest media company in the country by far, owner of 32% of the magazine revenue ($4,514 million) among the top 10 magazine companies, followed by Advance ($2,197 million), Hearst ($1,900 million), Primedia ($1079 million), Readers Digest ($835 million), Meredith ($808 million), Gruner & Jahr/Bertelsman ($788 million), International Data Group ($760 million), McGraw-Hill ($688 million) and Reed Elsevier ($601 million) ("100 Leading Media," 2004).[8]

In 2004, the changes again were minimal. Meredith climbed from number 6 in revenues to number 4, after buying a handful of magazines from Gruner + Jahr. That company, on the other hand seems to be largely leaving the magazine business, selling off the magazines to Meredith as well as *Fast Company* and *Inc.* It no longer resides anywhere in the top 100 media companies, according to *Advertising Age.*

The magazine industry ownership is peculiar in that the biggest of the big media companies are largely forgoing magazines as potential outlets. Of the top 10 media companies overall, only Time-Warner has invested seriously in the business, and that is largely an accident of history. The company is rooted in magazines; Time Inc. was started in 1923 as a magazine company, and then expanded into cable with HBO. That said, after mergers with Warner and later AOL, today only 15% of Time-Warner's media revenue comes from magazines. The multimedia giant Disney is a player in the magazine game, although in revenues it places only at number 20 among magazine companies. And although the media heavyweights Viacom and News Corp. do publish magazines, they don't even come close to making the top 25 magazine companies by revenue ("100 Leading Media," 2004).

News Magazine Owners

The biggest media companies generally steer clear of magazines and the biggest magazine companies generally steer clear of the news genre, with of course the exception of the biggest, Time-Warner. *Newsweek*, the number two news magazine by circulation and ad pages, is owned by the 15th largest magazine company, The Washington Post Co. *U.S. News and World Report*, the number three news magazine, is owned by the 25th largest magazine company, Zuckerman Media Properties. Neither of those two companies is a magazine company per se. The Washington Post Co. has its hands in many media, but is most heavily invested in newspapers, although it made an interesting play for online readers with its purchase of *Slate* at the end of 2004. And Zuckerman owns one newspaper and one magazine—the company was reportedly in hot pursuit of *New York Magazine*, which Primedia sold in 2003, but it didn't get it ("100 Leading Media," 2004).

Outside of the Big Three, ownership in the magazine world looks a bit more diverse, and there are opportunities for others to get into the field. *The New Yorker*, which has increasingly waded into the more "newsy" side of content, is owned by the magazine industry's second-largest player, Advance. That ownership has given the magazine some of the same advantages Time has: the ability to sell ads to an entire raft of books rather than simply one title and even the ability to consolidate things like office space—not a small concern in New York City. These advantages may be yielding dividends to the magazine, which is thriving. Still, it is hard to see *The New Yorker* ultimately as a model for innovation in the news sector.

The biggest change in newsmagazines in the last several years has been the launch of *The Week* by publisher David Bradley. The news magazine creates no new original content but repackages already published pieces in its weekly wrap-up. According to its 2005 publisher's statements, both circulation and revenues were climbing rapidly.

Aside from this newcomer, however, ownership among the news titles looks as stable as ever with little talk of any new titles emerging. There does not seem to be even the faintest chance that Time-Warner will unload its namesake magazine anytime soon. The Washington Post Co. considers *Newsweek* a critical part of its brand exposure, particularly overseas, where the newspaper carries less weight. And Mort Zuckerman, who has held *U.S. News & World Report* through a long down period, has shown no interest in selling it now that ad dollars seem to be creeping up and circulation has ceased dropping, at least for the moment. Neither the Washington Post Co. nor Zuckerman has been interested in creating new titles in news. They already have players in the game, and considering their small footprint in the industry they have not appeared interested in committing the serious money it would take to put forward a new magazine.

Time-Warner is the medium's 800-pound gorilla. In theory, it is well positioned to try to innovate and remake the news magazine. It certainly has advantages: a special knowledge of how the industry works, a vast stable of writers and editors, and a fat checkbook. The company is also showing an interest in launching new titles. After a few years of weathering the economic storm, and the absorption of AOL, the company for 4 years now has been entertaining new ideas focused on news niches. In 2004, it has revived and relaunched *Life* as a magazine inserted into 12 million Friday newspapers around the country, including some big dailies.[9] In March 2007, this venture also ended. Time Inc. shut down the print magazine and planned to keep the brand alive on the internet as a Web site (Tribune News Services, 2007). But for whatever reason—perhaps unease about cannibalizing *Time*'s readership or simply more interest in other, more profitable areas—the company seems uninterested in trying something new in news aimed at new demographics.

Moving Online?

News magazine owners have yet to take decided action in the medium that seems to be the future of the news media, the Internet. Nearly all good-sized metropolitan daily newspapers are on the web. The big names, like *The New York Times* and *The Washington Post*, have become large-scale national Web sites: nytimes.com and washingtonpost.com. The broadcast networks are all represented in the top 20 most visited Web sites. But the news magazines have yet to take off on the web. And magazine owners have not yet expressed great interest in it beyond the efforts of some of the bigger owners to establish a presence for their titles there. In fact, the Washington Post Co., in a push for a bigger presence on the web, did not expand *Newsweek* online; instead, in late 2004, it bought *Slate*—a title that was born and exists on the web.

The web has proven helpful to a handful of more niche titles, however. *The Nation* reports that its free Web site has proved "an invaluable source of paid subscriptions to the hardcopy magazine." Interestingly, a few of the more successful commercial startups on the web—those launched without old media roots—call themselves "magazines," outlets like *Slate* and *Salon*. But they do not fit with the traditional magazine model of new content every week or month. They are updated more regularly, or as news happens. If these kinds of Web sites end up being the model for magazines on the web, it is possible that the true innovation in the magazine sector will jump right over the old print form of the news magazine and take place online. Smaller owners could compete more easily in a world without the costs of paper and mailing and might be more able to try new ideas. It is not clear, however, how these small owners could pay for the newsgathering they would need to compete with older established titles. *Slate*, underwritten by Microsoft, was just sold to the Washington Post Co., and *Salon* is struggling to stay afloat.

ONLINE

Who owns the web? Has the people's medium been taken over by big media? Or does attention to the big companies and their popular sites, the most obvious part of the web, miss the energy and verve of the Internet? There seems to be, in effect, two Internets: one that aggregates large numbers of users around a limited number of sites, and another that was rapidly changing, full of energy and innovation. The success of blogs in 2004 suggests that this other, less visible Internet is even more robust now than a year earlier. But data on where people go on the Internet also suggest that at the same time the biggest sites are getting even bigger.

Big, Bigger, Biggest

Although the Internet allows for a nearly infinite choice of news outlets, in practice a few sites dominate, and in 2004 they appeared to be separating themselves even more from the rest of the web. Two organizations are the central sources for online traffic figures, Nielsen/NetRatings and Comscore MediaMetrix. Both show surges in 2004 traffic among the top sites, although their numbers for specific sites vary somewhat.

The top 20 news sites, as measured by Nielsen/NetRatings, averaged 69.6 million visitors per month from January to October 2004.[10] According to Nielsen/NetRatings data, CNN led the way during these months with 23 million, followed by Yahoo! News with 21.4 million and MSNBC with 21 million. AOL News, according to these figures, trailed behind with 14.4 million unique visitors, a figure quite different than that reported by comScore. ComScore reported Yahoo! in the lead with an average of 23.2 million, followed by AOL News at 22.6 million. CNN and MSNBC then followed with 21.3 million and 19.3 million, respectively.[11] After that there was a huge dropoff in traffic—an average of around 7.1 million visitors per month. Through October 2004, Yahoo! News showed the most growth among the top three sites since 2003, up 27%. CNN was up 15% and MSNBC 3%. Traffic figures from both comScore and Nielsen/NetRatings in the first half of 2005 showed these three sites to have held their lead. Yahoo!, MSNBC, and CNN took the top three spots each month.

"Traditional" journalism brands still seem to hold the most appeal. Of the top 25 Nielsen/NetRatings sites from January to October 2004, 17 were associated with traditional news companies—those that produce their own offline content for newspapers, television, or magazines. Five sites were unique and could not be linked to traditional offline news providers. The three sites that aggregated data from other news services and did not do any original reporting were AOL News, Yahoo! News, and Google News. AOL News and Yahoo! News use editors who post stories from other news services and do not do any original reporting. And although Google News also offers a considerable number of wire stories from other news sources, it does not use human editors but rather an algorithm to find articles, and a specific search often turns up material from outside the mainstream news pool. A second category among news sites not linked to traditional offline news providers comprises those that produce Internet-only content, such as *Drudge Report* and *Slate.* And the last category is those sites that produce local online news content: Advance Internet, World Now, and Internet Broadcasting Systems.

In 2005, there was some blending of these categories. Yahoo! News announced plans to begin creating some of its own original news content and perhaps adding original video or photos surrounding big story coverage (see *Fortune Magazine*, August 8, 2005). Google has not yet invested in original news content—even on the limited scale that Yahoo! has done. Rather, it has perpetuated its image largely as a software company rather than a media company.

Local News May Be Getting More Important

Looking further down the list, another distinction is worth noting. Several of the names are not actually single Web sites, but aggregations of various sites owned by the same company, such as Gannett's local newspapers, or those of Knight

Ridder. Of the top sites for the first 10 months of 2004, 12 were actually single sites. In addition to CNN, Yahoo!, and MSNBC, they include, in order, AOL, *The New York Times*, *USA Today*, ABC News, Google News, WashingtonPost.com, Fox News, CBS, and the BBC. Some of the strongest growth online was occurring at those sites operating at the local level. For the first 10 months of 2004, Gannett and Knight Ridder, which reported the traffic of all their newspapers' Web sites in aggregate, saw increases of 37% and 22%, respectively, over the previous year.[12] And Internet Broadcasting Systems, a company that operates local TV news Web sites, experienced an increase of 36%.

New Players in the Top 20

Still a third story line in online news growth in 2004 involved two other sites, the Associated Press' (AP) CustomNews service and Google News, which gained footholds in the top 20 news sites for the first time. AP CustomNews is designed primarily for smaller sites. The AP sends stories to a site in the newspaper's own design and structure, thereby requiring minimal investment in hardware, bandwidth or staff ("CustomNews," n.d.). Over the first 10 months of 2004, AP CustomNews received 6.2 million unique visitors a month.

Google News was unique among the news sites in the top 20 list in 2004. In addition to its famous search engine, which allows visitors to search for a seemingly endless supply of news stories on topics they are interested in, the site uses a computer algorithm that selects headlines based on how and where the stories appear in other places on the Web.[13] Although the headlines are often from major players like Reuters.com, there are frequently news stories from obscure and unorthodox news and political sites. The site grew from 4.1 million unique visitors in January 2004, the first time it was reported, to an average of 6 million unique visitors per month. In 2005, Google News remained a solid figure in the top 20 news sites.

Online Media Ownership Trends

Of the 25 online news sites in 2004, 8 in 10 are owned by media companies in the 100 largest in revenue as reported by *Advertising Age* ("100 Leading Media," 2004).[14] For example, CNN was the most visited news site and AOL News was fourth. Both sites are owned by Time-Warner, the leading media company in total U.S. media revenue in 2004. Nearly one third (32%) of the top 25 news sites in 2004 were owned by the top 10 U.S. media companies; more than half (56%) were owned by the 20 biggest companies. The next 80 media companies owned 24%, and all others owned 20%. Thus, the data point to heavy concentration of ownership of the news the public is getting online.

Acquisition of Non-News Sites

2005 saw a *new* trend emerge—that of large media companies acquiring smaller Internet properties. Over a 16-month period from early 2004 to mid-2005, Scripps bought Shopzilla; Dow Jones bought MarketWatch; The New York Times Co. bought About.com; and Gannett, Knight Ridder, and Tribune Co. acquired 75% of the local news aggregator Topix.net.

In late summer 2005, the News Corporation spent $1.5 billion on three Internet companies in just 7 weeks. According to *The New York Times*, Murdoch "has been particularly concerned about younger audiences spending less time reading newspapers and watching television, while spending more time online and embracing such features as interactivity and virtual communities." With their acquisitions, the News Corporation will rank fourth among companies in terms of monthly page views, behind Yahoo!, Time-Warner and MSN, but ahead of eBay and Google. However, this acquisition looks a little different from the 1990s: "Rather than taking small stakes in promising start ups, it is focusing on buying companies outright that are somewhat proven and generating operating earnings" (Siklos, 2005).

ETHNIC MEDIA

The "ethnic media" actually comprise hundreds of language-defined media outlets, with distinct differences both within and across languages. The Spanish-language press of New York is very different from that of Los Angeles. And what commonality is there between the Mandarin and Cantonese radio stations of greater Los Angeles and the Pakistani press of New York City, beyond the fact that they target non-English speakers?

For most of the ethnic press, ownership is not a complicated matter. Shares are not traded on the Nasdaq or the New York Stock Exchange. A person or a group of people or a family decides their community needs to be served by a news outlet and a "media organization" is born. Operations are often hand to mouth, and advertising revenues are usually generated by a handful of local businesses, many of which are owned by members of the local ethnic community.

But now some of the larger ethnic groups—again, particularly Spanish-speakers but also some Asians—are finding that larger scale advertisers are interested in reaching their communities. At the same time, some large owners have waded pretty deeply into the pool of the ethnic media. NBC's purchase of the Hispanic network Telemundo and Tribune's launch of its "Hoy" dailies are prime examples. In October 2004, Univision signed a $100 million television and radio advertising contract with Miller Brewing ("Univision, Miller," 2004). Even before the deal, in the first 6 months of 2004, the company's net income was more than double what it was in the first half of 2003—$115 million versus $54 million ("Univision, Miller," 2004). That follows steady increases over the past few years in net income, from $54.7 million in 2001 to $155.4 million in 2003 (Hoovers Company).

Beyond the Spanish-language media, however, there are three large barriers preventing large-scale national growth: lack of a single common language, thus reducing single audience size; lack of geographical concentration of many ethnic groups; and less interest on large-scale, national advertisers in smaller ethnic groups.

2004 brought some interesting changes in ownership in the ethnic news media. As the growth of audiences continued through the last decade, a question emerged about what the ethnic press of the future might look like. Would the various ethnic communities across the country grow to the point where large, English-language companies would begin trying to buy into the audiences that they missed with their mainstream outlets? That happened relatively early on with Knight-Ridder's creation of *El Nuevo Herald* to capture readers who might not read *The Miami Herald.* Or, as the ethnic populations grew, would competitors from outside the traditional big media companies emerge and develop their own foreign-language equivalents of the English-language chains? That has happened in television with Univision, which has evolved into the fifth largest television network in America.

Spanish-language papers in the three largest U.S. cities—New York, Los Angeles, and Chicago—were joined under one company, ImpreMedia in 2004. *La Opinion* in Los Angeles, *El Diario/La Prensa* in New York, and *La Raza*, a Chicago weekly, will now find out if a national corporate identity and presumably a more nationalized approach to some ad sales is a ticket to better times. Together *La Opinion* and *El Diario/La Prensa* have a daily circulation of more than 175,000 and *La Raza* reports a weekly readership of nearly 170,000.[15] In theory, the three could create a force in the Spanish-language print market in terms of ad sales and the ability to share bureau resources. But although the readers of all three papers speak Spanish, the similarities between their readerships end there. The papers serve very different populations—*La Opinion* largely a Mexican immigrant group, *El Diario* and *La Raza* a more diverse readership from a variety of Latin American countries. It seems thus far that there is no rush to create one national paper out of the three. In part, that may be due to respect for each paper's individual identity, but it may also stem from the belief that a true merging of the papers into a national daily wouldn't work.

At the same time, Tribune Co., which once co-owned *La Opinion*, has pushed forward with a version of *Hoy* on newsstands in New York, Chicago, and Los Angeles. A big question is whether the company, which may want to apply a one-size-fits-all approach to all its Hoys, will understand the differences in the populations it is dealing with. On a more local scale, Meximerica Media (funded by the Spanish company Recoletos Grupo de Comunicacion S.A. to the tune of $16.5 million in 2004) launched four daily newspapers in Texas in 2004.

Traditional English-language companies continue to step into the foreign-language marketplace, although as a rule gingerly. In the Dallas-Fort Worth market, Knight-Ridder and Belo have Spanish-language papers competing with each other, *La Estrella* and *Al Dia* respectively. In Denver, the Media News Group's *Post* and Scripps' *Rocky Mountain News*, which are run under a joint operating agreement, are working together on a Hispanic weekly that would take on the long-time weekly *La Voz*. In Detroit, another JOA city, the News (Gannett) and Free Press (Knight-Ridder) are reportedly looking into starting a weekly to compete with the local bilingual weekly *El Central*. Over all, mainstream news organizations own 46 (mostly weekly) Hispanic publications that have a combined circulation of 2.9 million, according to the Latino Print Network (Tim, 2003).

And it is not just Hispanic publications. In San Jose, Knight-Ridder's Mercury News has been running the weekly *Viet Mercury*, a newspaper aimed at the area's large Vietnamese population (110,000 living in a relatively small area with its own shops and businesses), since 1999. Interestingly, the *Viet Mercury* looks like a local Vietnamese-American newspaper rather than a copy of the *San Jose Mercury* in Vietnamese (see *San Jose Mercury News* Web site, jobs.sjmercury.com/about.html).

In radio, Clear Channel, the medium's giant, announced in September 2004 that it was launching a bigger push into the Spanish-language market. The company had only 18 Spanish-language stations when it announced the move, but planed to convert 25 more stations to the format in 2005 and early 2006. By 2006, Clear Channel Radio had 29 stations airing Spanish-language programming in 24 markets. But even with an additional 25 stations, less than 4% of all the company's stations would be broadcasting in Spanish (Jeff, 2004).

In television, where there are fewer outlets, there is a more clear-cut winner so far in the competition between the two ownership approaches: Hispanic-run Univision is ahead of NBC-run Telemundo. How successful has Univision been? It doesn't just run the most popular Spanish-language network in the United States, it also runs the second most popular, Telefutura, aimed at a younger audience. The question for NBC is relatively straightforward: Are Telemundo's struggles a function of bad management or poor programming or could it be that the network simply hasn't put the resources into Telemundo that it needs to compete? NBC and Telemundo haven't sat completely still. Like Univision, Telemundo has added a second network, Mun2, or Mundos, aimed at younger viewers, aged 18 to 34. But over all, Telemundo is something of a side-project for NBC, one of many networks the broadcast giant is using to reassemble the mass audience it had when it was one of only three options available to viewers.

As the nation's immigrant population grows, the universe of the ethnic media will continue to expand. Economically speaking, the ethnic media are operating on at least two tracks: the small-scale, local business model of most ethnic media and the large-scale national model of some of the Spanish-language press. One great unknown in the discussion of the ethnic media is how U.S. government policy will affect immigration, and ultimately the next generation of ethnic media readers, listeners and viewers.

ALTERNATIVE MEDIA

Ownership among the biggest players in the alternative press remained stable in 2004, with no major openings or closings. New Times, the largest owner, held on to its 11 weeklies, whereas Village Voice media kept its 6 (AAN Member Web site pages). There was activity in specific markets around the country, often involving battles between large companies and independent owners. The increasing revenue numbers and desirable demographics the alternative weekly publications offer has not gone unnoticed. In San Francisco in October 2004, the family-owned *Bay Guardian* filed suit against two New Times weeklies in the area, charging that New Times and the weeklies had sold ads below cost to put the *Guardian* out of business (Redmond, 2004).

In Boston there was a new owner when Metrocorp, the publisher of *Boston Magazine* and *Philadelphia Magazine*, purchased a majority ownership of *Boston's Weekly Dig*. The *Dig*, with a light approach to the news and a lad-magazine sensibility (think *Maxim*) is the self-proclaimed fastest-growing alternative weekly in the area with an unaudited circulation of 50,000 (AAN Member Web site pages). That number still places it far behind the long-time leader, the *Boston Phoenix*, which has an unaudited readership of 109,000 (AAN Member web site pages). But the influx of dollars and resources Metrocorp could put into the *Dig* may make it more of a competitor.

New Challenges

In addition to battling among themselves, alternative weeklies are tying to ward off competition from two different fronts. First is the arrival of free dailies, small ad-driven tabloids given out by daily newspapers primarily to mass-transit commuters. At the same time they are seeing the rise of what they call "faux alternatives," weekly ad-driven tabs started by the towns' dominant daily newspapers, aimed at competing with the alternative weeklies directly.

Those changes have all hit the alternative weeklies where they are most vulnerable: in advertising revenue, where the weeklies get all their money for operation, as they don't get any boost from subscription dollars. Yet the new competition, while potentially threatening, has not yet dented in any real way the bottom line of the alternative weeklies. After 2 or 3 soft years, industry revenues have begun to climb again, albeit not at the torrid pace of the 1990s. The alternative weeklies are arguably the most sensitive to economic problems. The question is whether the weeklies are simply more resilient than other media that have suffered against new competition—like newspapers that have suffered with the Internet's growth—or whether these alternative weeklies' challengers are still relatively immature.

The weeklies' new competitors for revenue bring some distinct advantages to the table. The commuter dailies and chain-owned free weeklies have more extensive distribution networks in place. They can sell ads at special rates for both the daily paper and the free paper. Their local marketing muscle can overwhelm smaller competitors. And the fact that they are affiliated with or owned by the local daily newspapers means they often have the advantage of lower printing costs. Where classifieds are concerned, Internet services can offer search engines that make the task easier for those seeking goods and services, along with the ability to update with new listings regularly. Most alternative weeklies have their own Web classifieds, and many of them have been quite successful in generating revenues. But services like Craigslist are free, and that is hard for any pay-for-ads Web site to compete with.

NOTES

1. Data gathered from *Broadcasting and Cable Yearbook* and back issues of *Broadcasting and Cable* magazines.
2. In spring 2004, NewsSource became briefly notorious when it became a conduit for retransmitting a video news release praising the administration's Medicare bill. For more on the system, see Kerschbaumer, (2003). CNN heats up with Pathfire. *Broadcasting & Cable.*
3. According to Dave Devoe, News Corp. CFO, in a conference call with investors published by business news wire FD Wire on May 6, 2004.
4. Smith Barney Citigroup report, June 17, 2004.
5. NBC press release, June 10, 2004.
6. For example, titles of early programs included *Internight* and *The Site.* See Thompson (2004, Oct 27). MSNBC and the cable news war. *Palm Beach Post.*
7. For a comparison of how BBC World and American cable news covered the war in Iraq, see (2003). Bercovici, J. (2003, March 31). BBC World, for those in the know. *Media Life Magazine.*
8. Companies are ranked by their total media revenues collected in the United States in 2002. The list is available at http://www.adage.com/page.cms?pageId=1084.
9. Figures from *Life* magazine. http://www.life.com/Life/lifeabout.html
10. Nielsen/NetRatings. The top 20 sites from January to October 2004 include CNN, Yahoo! News, MSNBC, AOL News, Gannett Newspapers and Newspaper Division, Internet Broadcasting Systems, Knight Ridder Digital, NYTimes.com, Tribune Newspapers, USATODAY.com, ABCNEWS Digital, Associated Press, Hearst Newspapers Digital, Google News, Washingtonpost.com, Fox News, CBS News, WorldNow, Advance Internet, and BBC News. The next five (referenced later in the report) are MSN Slate, McClatchy Newspapers, drudgereport.com, Cox Newspapers, and Boston.com.
11. Beginning with September 2004 monthly data, Nielsen/NetRatings improved its NetView U.S. home panel sample weighting methodology by updating projections for key segments of the Internet universe, improving the accuracy of audience and reach

projections. As a result of the data reporting changes, variances with the AOL data exist prior to September 2004. Subsequent to the methodology update, AOL News' traffic garnered 17.6 million unique audience during September 2004.

12. 2003 Knight Ridder data did not include January through March.
13. The algorithm Google uses is called PageRank, a system developed by Google's founders at Stanford University in Palo Alto, California. Google explains how PageRank operates: "PageRank relies on the uniquely democratic nature of the web by using its vast link structure as an indicator of an individual page's value. In essence, Google interprets a link from Page A to Page B as a vote, by Page A, for Page B. But Google looks at more than the sheer volume of votes, or links a page receives; it also analyzes the page that casts the vote. Votes cast by pages that are themselves 'important' weigh more heavily and help make other pages 'important'." (http://www.google.com/technology/). For a critique of PageRank, go to http://www.google-watch.org/pagerank.html.
14. Companies are ranked by their total media revenues collected in the United States in 2003. The list is available online at http://www.adage.com/page.cms?pageId=1084.
15. Information from *Audit Bureau of Circulations*, *El Diario*, *La Opinion* and *Impremedia* website, www.impremedia.com/newspapers/.

REFERENCES

100 Leading Media Companies. (2004, August). *AdAge.com.* http://www.medialife magazine.com/news2003/mar03/mar31/3_wed/news4wednesday.html

Angwin, J. (2004, May 20). The right price: For Fox News, ad-sales market isn't fair, balanced. *The Wall Street Journal*, p. A1.

Bercovici, J. (2003, March 31). BBC World, for those in the know. *Media Life Magazine.*

BIA Research, Inc. BIAfn Media Access Pro online database.

BIA Research, Inc. (2005, May 12). *BIA Financial Network reports on revenue growth of top television station owners over the past four years.*

Burt, T. (2004, November 30). A new media model? *The Financial Times*, p. 17.

Burt, T., & Yeager, H. (2004, April 7). Disney ponders changes at ABC Network. *The Financial Times*, p. 24.

Carter, B. (2004a, November 24). Courage, CBS news. *The New York Times*, p. C1.

Carter, B. (2004b, May 13). Deal complete, NBC is planning to cross-market. *The New York Times*, p. C1.

Carter, B., & Steinberg, J. (2004, November 25). Improved numbers shelter CBS from the fallout. *The New York Times*, p. C1.

Chunovic, L. (2004, October 25). A "fierce warrior" learns the ropes. *Broadcasting & Cable.* http://www.broadcastingcable.com/article/CA474533.html?display=Special +Report

Clear Channel chooses Fox for news deal. (2004, December 6). *Associated Press.*

CustomNews: Turn your Web site into your audience's online news source. Available online at: www.apbroadcast.com

Eggerton, J. (2004, August 9). Parsons takes shot at Fox. *Broadcasting & Cable*, p. 2.

Emmis may sell TV stations (2005, May 10). *Austin Business Journal.* http://www. bizjournals.com/austin/stories/2005/05/09/daily18.html

Freedom du Lac, J. (2004, December 7). Viacom move puts local TV folks on edge. *Sacramento Bee*, p. E1.

Frew, W. (2004, November 3). Murdoch packs his kitbag and heads off. *Sydney Morning Herald*, p. 23.

Gannett press release. Gannett to acquire Home Town Communications, (2004, November 19).

Gibson, C. (2004, September 27). BBC World plugs into US cable talks. *The Evening Standard*, p. A34.

Ginocchio, P. (2003, August 11). *Newspaper publishers keeping cash in circulation.* New York: Deutsche Bank Securities.

Higgings, J. M. (2004, December 13). Still strutting after all these years. *Broadcasting and Cable.* http://www.broadcastingcable.com/article/CA487193.html?display=Special +Report

Hoovers Company. *Univision Corporate Profile.* www.hoovers.com

Humphreys, J. M. (2004). The multicultural economy 2004. *Georgia Business and Economic Conditions*, *64*(3), 6.

Jeff, L. (2004, September 17). Clear Channel ready to push Spanish radio. *New York Times News Service.*

Kerschbaumer, K. (2003, February 3). CNN heats up with Pathfire. *Broadcasting & Cable.* http://www.broadcastingcable.com/article/CA274610.html?display=Technology.

KVI, Fox News partner for first Fox-branded radio station. (2004, September 20) *Radio Ink.* http://www.radioink.com/HeadlineEntry.asp?hid=124946&pt=archive

Matthews, S. (2005, October 11). *BBC hints at launch of 24-hour US news station.* Brand Republic.

Melvin, D. (2004, August 26). More Americans turn to BBC for news. *Cox News Service.*

Mindlin, A. (2005, September 19). Browsing the site, not watching the show. *The New York Times*, p. 3.

Murdoch's News Corp sees earnings jump (2005; August 10). *Agence France Presse.*

Nielsen/NetRatings. (2004, October). *Top 20 online current events & global news destinations.*

Project for Excellence in Journalism. (2007, March 12). *Online Ownership.* State of the News Media 2007. www.stateofthenewsmedia.orf/2007/index.asp

Ranii, D., & Biesecker, M. (2005, January 4). Herald-Sun fires executives, others. *Newsobserver.com.*

Radio & Television Business Report Newsletter. (2005, August 29). www.rbr.com

Redmond, T. (2004, October 20-26). Bay Guardian sues New Times chain for predatory pricing. *San Francisco Bay Guardian.* http://www.sfbg.com/39/03/news_newtimessuit .html

Romano, A. (2004, July 12). Behind the cameras. *Broadcasting & Cable.*

Siklos, R. (2005, September 9). News Corporation with IGN in its stable, backs up promise to be bigger web player. *The New York Times.*

Steinberg, J. (2005, August 16). News Corp. starts to fill shoes of departed Murdoch. *The New York Times*, p. C1.

Sutel, S. (July 18, 2005). News launches Fox interactive–a new net division. *AP Financial Wire.*

Thompson, K. D. (2004, October 27). MSNBC and the cable news war. *Palm Beach Post.*

Tim, P. (2003, October/November). Dismantling the language barrier. *American Journalism Review.*

Top news sites for December [2004]. (2005, January 12). *Cyberjournalist.net.* http://www.cyberjournalist.net/news/001831.php.

Tribune News Services. (2007, March 27). Life Magazine's 3rd Life Coming to an End. *Chicago Tribune.*

Turner, T. (2004, July/August). My beef with big media. *Washington Monthly*, p. 30.

U.S. policy: Telecommunications Act of 1996, Federal Communications Commission. *Encyclopedia of television.* http://www.fcc.gov/telecom.html

Univision, Miller toast $100 million ad deal. (2004, October 8). *Los Angeles Business.* http://www.bizjournals.com/losangeles/stories/2004/10/04/daily50.html.

Van Klaveren, A. (2004, November 4). Newsgathering. *BBC News.* http://news.bbc.co.uk/newswatch/ifs/hi/newsid_3960000/newsid_3964600/3964687.stm

11

A CRISIS IN THE NEWSROOM

A. L. Bardach

TOP EDITORS AND JOURNALISTS DISCUSS MEDIA CONCENTRATION AND MEDIA BIAS

On May 7, 2005, The Media Project of UCSB's Center for Film, Television, and New Media (CFTNM) brought together three of the most influential newsmen of our time: Bill Keller, the executive editor of *The New York Times*; Jacob Weisberg, the editor of *Slate*, the number one online magazine, started by Bill Gates and now published by *The Washington Post*; and Lionel Barber, the managing editor for the *Financial Times* (FT) in the United States and presently the editor for all the FT editions.

Entitled "A Crisis in the Newsroom," the debate zig-zagged between the two hot button issues of the day: media concentration and media bias. Moderating the 90-minute debate was Virginia Postrel, author and business columnist for *The New York Times* and me. So although, an academic institution, UCSB, was the host and producing entity, all the participants were working professionals.

Constance Penley, co-director of the CFTNM, introduced the debate and its participants. She spoke of a general sense of alarm about the ever-increasing levels of media concentration and "the effects of conglomerations on creativity, competition, and democratic access to the marketplace of ideas." She pointed out the there is, minimally, "a strong anecdotal sense" that having six companies own most of the media outlets in the United States today has had a deleterious effect on news gathering, news dissemination, and indeed, democracy.

Like most news stories, the debate defied the anticipated story line. The story turned out not to be a black-and-white tale of tyrannical owners and terrified reporters, but a nuanced portrait of a troubled and demoralized profession, told in shades of grays. We soon learned, for instance, that our three editors—working for three of the most prestigious media companies today—did not share the burdens of say, editors working for most publicly held companies such as the Tribune Company—which owns the *Los Angeles Times, South Florida Sun-Sentinel, Chicago Tribune*—or say, a newspaper owned by Gannet or Knight Ridder.

Two of the editors—Bill Keller and Jacob Weisberg—work for family-run companies: *The New York Times* is owned by the Sulzberger family, and *Slate* is now owned by *The Washington Post,* which is controlled by the Graham family. Although there may have been a degree of disingenuousness here, both said that the deep involvement of their respective families in media issues had insulated them from many of the bottom-line pressures that burden most editors today. Lionel Barber's *FT* is owned by Pearson, the behemoth that owns Penguin Books, *The Economist,* and other entities. Barber claimed, however, that despite the FT's stunning losses in 2004 and 2005, neither blame nor undue pressure had yet to be felt in the newsroom. Hence, it soon became apparent that the three editors enjoyed relatively buffered jobs—something not shared by say, Jon Carroll, the now departed editor of *The Los Angles Times.*

The other big surprise of the debate was that media concentration seemed not to be nearly as threatening to the panel as is media fragmentation. The uncertainty and anarchy caused by the influx of thousands of bloggers, Web sites, and alternative media sources—such as satellite radio—seemed the greater threat to the panel than the dramatic consolidation of media ownership. The fact that some 80,000 blogs are created every day seemed more problematic to them than such media monopolists as Rupert Murdoch or Viacom's Sumner Redstone.

I kicked off the discussion with a cautionary tale about the most demoralized newsroom in America today. *The Miami Herald* is an especially sad story because the *Herald* was once a terrific newspaper that broke big stories, boasted international bureaus, and even stopped presidential candidate Gary Hart in his tracks when it broke the Donna Rice scandal. However, around 10 years ago, its corporate owner, Knight-Ridder, decided the *Miami Herald* was not making enough money, despite the fact it was making about 18% profit at the time. But Knight Ridder decided it had to make 22% to 25%. To achieve its goal, Knight-Ridder eviscerated the newsroom, closing the foreign bureaus—with the exception of Latin America—prompting a flight of fantastic talent out of the *Herald* newsroom to *The New York Times*, *The Washington Post,* and others. In 2006, Knight-Ridder sold all its newspapers to the McClatchy Co.

Now we hear from staffers at other Knight-Ridder newspapers such as the *Philadelphia Inquirer* and the *San Jose Mercury News* that the same model is being imposed on them. So I began our debate asking whether having a free press means that media owners are permitted unfettered freedom to optimize profits. Was that really the intent of our founding fathers? And is there a corporate model that guarantees profits but also preserves the integrity of the newsroom and the news-gathering process?

What follows are the central arguments of what proved to be an engaging, often blunt, and almost always surprising 2-hour debate. It has been edited only for clarity and, in some instances, brevity.

THE AGE OF MEDIA CONGLOMERATES AND ITS CONSEQUENCES

Lionel Barber: Well, obviously I speak largely from my experiences as a British journalist, admittedly somebody who has spent a lot of time in America, but I think that something important happened in American journalism in the early 1970s, late 1970s, when a number of companies decided that they needed to go public, to tap the capital markets to raise money. And this changed the calculations made in the executive board rooms. Previously, private companies were under nowhere near the same kind of pressure to generate the kind of returns that they are now expected to generate and that certainly applies to a company like Knight-Ridder.

I'll just say one other thing about the *Financial Times,* which is indeed owned by a larger organization, a publishing group called Pearson. Its origins are actually in engineering, not in publishing, and Pearson expects us to be profitable but it does not expect us to make the kind of rates of return that are expected today in this country that can range up to 25%. And, indeed, in the full spirit of transparency, I should say that the *Financial Times* group, largely as a result of the steep drop in advertising revenue, has actually lost money for the last 2 years. But we haven't had the direct proprietary interventions to take away from our quality journalism.

A. L. Bardach: Is there a corporate model that can generate profits to the owners and also guarantee integrity of the news-gathering process?

Bill Keller: I speak from a position of a somewhat lucky man. Certainly, one answer to that question is to have your corporation owned by a family that controls the voting stock and that takes a legacy of serious journalism seriously. One result of that is that our profit margins are considerably lower than Knight-Ridder or Scripps Howard or the Tribune Company, but so far anyway, the company has been willing to put up with that. Lionel is absolutely right that there was a turning point when big media companies went public.

But the interesting thing now is that if you walk into most of the big news organizations in America or to places like the American Society of Newspaper Editors, or the other cabals where editors and publishers hang out and ask them if there was a crisis, the crisis would have nothing to do with concentration, it would be about survival, because they all now sense that they are seriously endangered, not by concentrations but by quite the opposite; by the fragmentation of both the sources of news and information and the places that advertisers can go, on the Internet, the niche cable, satellite radio, you name it. There are lots and lots of places where I mean advertisers have a lot more choices and to some extent they are deserting newspapers.

Jacob Weisberg: Five months ago [when] we were owned by Microsoft, everyone was suspicious that we would be subject to some sort of nefarious corporate control. In my experience actually both [Microsoft and *The Washington Post*] are identical in that they are totally hands off editorially, which is great from my point of view but I think also points to the fact that corporations—for whatever reason they want to own media properties—can be very good owners with a great deal of integrity.

There are three great newspaper companies still run by three great newspaper families. *The Wall Street Journal, The Washington Post,* and *The New York Times.* Each of which, although a public [company], is controlled by a family. And those three companies honestly do not profit-maximize the way just about every other publicly held company has to. One of the concerns that I have is that it may become less and less possible for those companies to do that. That is, for them to say, "We are willing to pay a price in terms of reduced profitability because the values most important to us are the journalistic ones."

But the distance between the best newspapers in the country and the rest has gotten bigger and bigger. There used to be a large number of second tier, and I don't mean that critically. Not the leading newspapers in this country, but very good newspapers like the *Miami Herald,* like the *Philadelphia Inquirer,* like the *Chicago Tribune.* I think if you went back 30 years and picked up most of those papers and put them alongside *The Washington Post* and *The New York Times,* there was a gap but it's nothing like the gap there is now. And many of those [papers] really are in decline and I think it's a concern. I think it's a shame that when you go to other cities around the country and pick up the morning newspaper, which some of us above the age of 40 still like to do, many of them are extremely disappointing in terms of their basic journalistic quality.

MISSING THE STORY

Virginia Postrel: Lionel talked about a change in the nature of the media business in terms of ownership, that is in going public in the 1970s. There has also been a significant change in the relationships of the media business, not just news but all media to the economy as a whole. . . . And [one] result is that journalists wind up covering their own industry. . . . The conventional wisdom quote from 2000, talking about the AOL/Time-Warner deal, is "What also drives the deal was the clear sense that the only way to survive at that level of competition was to unite every possible spectrum of electronic distribution and every possible source of content that said we're under a single banner." And that was in *The New York Times*. That was the way most journalists covered the AOL/Time-Warner story. I went back and did a Nexus search, and in the world of economists and business school professors, people were going, "Huh? What is the justification for this merger?"

Barber: Well Virginia, it's a great theory, but it doesn't quite stand up does it.

Keller: If you're talking about the media industry, you need to be careful about talking about newspaper journalists writing about the Federal Communications Commission [FCC] or about the telecom industry. That we don't see the telecom industry or the cable industry in this country as somehow intrinsically bound up with our business. And the third point is, you talk about media concentration and the great AOL/Time-Warner deal and we all remember that wonderful image on the picture of Steve Case hugging poor Gerry Levin and he knew he [had] just screwed him. And what happened—the whole theory you've eloquently outlined this afternoon—fell apart. That integration deal hasn't worked.

Postrel: No, that's my point. My point is that the journalists had this theory, which if they'd taken the trouble to interview people who are not Wall Street analysts, but who are people like business school professors and economists, who think about vertical integration in a non-media way, they would have heard a lot more skepticism. . . . And now that [AOL/Time-Warner] merger is seen as, to quote David Carr in *The New York Times*, "one of the worst mergers in American business history." So why [were] journalists not skeptical?

Weisberg: Journalists weren't skeptical of Enron either, I mean there were endless pieces of hyperbole in the business press, about how this was a great company, about how this was the greatest company ever built. Only at the very end, [was there] the slightest note that maybe . . . the balance sheets weren't quite as persuasive as they [should have been].

Barber: Well, journalists *were* skeptical of AOL/Time-Warner. But they were skeptical that it would be too profitable and too powerful, not that it was a really stupid deal for Time-Warner's shareholders.

Keller: Just to direct a hit at myself, journalists were insufficiently skeptical about the existence of weapons of mass destruction in Iraq too, but what have you established aside from the fact that sometimes journalists are credulous. This is true. Journalists screw things up. My own paper sort of took a pass on the Holocaust.

Look, I didn't go to journalism school. I've only ever had one good idea about journalism education: that nobody should go through journalism school without the experience of being written about. I said this to the recent current dean of the Columbia Journalism School and he didn't buy it. I had it in my seminars in which journalists would subject themselves to being profiled by their classmates and sort of taking turns—maybe UCSB will take this up—but Nicholas Lemann didn't take it up, but the whole rest of the world has. There is such a profusion of critical voices writing about our industry nowadays, and it sometimes feels as though they're all writing about my paper in particular. You know, I think if the AOL/Time-Warner deal were proposed today I expect you would get a much more diverse range of critical [writing].

Postrel: One thing I learned in my days as a business reporter, business reports have amazing 20/20 hindsight. They'll always tell you what went wrong after it goes wrong.

Weisberg: There are, I think, a couple of tremendously positive transformations going on in journalism that are perceived by many journalists as threats, and as negative. And the primary one is the concentration. We're not living through any media concentration, we're living through an age of radical de-concentration, when because of the proliferation of forms of all kinds of media, for the first time in human history, essentially anyone [can] publish their views, their version of events.

We have seen this massive leveling of the barrier to entry in journalism, tremendous democratization of journalism, and surprise, a lot of professional journalists don't like that. It's very threatening because their monopoly, not their monopoly writ small, but their monopoly writ large, is vulnerable. And it is very vulnerable, but I think from the point of view of society, it's a very good thing, and ultimately from the point of view of journalism it's a very good thing.

The other thing that's going on now that's misperceived is that you have a revolution of rising standards in journalism. Journalism now is better than any journalism that has ever been done anywhere in the world. You have better journalists who are more accurate, who are under more pressure to be accurate, to be fair, and in large measure they're rising to the occasion and being more fair and more accurate and more ethical. But what happens when you have comparatively rising standards like this is that there is a tremendous focus on the bad apples, on the mistakes. People get caught; people are getting caught and fired for things now in journalism that didn't used to be thought of as wrong.

Postrel: So basically this is the car industry, circa, 1980.

Barber: I worry, he's absolutely right, that definitely there is more emphasis on training, there is more emphasis on accountability, that organizations, including our own, have tightened up a great deal, but when someone, which happened the other day in Milwaukee—gets fired for actually having a beer at the pub, because he's doing an article on binge drinking, I would say the new Puritanism has grabbed hold of America.

BIAS IN THE NEWSROOM

Bardach: I want to go back to the issue of one of our topics, which is bias and I'll address this to Bill Keller. My sense of *The New York Times* and frankly all of what we call the mainstream media is that when they discuss religious faith or people of belief, it's done with a kind of shuffling embarrassment. I'm not talking about the politics of religion—but it's [coverage of religion] is almost as if [one's] talking about an exotic bird species. Now I know that most reporters are godless intellectuals, but some of us are not, and the vast majority of readers are not and I'm not just talking about only about Christianity or what is familiar to us. I remember reading a couple months ago a piece on Hinduism in *The New York Times* and it really did have a tone of dismissiveness and condescension toward the idea of belief, never mind, miracle. This is a big beef among conservatives. How do you respond?

Keller: I think there is a large element of truth in what you say. I think it's changing, we're certainly trying to change it. We have a public editor, Dan Okrent, he wrote the most widely quoted column of all his 18 months on the job under the headline, "Is *The New York Times* a liberal newspaper?" And he answered it, which is the only sentence anybody that anyone ever quotes, which was "Of course it is." But he went on to say he wasn't talking about a bias toward the Democratic Party or the Republican Party, he was talking about a bias to essentially urban, secular, values, and I think it's generally true that *The New York Times* is a Liberal Newspaper in the sense that a liberal arts college is liberal.

We have more reporters covering science than we have covering religion, although I think if you read the coverage of the papal succession, you will find there was not a drop of condescension in it. It was extraordinarily probing and thorough. But you know we have critics writing about entertainments that a lot of Americans would find horribly offensive and they write about them without passing moral judgment on the sexual content. We cover gay unions on our weddings page, and I would not back away from all of that for the sake of being all things to all people.

But I do think newsrooms go through waves of being obsessed with the idea of diversity. We certainly agonize over that a lot, and it's usually focused on the need to get racial diversity and gender diversity into the newsroom, which is important. But if the notion of diversity…is about how you serve a diverse audience in a diverse country, then we should be looking beyond race and gender, we should be looking for a diversity of background of class, experience, not trying to pick people based on their religious views or their political views. For me, at least a diversity hire would be somebody who had grown up having regular contact with people who went to church, owned guns, and voted Republican.

Bardach: Not necessarily the feeding pool out of Harvard and Yale? Jacob, how do you respond to that?

Weisberg: Well, all of us come to whatever we do in life with all of our experiences, beliefs, prejudices, call them biases if you want. The exercise of being a journalist is among other things an exercise in trying to put your prejudices, your views aside, and look at things in an open minded way, which doesn't mean you don't have a hypothesis about it, and it doesn't mean you're a blank slate. *Slate* has a different way of dealing this problem. Newspapers deal with this problem basically by saying, let's pretend we are a blank slate. Let's try to act like blank slates and maybe we'll become slates.... I actually thought it was quite striking that Dan Okrent, who was the public editor of *The Times,* made that acknowledgment and that comment.

Slate, because we're an opinion magazine, we adhere to an idea of fairness that is slightly more abstract, having to do with intellectual fairness rather than balance. We say, let's put all of those things on the table. We're not pretending that we can put them aside in some way. So, for example, we've done this for the last couple of presidential elections, we have asked everybody on our staff who they were intending to vote for. We asked the business people, we asked the copy editors, we asked everybody on staff and most people did participate. And unsurprisingly, the vast majority were for Kerry this time and had been for Gore the previous time, although we did have several people who were for Bush both times.

Again, the buzz word is transparency; let's be transparent about it, let's not hide who we are. But it doesn't mean because we're telling people who we are, that we're not just trying to be fair. What we do every day is scrutinize our own prejudices and our own assumptions and sometimes we end up changing our mind about things.

Barber: I just had one comment on *The New York Times* as a liberal paper.... There was one thing that I found very odd, the decision to appoint a particular correspondent to cover the conservative movement almost as an exotic species. I wonder why this wasn't assigned to the general reporting beat, as part of the general political coverage, because otherwise you do tend to see a coverage of conservative ideas or the conservative movement ghettoized within the paper.

Keller: First of all the idea wasn't to ghettoize it, the idea was to force it on to the front page on a more regular basis and also to serve as an impetus and a go to the people who cover the routine politics of Washington to be better informed about the people, the ideas, the groups that are, let's face it, run the country. And I don't think there is anything particular eccentric about assigning a reporter to cover the movements and the sentiments and the ideas of the people who voted the current government into office. In fact, I've replaced the reporter who was doing that. We're continuing that as a beat because it was so successful and you know it's easy and it would be comforting to say that "yes, we will get all of the viewpoints reflected because we're covering the administration after all" but it's one thing to get them coming out of official government mouthpieces and another to make an active effort to go into the lobby groups that work these issues, to go into the communities that support those lobby groups, to look at the funding of them. I think it's produced a lot of really good journalism.

Weisberg: You get it from both sides here, but it almost seems to me in the paper [*Times*] recently that you're bending over too far in the other direction to be fair to conservatives. I've been following your coverage of this case in Kansas, and the debate about teaching evolution in the schools, and the argument from the people who believe in intelligent design and God. There was a story that was so fair to these people, who you know are offensively wrong. It wasn't that they were being treated as an exotic species, it was as if they held legitimate views, and they don't.

Keller: Well, I said at the paper, that giving more serious three dimensional coverage to conservatives did not mean equal time for creationism and I don't think that if you read the paper you will have the impression that *The New York Times* has come out for the literal interpretation of Genesis. I really don't. But I do think it's important if there are people in the state of Kansas who control the school board and who are writing the stuff into text books that you let them explain what it is they're trying to do.

That doesn't count? That was actually the type of attitude we were trying to root out by having a conservative . . . tasked to write about conservatives. Because when I went around the reporting staff, trying to pick somebody to cover that beat when we first assigned it, I had somebody—a serious, really smart reporter—who said "no thanks, the yuck factor is too high." And I think you do sort of sense that there is a predisposition to cringe or sneer at ideas that we find unacceptable.

If you read our paper, especially if you read our weekly science section, I don't think you could possibly argue that the creationists have gotten a foothold in *The New York Times*. But yes, we do try to put their voices into the paper and explain what their argument is on behalf of rewriting the textbooks in favor of [creationism].

Weisberg: But you don't put the points of the segregationists in the paper. I mean at what point do you, and I'm not arguing that you have crossed this line. But there is a fine line, wouldn't you agree, between portraying the views of people who are either wrong or reprehensible or both, and on the other hand, lending some legitimacy to that point of view.

Keller: In fact, during the coverage over the debate over the Civil Rights Act, that was before you were born, . . . when Johnson was pushing through the Civil Rights Act and the Voting Rights Act, you would have seen that we in fact did put the voices of segregationists into the paper, because some of them were Senators of our United States. And I don't think quoting somebody constitutes endorsement. I think it constitutes an essential step towards understanding.

Bardach: This whole discussion raises a very interesting challenge to the traditional notion of the daily newspaper as objective and all things to all people. I would argue that *The Times* is struggling with this legacy of an ideology [of] writing for the Upper West Side or its equivalent throughout the country

Keller: Well, I don't think we're writing for the Upper West Side only, although I live on the Upper West Side so I could be wrong about that. We actually are privy to an advantage economically different from most newspapers which are confined more or less to a fixed geographic area, and we do know quite a bit about who our audience is. And we know that they tend to be well educated, fairly successful, they tend to be pretty interested in stuff that happens beyond the boundaries of their own community or their own country. They have a very big appetite for coverage of culture that includes both popular and high culture, and we know this. This is our constituency, and we tend to write to those people, when we set the throttle and the tone and the vocabulary.

THE BOTTOM LINE

Postrel: Isn't there an economic imperative to segment your market that way? I mean, not just for *The Times* but for newspapers in general, I mean, they're losing classified ads to the Internet. The retailers are consolidating, . . . I mean, the readership that you just described is a very desirable one for advertisers.

Keller: That's right.

Postrel: It would be less desirable if it included everybody in New York City.

Keller: Well, when we started out our market was New York City and we sold mostly local advertisements to retail stores that wanted to sell to people in New York. Now, we sell mostly national ads to people who want to reach an audience of some influence and above average level of income and education

Postrel: And I guess my question is—isn't that the newspaper of the future in some sense?

Keller: I sure hope so.

Barber: This does point out the very important debate, which we're all having in our newsrooms and executive suites, which is, "What is the future of newspapers?" I mean, if you have huge numbers of different outlets which are reporting around the clock, you've got more than 260 cable channels, you've got the Internet, you've got bloggers, you've got some of the traditional media, it's very difficult to figure out how you distinguish yourself.

Now, in the *Financial Times*, we are purveyors of business and financial information as well as being as a kind of window on the world. So, and we also happen to be pink [paper], which helps. It's salmon in Washington State. We're peach in

Georgia. A very national newspaper. But I think in general it seems to me the business newspapers have great challenges, really serious challenges, with the likes of Bloomberg and electronic news, instant information.

THE INTERNET: FREEDOM OR MISSED OPPORTUNITY

Bardach: I want to get back to this Internet conversation. The idea or canard that the Internet was going to be the answer to media concentration. Somebody said that the Internet was going to make us free!—give us access to information and everybody could be a journalist, and everybody could be a blogger. . . . But wasn't the idea that the Internet was going to fill in some of the deficiencies of mainstream journalism? For instance, we know with corporate cost-cutting that the first thing that goes is investigative reporting. So why have not Internet magazines, like *Slate* and other ones, not jumped in to pick up the slack. Why don't we see more investigative reporting?

Weisberg: Because it costs lots of money.

Bardach: It seems to me that a lot of Internet magazines offer a lot of opining and whining and not much reporting. You would think that they would want to seize the opportunity that is available to them.

Weisberg: Investigative reporting is difficult for Internet publication for a bunch of reasons and probably one of the biggest ones is resources. It's expensive to do, and the business is not developed to the extent that you can really find elements, but I think it will. But there are more tedious issues. I mean, people don't like to read long stories on a computer. Investigative stories tend to be long. It's not the kind of thing that works online and *Slate* in particular, things have to be boiled down, be pithy, like the questions we're going to have at the end of the session.

I spoke to a journalism class at New York University several weeks ago, a class of seniors, journalism students, really smart group of about 20 or 25 college students. And I asked them because I was curious, "How many of you read *The New York Times* newspaper?" And I'm sorry to say not a single hand went up. I was a little disturbed by that. When I was in college I read *The New York Times* every day, and most other people I knew did too. When I said, "How many of you read *The New York Times* on the web?" And every single hand went up, and I was very relieved. All that said, is that the audience for your product is not going away, it's just migrating to a different form. And there are some very big practical hurdles in making that transition, perhaps the biggest one being this economic problem that it's hard to see right now how even a very good online business generates the kind of revenue needed to support the robust reporting operation you have, but I firmly believe it will happen, and you know, whether it takes 15 years, or 20 years, or 25 years, *The New York Times* may end up with a much bigger audience than it's ever had because of the Web. It is a question of figuring out, and hopefully this is the job of our publishers, not our job, but someone has got to figure out how to monetize this for us so we can continue to do the kind of journalism we want to do.

Keller: It's an eccentric view in old media, which I guess I represent here, but I actually think Jacob is right. I don't see the profusion of new voices on the Internet as a threat. First of all, I read them, I love them, I think they're good for us, they provide a stimulus. Sometimes it feels like a cattle prod, but they keep our feet to the fire, you know, a lot of them aren't even news, they're sort of resources that all of us use, and a lot of gossip and a lot of propaganda, and a lot of polemic and all of that is great fun.

But, I'm going to slip into economic jargon here and afterward Lionel is going to tell me I've made a complete fool out of myself, but the barriers to entry in the media market have essentially fallen to almost nothing—and what do you do in an easy entry market? You move up the value chain. Isn't that what they say? You offer some assets that the new competitors can't offer and first and foremost the asset that the old media offers is reporting. And by that I mean the whole complex of, we have 350 reporters at *The New York Times*, and they go and witness stuff and they write it up and we have editors who put it in place, and we have photographers who bring back pictures. The blog world is wonderful but it isn't enough, and I do think that there is enough of a demand for actual reporting, that we will make that transition. A few of us will anyway.

VERACITY/OPINION/FACT-CHECKING

Barber: I share one thing in common with Bill, that I do represent old media. I also represent old Europe. We in Britain have a kind of "news with attitude." We're quite comfortable with partisan journalism, and I think it's done quite well for us. We admire what's been done in the United States, the standards of journalism. The one thing that strikes us is that you all are a little bit too nice about the new journalism on the Web. I mean, quite frankly, I think some of it is totally irresponsible. Some of these bloggers who have no qualifications, who come out making accusations about people, demanding scalps. I slightly envy their ability to write a wonderful prose at three o'clock in the morning, but they're not in any way regulated, or watched over and they've done a lot of damage.

Weisberg: Compared to *Geraldo*, compared to the editor of the *Manchester Union Leader*, there is good journalism and bad journalism in every form and there is a lot of attention paid to the loudest voices I suppose on the Internet. But it's a perfect marketplace. Only the people who are persuasive and who attract an audience thrive and get attention.

Barber: We need to take it all with a little bit of a pinch of salt. Some of the stuff that is put out on the Web is not fact checked. It's just put out there and the assumption is that somehow the reading public is wary enough, informed enough to be able to check whether this is true or not—and that's I guess the problem

Bardach: Newspapers don't fact check either.

Postrel: And as Eugene Volokh, a very fine blogger who is a UCLA law professor . . . [who] blogs, pointed out, "generalizing about blogs is like generalizing about books." It's saying, "Well, I don't read blogs because some of them are irresponsible and gossipy," is like saying "I don't read books because some of them are irresponsible and gossipy," which is true.

Weisberg: Just as an illustration, about a month ago, when things were happening in Syria, we didn't know very much about Syria. Someone who knew about it told me that if you follow Syria, there are two essential sources of information. What are they? Are they newspapers, are they magazine correspondents? No, they're two blogs that are written, one by an American academic here and one by someone in Syria, that have become these sort of grand collections of everything interesting and important that happens in Syria. If you want to know about what's going on in Syria, you absolutely will be using these sources. Now, [there's also] The Drudge Report. They're both on the Internet. They [both] use words. They have nothing more than that in common.

Barber: I'm kind of defending a tradition and a practice with which I've been involved for more than 25 years. Even as a British journalist, in which reporters report the news, they come back to an office, or, now, a mobile phone, or a computer. They send a story and it goes through two or three layers of checks and editing before it's put out to the reading public. And I would suggest that that is a different process to the one of the instant publication of news with attitude.

Keller: This does not characterize the entire blog world, but there is within the blog world a kind of cult. A sort of theology that the kind of journalism that Lionel was talking about is a bad thing. I mean Jill Abramson [assistant managing editor] went to a blog conference up at Harvard and was sort of beset because of people [whose] concept of how truth emerges from blog journalism is that everybody throws their facts up into the air and all their opinions into the air and through some sort of friction, the truth emerges organically.

Postrel: Although I spent 10 years as an editor, there weren't a lot of people fact-checking me. But the idea is that instead of having two or three editors—you have thousands [of editors]: people will check every thing! Because I am a traditional journalist, I'm very careful what I write on my blog, but people will check everything from my misspelling of the word Studebaker to knowing more about some [topic] . . .

Weisberg: Newspapers are filled with mistakes, [but] they are much less filled with mistakes now because they are, after publication, essentially, peer-reviewed by millions of people around the world, collectively, who collectively know everything. And will let you know in no uncertain or polite terms when you have erred. And as a result, they are making newspapers much more accurate.

Now the fact that there are lots of blogs that are published that are inaccurate and aren't corrected, is largely because those blogs aren't significant enough for anybody to bother to correct them. If you make a mistake in a prominent blog, you will be roundly denounced, ridiculed, corrected by other bloggers, by readers, there is an immediate feedback mechanism, which does constitute this kind of massive postpublication peer review, and I do believe that the result is greater accuracy both for them and in particular for old media.

Keller: I'm not arguing with you about whether blogs constitute a kind of reality check, an extra line of defense. I'm quarreling about whether they're a substitute, and I don't think [so] yet. I'm straying out on a speculative limb here, but what I think is happening, we're beginning to see the beginnings of a convergence. We're seeing a lot more newspapers beginning to operate in something closer to real time. I mean, *The New York Times* has for 6 or 7 years had a continuous news desk, update news during the day, *The Washington Post* does that. A lot of newspapers are doing blogging. Not quite in the fashion of the wild west bloggers, but they are doing blogging, and on the other hand, the biggest news story in the past couple of months in the Internet world has been that Yahoo! news has hired editors because unlike Google news, where all you get is an aggregation of information that's done by a computer, they think that people would actually be interested in having some human judgment exercised over the selection of the material they get when they do a news search on Yahoo!

Weisberg: Actually [at *Slate*], most of it is edited but some things aren't edited. We believe the trade off in terms of speed and the writer's freedom to express himself or people who have a record of being careful and who are capable of not being edited. But, the point is that because of the Internet, and partly because of bloggers and partly because of readers who don't blog, but are engaged electronically, the cost of error has risen dramatically.

I know as a journalist, that I used to make mistakes, not check things. I would like to think I was not sloppy, but I know that I've made mistakes that were never caught. They would now be caught. And because they would now be caught, I am much more careful that I ever was to check everything. . . . That's the way that the Internet, of which bloggers are a phenomenon at the moment, have made all of us more accurate.

Keller: They've speeded up the process of "letters to the editor" about a thousand-fold.

Weisberg: And publicly so, they can't be buried. . . . I remember the first time I was written about, I was about 15 and it was a caption in my local newspaper and it must have been less than 50 words, and there were two factual errors in it. I think that probably happens less today, not never, it certainly does happen, but less.

MISUSES OF INTERNET/ CENSORSHIP/BIAS

Bardach: I want to pick up on the Internet again, making us free. We know that China, Saudi Arabia, Iran have imposed draconian restrictions on Internet access. In Saudi Arabia, you cannot publish anything that is deemed critical of the state. Now, we don't know how this is all going to play out, but three countries have already put in very severe controls.

We think that is very shocking, but even if the rules aren't posted on the wall, [as to what] you can do or can't do, what about the rules here in the U.S. media? What are the limits and bounds? What about self-censorship? What about incurring the wrath of the powers that be or the administration? And what are the consequences of the failure to stay within those guidelines? We all have a certain amount of self-censorship. You [Bill Keller] were talking earlier that you weren't happy with your WMD story. Did that come out of pressure [that] we're going to war and should be doing stories about this, that the administration is putting this out?

Keller: Look, there clearly is distortion in the press because everything that is put out over the airways or in a newspaper comes through, is filtered through, the human being who has got experiences and attitudes, some of which you can put aside. And some of which you can't. I think the notion that is implicit in your question and explicit . . . that this is all kind of ideologically governed is largely, greatly exaggerated. I would say that's even true of Rupert Murdoch, if you want to come down to it.

I think most of the distortion that happens in *The New York Times* where I work, has to do with people trying to take information and turn it into a compelling story. They want a story that will get put on the front page, they want a story that will make their competitors envious, they want a story that people will read. In its most benign form, that consists of trying to order the story in such a way that you draw readers in and keep them reading. In its more malign form, it comes from being insufficiently skeptical of something that somebody has told you because, damn, that would make a good story.

Weisberg: Too good to check.

STORIES TOO GOOD TO CHECK OUT/ THE ACCESS TRADE/BIAS

Barber: The earlier question is extremely important I think. I disagree fundamentally with you putting China and Saudi Arabia and their limits on the press or with the Internet with what goes on in this country because it's really two very, very different situations.

The problem I think in my experience with American journalism is it is even more reactive than British journalism. It's amazing to me how many people think that the press or the television is somehow setting the agenda. The fact is, that intelligent, savvy politicians are extremely good at encouraging journalists without putting them under coercion but just seducing them into writing about what their agenda is, and we've seen this throughout history.

We look at the Reagan administration in the first term, look particularly at what happened after 9/11, some of which I can fully understand given the tragedy in New York but the fact is that the press in America was very gullible and too uncritical and too unquestioning in its reporting of the Bush Administration's war on terror. It was very uncritical in examining what that meant. A War on Terror. It was very uncritical in terms of not looking at, say, Europe's experience of dealing with terror and whether there were any lessons to be learned and second, in the run up to the war in Iraq, many newspapers, including the best, were solely focused on the case of whether, the likelihood that Iraq possessed weapons of mass destruction. There was nowhere near enough, what I would call anticipatory journalism, . . . given that the great probability was that the American military campaign would depose Saddam within weeks, if not in a shorter period. So everything then was a terrible surprise when there was utter chaos. And I think therefore that the America press fell down in their duty. They would say, "Well we couldn't report on that because it hadn't happened or nobody was telling us about this." And I think there, the fact is you need to cross a certain threshold here in terms of your skepticism toward authority.

Weisberg: I do think though that when journalists fail in some way there is an immediate instinct to assume that the bias is ideological, when in fact it is almost never ideological. It is first of all, very often, the result of a herd mentality.

The moment of patriotic correctness after September 11th, and in the run up to the Iraq war on the weapons of mass destruction issue for example. Nobody questioned that in any serious way, because nobody questioned it in any serious way. There was a collective belief—and it was really an assumption that was common, not just to our government, but to governments, countries in Europe, that ended up opposing the war—one of the main arguments they made was that Saddam would use these terrible weapons against us in the war. Everyone thought that and it was a journalistic failure, but it wasn't an ideological failure. It was a failure of imagination. It was a failure to challenge this assumption.

Another place where I think you have what sometimes appears as a common political bias but isn't is in the basic bargaining that goes on in the trade off and trade for access and positive coverage. You can call this the *Vanity Fair* effect [as] it's not as extreme anywhere other than *Vanity Fair*. But if you look at whatever movie star is on the cover of *Vanity Fair*, this month, he will not be a negative story. And the reason it will only be a positive story is because the magazine needed access to that person. They needed to take their photograph in a posed setting, they wanted the interview and the person wouldn't be there otherwise. And if the person didn't, if the subject didn't feel fairly confident that he or she was not going to be written about in a critical way, they wouldn't have cooperated.

And you have a much milder version of that that goes on all the time, particularly in Washington with the beat reporting system. And particularly around the White House, where if you want access, which journalists feel they need because they cover things as reporters, they want interviews, you want an interview with the president, you want to be the first to know, you want leaks. Those don't tend to go to the newspaper that is most negative. They tend to either go to the newspaper that is most positive, or the good stuff is divided up according to some infinitely complicated formula that does incorporate one of the two components, how nice you are to those people. And actually in a certain way the Bush Administration has been liberating for a lot of journalists because they don't really give much access to anybody. So you aren't going to get the access, even if you do suck up to them. So you don't bother to suck up to them.

Bardach: But invariably officials from the administration will go on Fox News where they presume they're going to get a better treatment, and it is true. I did spend ten years at *Vanity Fair* and there is some truth to what you just said.

Postrel: Re: Fox News, one of the things that is most striking about their coverage is they get more interviews with people from Hamas, and . . . all these Palestinians, because I guess they just want to see their spiel, and they don't care that it's not framed in positive way.

THE NEW YORK TIMES AND WMD: WHAT WENT WRONG?

Keller: I'd like to react to a couple of things that Lionel said, first of all, just as a generic thing. It is true that savvy politicians are very good about getting their agenda set, I think it's also true that savvy reporters are pretty good at reporting against that agenda, and if you doubt it, you can ask Tom De Lay, John Bolton or the people who are trying to promote the president's social security program, on a larger question of WMD in Iraq.

I would guess that I'm the only here who has had the fairly recent experience of going back and reading all of *The New York Times* coverage leading up to the war in Iraq, which I did because when I came in to this job, it was already a subject of some controversy, and I had to decide, did we want to do a mea culpa or not,? And we did end doing a mea culpa, because we ran some really bad, really credulous stories.

But there is a kind of mythology that the mainstream press, led by *The New York Times*, took us to war, as if the administration had nothing to do with it, in some versions. And, to put this in some context, the notion that Saddam Hussein had weapons of mass destruction was not a mainstream press notion, it was not a Bush administration notion, the Clinton administration believed it, the former weapons inspectors without exception believed it, the UN Security Council believed they had it, there was dispute about it. First of all there was a considerable dispute about whether he had anything in the way of a serious nuclear program, but that he had chemical and quite possibly biological weapons was the universal point of view. The Democrats in Congress believed that too. The debate in those months leading up to the war was partly about . . . the question of what would be the consequences if we go in and wage war and there was a robust debate. We had a number of stories quoting Generals, poking fun at Rumsfeld and Cheney, for being overly optimistic about how Baghdad would fall in our hands and the people would come out with flowers in their arms. But most of the debate was about whether or not, whatever Saddam Hussein had, constituted an imminent threat that justified going to war. Or whether we should wait and let the inspectors go in and find out what he actually had before hand. And that debate, which was the central debate in the lead up to the war, was covered extensively, not just in our paper, but all over the place, even TV covered that debate. So the sense that everybody was herded sleeplike to war without a dissenting bleat is a bit of an exaggeration.

Bardach: I think it speaks to the prestige of *The New York Times* and the scrutiny that is given to *The New York Times* because no matter what is going on with the Internet, and blogs, the *Times* very much feeds the news cycle. . . . So it gets the credit and it gets the blame, it goes with the territory.

Postrel: Which raises the question, given the influence of the *Times*, which is called after all *The New York Times*. . . . I'm very struck by the fact that we're sitting here in California, and we're going to talk about media. So we get these three guys to fly in from New York because God forbid anybody in California would know anything about the media, and at least you're not all native New Yorkers, it could be worse. None of you is a native New Yorker, I believe.

Weisberg: I think of myself as baked in the ether.

Postrel: New Yorkers don't live like the rest of Americans, including Americans in blue states. It's not strictly an urban ideology, they take the train more than they drive, they live in small apartments, they think that no-one could possibly survive on less that $100,000 a year, the idea that people own houses on a median income is very hard for them to grasp. You're in New York and we look to you for leadership. How do you avoid these natural, every day lifestyle biases from distorting the lens of your reporters?

Keller: Most of the reporters—who do the witnessing and interview the people and send their impressions—in don't live in New York. The business section and the metro section together, the majority of them do, but the people who are reporting about the state of affairs in the country and the world are mostly in the country and the world.

Postrel: Okay, take out the foreign bureaus, what is the division?

Keller: Well, we have as many national correspondents around the country, not counting Washington, as we do foreign correspondents, and we have 40 people in Washington, too.

Postrel: So you have roughly 40 non-Washington people versus 350 in New York. Is that what you said earlier?

Keller: We've got 350 reporters, of whom 150 of them are outside of New York regularly working, roughly.

Weisberg: My point is only that we all have to fight the bias, and in New York it is probably particularly strong. An example that I ran into was actually after September 11, the bulk of our staff has only recently sort of moved to New York. Now the majority of our staff is in New York. And in New York everyone was obsessed with what they were going to do on Ground Zero, what they were going to build there. And I happen to live nearby so I'm particularly obsessed, but this story's in *The New York Times* constantly about the politics of it, the changing plans, the debate about it, has been going on since September 11th. Whenever it comes up in one of our editorial meetings, . . . we have on the phone, all the people who aren't in New York just start snickering, because they think the most ridiculous thing to be obsessed with. It's like if we were obsessed.

They recognize that there are some larger implications, but they don't live in New York, they don't care what's built at the World Trade Center, maybe they'll go see it someday. You know, but it is, it was sort of a lesson to me in how you can become obsessed.

Bardach: Well there's the idea that somewhere between New York and Los Angeles is a country called America.

Barber: Our correspondents are based outside New York. We've got a very large Washington bureau and we have people who actually write and report on the midwest, and I certainly make sure, that the reporters who are based in Washington, who are covering political stories, travel regularly. But not to Santa Barbara because they might not come back.

Audience member: My question relates to the role of consumers in a sort of market-driven economy in the media. I witness a reduction in the amount of foreign reporting . . . we talk about are people watching more television and reading less newspaper. I only read *The New York Times*, I watch less television because I find that the stuff I see on CNN, which I used to turn to, being sort of infatuated with pop culture. I wonder, in a market economy where consumers dictate the type of goods that are provided by companies, the media companies included, is there a need to balance this sort of, providing the consumers what they want versus having some sort of a public service component to the news.

I understand that when Congress originally granted license for the airwaves to the radio an television companies, there was an explicit public service sort of rule—that you were there to provide good for the country. I'm wondering what are your thoughts are in terms of how consumers direct the news based on their market preferences?

Keller: That's a very good question. . . . Virginia mentioned earlier that all these blue bags full of *New York Times* popping up in Dallas. I actually used to work for the *Dallas Times Herald*, which went out of business, but *The New York Times* didn't drive it out of business. It was already out of business when the *Times* got there.

But the point I wanted to make is . . . newsrooms tend to be nervous about that because we don't really want to know. But one thing we do know is that the reason that people in Dallas and Baltimore and Philadelphia, and Denver will pay $500 a year to get one delivery of *The New York Times* is because their local paper stopped covering foreign news. The two big things that they say is "Our local newspaper doesn't cover foreign news anymore," and "they don't cover culture." There are other things too: sometimes national [news]. Some don't like their local political coverage. . . . So there is a market there.

Bardach: How many papers are covering foreign news now? And doing investigative reporting?

Keller: Actually the question about the public service component, which applies to broadcast, not to print is a good one, because television stations have discovered they can satisfy that nominal requirement very cheaply. If you put on a news show, it doesn't matter if it's all lurid crime and the sort of faux investigative stories brought to you by the local spotlight team. It doesn't matter if it's crappy, the government hasn't set the definition of news coverage and public service that requires them to actually send reporters, hire reporters and go out and do serious work.

Postrel: But your *Times* example I think illustrates there is a small niche market, potentially quite lucrative, for people who really care about these things and the other people are not going to read it. It's like if a tree falls in the forest, you can say it's public service—but is it service, if nobody pays attention?

Bardach: Well, if everybody cuts back on investigative reporting, someone will say, "My God, there is a market, we could get in there." But it doesn't seem to be happening.

Barber: I will say our paper has a foreign correspondents network almost as big as *The New York Times* and clearly we think, and the growth and circulation of the *Financial Times* over the last seven years, which is fourfold in this country, shows that there is an informed globally minded audience of Americans who are interested in reading the paper.

The trick is to find ways of presenting news, foreign news in a way that is relevant to the audience in America. And I think that there are two very important phenomena in the last seven years: first, the Asian financial crisis, which brought home to many investors . . . a crisis in the markets in one area, across the other side of the world—and almost knock out a highly leveraged hedge fund that then could really have disrupted the U.S. financial system. That was one very important event. And the second was clearly 9/11—that somehow a country that appeared to be far away, of which Americans knew and cared nothing, could actually be a breeding ground for terrorism. So it's in everybody's interests that we understand what's going on there and that it's reported. And I would draw a distinction between that kind of foreign news reporting, which is explanatory, and what Annie has referred to several times this afternoon, which is investigative journalism, which is important, but is at an even higher level and requires even more resources.

Audience member: I would like to hear you speak more about legitimate criteria that are used by newspapers,

Keller: We certainly apply criteria to what constitutes something, particularly what's worth devoting a lot of space to. What's worth putting on the front page. I'm not sure that I could devise a set of legitimate criteria that even a lot of reasonable people would agree with. There is "different people will disagree" if you look at just the five or six desk papers in the country on a given day. . . . The sort of subquestion to your question was how do you get people to read it, and if only people read all of these stories, they'd be a much better informed electorate, and that's probably true but I wish I had an answer to that one.

Barber: I don't particularly feel comfortable with the word criteria, when it comes to news or what we put in the paper on the front page or inside. We're constantly operating on what, say the events of the day are, and analyzing them, but also operating on different time scales, so we are trying to anticipate what may happen in the markets, in countries, or whatever. And we're trying to do some journalism, which we know is not going to have bear fruition.

This is not rocket science. . . . We're setting different people on different stories, and we're also trying to make sure that the people we hire to work for us, also are diverse, and I don't mean racially diverse, I mean nationally diverse, so they bring different points of view about what they think is important.

We have definitely tried to have, two or three Arab speakers for example, in the last 5 years, who will come into the newsroom and because it's not a totally democratic place, that would be very dangerous, but people are allowed to speak up and offer their points of view on what they think is important in the news.

Weisberg: *Slate* has the luxury of not being a primary news source so we don't have all the responsibility really that goes into the kind of news judgments that Bill and Lionel have to make. But we take the approach that's very much driven by argument, analysis, novelty: Is there a fresh point? Is there a fresh idea here? Is it entertaining? We tend not to think so much in terms of what would our readers want. . . . We, the editors, think of ourselves as consumers of the magazine as well and the standards we tend to apply is: Do we find this interesting? Do we find it smart? Do we find it fresh? And if we're interested in it, it's a pretty good chance that enough of our readers will be as well.

12

NETWORKS OF INFLUENCE

The Political Power of the Communications Industry[1]

John Dunbar

The Center for Public Integrity (www.publicintegrity.org) is a nonprofit, nonpartisan, tax-exempt organization that conducts investigative research and reporting on public policy issues in the United States and around the world. Created in 1989 by former network news producer Charles Lewis, the Center's core work has been the examination of the nexus between political power and corporate greed.

In 2002, we created the "Well Connected" project. The goal was to monitor the corporations that control the flow of information to the public and the government overseers whose job it is to regulate them. From the beginning, we decided to focus our research on the "pipeline" companies: Those corporate giants that control valuable, publicly owned slices of the airwaves, telephone lines, and cable television systems—in short, all the modes of electronic communication that exist

[1]Material in this chapter is extracted, summarized, and updated from a lengthy and detailed report originally published October 28, 2004 (see "Networks of influence: The political power of the communications industry," www.publicintegrity.org/telecom). All quotations here were provided during interviews with the Center for Public Integrity.

in the economy. The communications industry is in a unique position. It spends hundreds of millions of dollars to affect government policy but at the same time asks for the public's trust in delivering information on those activities precisely because the public frequencies have been provided for its commercial use. The influence of these industries is particularly important given that they control the information that helps all Americans formulate their views on everything from whom to elect for president to what movie to see this weekend. It was clear to us that the communications industry was ill-suited to monitor itself, and the Center, by virtue of its nonprofit status and reputation for independence, has become an effective watchdog.

In October 2004, we released "Networks of Influence: The Political Power of the Communications Industry." The Web-based report, later distilled into a 318-page reference work, examined the political influence of the communications industry on Congress. The report focused on lobbying, campaign contributions, industry-funded trips ("junkets") and government officials who routinely travel back and forth between public and private service (the "revolving door").

The project was an ambitious one. The targets of our analysis included radio and television broadcast companies like Clear Channel Communications Inc. and News Corp.; phone companies like Verizon Communications Inc.; and cable television operators like Comcast Corp. One would normally expect these conglomerates to have little in common. But as communications technologies converge, companies that were once in segregated lines of business are becoming direct competitors as well as strategic partners. Although only a small percentage of Americans receive their television signals over the air, giant broadcast companies still control most of the programming seen every day. Cable television companies now provide telephone service; and traditional telephone companies are spending billions on fiber optic lines to deliver video programming to homes. These companies have another major characteristic in common: They are all regulated by the Federal Communications Commission (FCC).

For each of these sectors, the Center calculated how much the major companies within each spent on lobbying, campaign contributions and industry-funded trips for federal lawmakers and their staffs. This chapter provides updated figures. For the overall lobbying and contribution overview, the time frame was 1998 through year-end 2004. For some of the more detailed analyses, we relied on the original time frame published in our first report. (To search the database that made this report possible, visit the Well Connected Web site, at www.publicintegrity.org/telecom.)

This chapter first provides total amounts of lobbying, campaign contributions, junkets, and revolving door instances. Then for each of the three separate sectors, the chapter provides two analyses. The first provides an overview of total amounts and then explains the numbers in each of the areas of influence. The second looks at one influence-related issue in depth, to provide a way to comprehend the significance behind these daunting numbers.

OVERVIEW

The communications industry spent $1 billion from 1998 to 2004 to affect election outcomes and influence legislation before Congress and the White House. (Rounding out total spending was another $122 million that came from telecommunications companies that do not fit easily into the three sectors.) A sector breakdown shows the following:

- Broadcasters spent $204 million on lobbying and $30 million on campaign contributions and $160,000 on trips, for a total of $234 million.
- Telephone companies like Verizon Communications Inc. and AT&T Corp.* spent $436 million on lobbying, $65 million on campaign contributions, and $280,000 on trips, for an overall total of $502 million.
- Cable television providers spent $116 million lobbying, $26 million on contributions, and $230,000 on trips, for a total of $142 million.

Lobbying

Total lobbying expenditures from 1998 to 2004 by the industry were more than $840 million. The $840 million affected legislation ranging from local telephone competition to media consolidation to the deregulation of the cable television industry. Spending on lobbying by communications companies rose from $119 million in 1998 to $142 million in 2004, an increase of 18.7%.

General Electric Co., which owns 80% of NBC Universal topped the lobby spending list at $94.1 million. (The total includes all lobbying by the giant conglomerate, even though it draws only a portion of its revenue from broadcast operations. Federal disclosure rules do not require companies to separate lobbying expenditures by subject.) Second by a small margin was Verizon Communications Inc., the nation's second-largest phone company. The former regional Bell operating company spent $81.9 million from 1998 to 2004. Verizon has key financial interests in local and long-distance phone regulation, spectrum allocation for its wireless division and a multitude of other issues. Third on the list of top lobby spenders was SBC Communications, Inc.* at $58 million. SBC is followed by competitors AT&T Corp., at $53 million and Sprint Corp., with $47 million.

Contributions

Campaign contributions from 1998 to 2004 were $160 million, including both hard and soft money donations from industry employees, labor unions representing employees in the communications industry, and political action committees.

*Since this report was first published, SBC Communications, Inc. purchased AT&T Corp. and renamed the combined company AT&T, Inc.

The two largest contributors over the study period were both labor unions. The International Brotherhood of Electrical Workers (IBEW) spent $13.4 million since 1998. Second was the Communication Workers of America (CWA) at $13 million. A significant portion of the union total (roughly 24% from the IBEW, 43% from the CWA) came from unregulated "soft money" contributions. Such contributions are now banned by the McCain–Feingold Bipartisan Campaign Reform Act that went into effect late in 2002. Third on the list of top contributors was Verizon at $11.4 million. Verizon is followed by SBC at $11.3 million. Media conglomerate Time-Warner Inc., with holdings in publishing, film production, the Internet, and cable television was fifth at $9 million.

The partisan preference of the communications industry as a whole has leaned toward the Democratic Party. Total contributions were split 56.2% for Democratic candidates and party organizations and 43.2% for Republicans through September of 2004. Bush's top communications patron is SBC, the San Antonio-based former Bell operating company that has waged a long battle with the government over the Telecommunications Act of 1996, which was supposed to create competition in local phone service. SBC donated $189,510 from 1998 to 2004. Bush's second biggest booster from the communications business is Time-Warner ($175,600) and third is General Electric ($166,000). Kerry's top communications patron is Time-Warner, which donated $260,840 over the same period. Second is broadcast giant and owner of CBS, Viacom Inc., at $127,260. Verizon is third at $118,820.

Behind Bush and Kerry on the list of recipients of the most communications dollars is former Senate Minority Leader Tom Daschle (D-S.D.), who received $1.1 million. He is followed by New York Democratic Sen. Hillary Rodham Clinton, who despite her status at the time as a freshman legislator collected an impressive $1 million. Former Rep. Dick Gephardt (D-Mo.) was next on the list at $974,000. Senator John McCain (R-Ariz.), chairman of the Senate Committee on Commerce, Science and Transportation, came in fifth at $926,000.

Then there were the bundlers. At least 10 figures in the communications industry raised $100,000 or more for President Bush's re-election, including the top executives at the largest and second largest telephone companies in the nation. On the list of "Pioneers" were Salem Communications CEO Edward Atsinger; IDT Corp. CEO James Courter; lobbyist Ronald Kauffman, senior managing partner of the Dutko Group; Liggett Communications CEO Robert Liggett; lobbyist and former member of Congress Bill Paxon of Akin Gump Strauss Hauer & Feld; Univision Communications CEO Jerry Perenchio; Edge Wireless CEO Wayne Perry and Verizon CEO Ivan Seidenberg. SBC Communications CEO Edward Whitacre was named a "Ranger" for hitting the $200,000 mark.

Junkets

Closely related to lobbying was the much criticized Washington practice of "fact-finding trips" by lawmakers and staff, which are also referred to more derisively as "junkets." From 2000 to March 2004, the communications industry sponsored 450 trips for members, family, and staff of the Senate Committee on Science, Commerce and Transportation and the House Committee on Energy and Commerce, valued at $700,000.

The top sponsor marks the first entry of the cable industry into a top five category of influence in the Center survey of communications companies. The National Cable and Telecommunications Association funded 102 trips worth $198,727. Second was another lobbying group, the United States Telecom Association, with 62 trips at $106,689. Third was SBC with 47 trips valued at $76,327. Fourth was the National Association of Broadcasters with 43 trips at $99,084 and fifth was the Cellular Telecommunications & Internet Association with 41 trips valued at $57,044.

In May 2003, the Center did a survey on members of the FCC and staff who accepted free travel from industry. Over the 8-year period covered in the survey, businesses paid for $2.8 million in free travel for FCC commissioners and staff. Since the publication of that report, the practice has largely ceased. The frequent flyer award among members of Congress was no contest. The top recipient was former Louisiana Rep. Billy Tauzin, who was chairman of the House Committee on Energy and Commerce. Tauzin and members of his office staff went on 69 trips valued at $107,424. Tauzin's office was top traveler in every communications sector examined by the Center. A distant second is former Senator Conrad Burns (R-Mont.) at 31 trips valued at $41,529 and in third place is Rep. Fred Upton (R-Mich.), with 14 trips worth $27,398.

Revolving Doors

Center researchers also undertook an unprecedented study of former key FCC and congressional officials who left their government jobs for positions in the industries they used to regulate. The practice raises concern for a number of reasons, the most serious being that a federal employee may be tempted to show a regulated company favorable treatment in hopes of getting a job at the end of his government service.

The U.S. Office of Government Ethics has rules on postemployment activity by former federal workers, but they are narrowly drawn and easily sidestepped. Although the Center study did not find any former FCC or congressional employee who appeared to violate those rules, researchers found that an extraordinary number of them—398, in fact—have gone to work for companies they used to regulate. For example, all eight previous chairmen of the FCC have either worked for or represented corporations that are regulated by the agency. Two are registered lobbyists.

PART I: THE BROADCASTERS

Overview

The broadcast industry spent more than $204 million lobbying the federal government from 1998 to 2004—a period of increasingly intense battles over ownership rules. In addition, television and radio companies contributed more than $30.1 million to federal candidates and lawmakers during the same period. The companies and their principal representative organization—the National Association of Broadcasters (NAB)—also sponsored 84 trips for lawmakers from 2000 to March 2004 at a cost of $160,000, bringing total spending to affect policy and elections by the industry to a minimum of $235 million. The volatile political climate also saw 24 individuals with close ties to both the industry and its regulatory overseers make lucrative moves back and forth between the two.

Lobbying

From 1998 to 2004, lobbying expenditures by the broadcast industry rose 80%, from $20 million to $36 million. The spending frenzy peaked in 2003, the same year that the FCC proposed significant relaxation of the ownership rules, which would have allowed corporations to own more media outlets than ever before and reach a greater percentage of the national audience.

General Electric topped the lobby spending list for broadcasters at $94 million through 2004. Second in broadcast lobbying is NAB, an influential trade group that represents the interests of free, over-the-air radio and television broadcasters. NAB spent $42.9 million lobbying, according to records. Walt Disney Co., owner of the ABC television network, was a distant third at $23.6 million.

The steady rise in lobbying dollars followed the passage of the Telecommunications Act of 1996, which stipulated that both caps on station ownership and the percentage of the national audience a broadcast company may reach directly be reviewed every 2 years. Since 1998, the industry has responded by steadily increasing spending on lobbying.

Contributions

As for contributions, in the three election cycles between 1998 and 2004, broadcasters contributed more than $30 million to federal candidates and sitting officials. President George W. Bush and his Democratic challenger, Senator John F. Kerry, ranked first and second through 2004, with donations of $570,848 and $422,621, respectively. Senator Hillary Rodham Clinton, the sophomore from New York and former first lady, ranked third with $364,175.

The last election cycle was a boon to both Bush and Kerry, who raked in unprecedented amounts of contributions. Broadcasters gave nearly identical amounts to Republicans (50.1%) and Democrats (49.6%) during the period analyzed. Almost half of the contributions went to the national party committees, which then redistributed the money to affect local elections. Contributions made directly to individual candidates, such as Kerry and Clinton, made up the relatively minor difference between the two parties.

Once again, the industry's top donors of federal campaign dollars were major broadcast owners, with General Electric leading the way ($7.8 million), followed by Viacom ($4.3 million) and Disney ($3.7 million). Broadcasters also sponsored $165,474 in trips and junkets for members, family and staff of the Senate Committee on Science, Commerce and Transportation and the House Committee on Energy and Commerce from 2000 to March 2004.

Junkets

First on the list by a large margin is former Rep. Tauzin and his staff, who accepted 18 trips from the broadcast industry, valued at $25,006. The Louisiana Republican was chairman of the House Committee on Energy and Commerce. The NAB has been by far the greatest sponsor of such outings, having paid for 43 of the 84 industry-sponsored trips. During that time, the trade association accounted for two thirds of all broadcast-industry junket and trip spending. News Corp. and Disney followed the NAB with 15 and 13 trips, respectively.

Revolving Doors

Some two dozen individuals have moved through the lucrative "revolving door"—that is, between the broadcast industry and the government offices and agencies that regulate it. Marsha MacBride, currently the NAB's executive vice president for legal and regulatory affairs, has ping-ponged at a dizzying clip over the last 13 years: She went from the industry to the FCC, back to the industry, and then back to the FCC before landing—at least for now—at the NAB. Clear Channel Communications Inc., which had virtually no Washington lobbying presence for years, got into the influence business in a big way when it hired Andrew Levin. Levin was minority counsel for the House Committee on Energy and Commerce, where virtually all broadcast legislation is heard, before moving on to become senior vice president of government relations for the radio giant. Finally, David Goodfriend, interim executive vice president and general counsel for liberal radio network Air America, was at one time deputy staff secretary at the White House under President Bill Clinton, director of legal and business affairs at satellite television broadcaster EchoStar, legal advisor to Commissioner Susan Ness and a telecommunications lawyer with Willkie Farr & Gallagher.

BROADCAST CASE: SINCLAIR BROADCASTING

When it comes to politics, the broadcast industry does not play favorites: Between 1998 and September 2004, records show, broadcasters donated $13,528,000 to Democratic candidates and party organizations and $13,391,000 to Republicans. But a glaring exception to this evenhanded approach can be found at Sinclair Broadcasting Group, which endured a firestorm of negative publicity for ordering its 62 television stations to air a documentary—just days before the election—critical of Democratic presidential nominee John Kerry.

A little more than 95% of Sinclair's $334,000 in contributions went to Republicans—a lopsided record of giving unmatched by any other major television broadcaster. News Corp., for example, whose Fox cable-news operation is often criticized for its perceived right-wing bias, actually gave 64% of its contributions to Democrats from 1998 to September 2004 and 36% to Republicans. (Contributions by longtime Democratic fundraiser Haim Saban, who owned half of Fox Family Worldwide, were excluded from that ratio; when included, News Corp. contributions skew in favor of Democrats by 81% to 19%.) And Clear Channel Communications Inc., whose board of directors includes the businessman who bought the Texas Rangers baseball team from President George W. Bush, only favored Republicans by about 63% to 37%.

Reporters, editors, and news executives insist that their employers' campaign contributions have no bearing on how the news is reported. But recent events involving Sinclair and other large broadcasters, such as Viacom, coupled with an escalating lobbying effort to remove government controls, raise questions about whether the broadcast industry's corporate goals may be affecting news coverage.

The controversy involving Sinclair was kindled when management ordered the company's television stations, several of which are in so-called swing states, to pre-empt regular prime-time programming for an unflattering documentary, *Stolen Honor: Wounds That Never Heal,* about Kerry's Vietnam service. But if Sinclair's goal was to help itself by helping President Bush, whose administration consistently supported deregulation of major media corporations, the move clearly backfired. By ordering the airing of an overtly political broadcast so close to the election, the company managed to cast a negative light on the media ownership issue in general, and a harsh spotlight on itself in particular. FCC Commissioner Michael Copps called the Sinclair flap "an abuse of the public trust."

It wasn't the first time Sinclair became the subject of a newscast, as opposed to the producer of one. Six months earlier, for example, the company earned an unwelcome spotlight for instructing its ABC affiliates to pre-empt a *Nightline* program devoted entirely to the reading of a list of American soldiers who had died in the Iraq war. In explaining its decision, Sinclair claimed that "Despite the denials by a spokeswoman for the show, ABC appears to be motivated by a political agenda designed to undermine the efforts of the United States in Iraq." Two months earlier, Sinclair had dispatched news crews to Iraq to cover the "good news" that it believed the rest of the media had missed.

Sinclair was founded by Julian Sinclair Smith in 1971 with a single Baltimore television station that is now the company flagship. Led by Smith's son, David, Sinclair grew into the nation's largest owner of television stations—a collection of mostly small-market broadcasters that reach nearly 25% of all households.

The company has made no secret of its conservative leanings. In 2002, it created its "News Central" production to centrally manage news operations for all of its stations. Daily broadcasts include commentaries by Mark Hyman, the company's vice president for corporate relations, whose minute-long segments have often included questionable claims. In one segment, for instance, Hyman said that Kerry enlisted in the U.S. Navy to avoid being drafted into the U.S. Army. Overall, however, Sinclair has shown its partisan colors with impunity—at least until now. When news surfaced of the company's plans to air *Stolen Honor*, advertisers threatened to boycott, Wall Street analysts criticized the company, and the stock dropped to a 52-week low. Shareholders threatened lawsuits over the broadcast of the anti-Kerry documentary, prompting Sinclair to issue a statement saying it would not air the film in its entirety, but rather broadcast segments from it as part of a news program.

PART II: YOUR PHONE COMPANY

Overview

The four former "Baby Bell" local phone companies and their rivals spent nearly $200 million lobbying Congress and the executive branch from 1998 to 2004 in an effort to influence federal telecommunications policy. Phone companies have contributed some $65.5 million to federal campaigns and political parties and paid for $276,072 worth of travel for members and staff of Congress, bringing total spending to affect policy and politics to more than $502 million.

Lobbying

Led by Verizon Communications Inc. at $81.8 million, the former Bells collectively have far outspent former chief rival AT&T in an attempt to win favor with Congress, the FCC, and other government agencies. All traditional telephone companies combined spent $436 million lobbying from 1998 to 2004, making the industry among the most prolific spenders in Washington, DC. SBC Communications Inc. was second, spending $58 million. Third behind them in lobby spending was AT&T Corp. Despite experiencing tough economic times over the period, the former phone giant nevertheless spent $53 million on lobbying from January 1998 to 2004.

Contributions

The three companies that led the pack in lobbying also topped the list of campaign contributors: Verizon shelled out $11.4 million, while SBC spent about the same, and AT&T spent $7.9 million.

Telephone companies prefer Republicans to Democrats. Telephone companies gave roughly 59% of contributions to Republican parties and candidates and 41% to Democrats. The top recipient of telecommunications industry campaign contributions is President Bush with $871,000; Senator John McCain (R-Ariz.), former chairman of the Committee on Commerce, Science & Transportation, is second at $487,000; and Rep. Dennis Hastert (R-Ill.), former speaker of the House, is third at $486,000.

From 1998 to 2004, Bush out-raised Democratic Kerry in phone contributions 5 to 2. Bush's top four phone supporters were the four Bell companies: SBC ($189,510), Verizon ($135,480), BellSouth, which is now part of AT&T Inc. ($122,950), and Qwest ($68,800). Together, the Bells gave Bush 62% of the money he received from telephone interests and favored him over Kerry almost 3 to 1. Kerry's top four phone contributors were: Verizon ($118,820), AT&T ($37,200), SBC ($33,130) and MCI ($24,980). Together, the Bells gave Kerry a bit more than half of the money he received from phone interests.

Junkets

The telephone industry is also a frequent sponsor of trips for members, family, and staff of the two congressional committees that regulate the industry—the Senate Committee on Science, Commerce and Transportation and the House Committee on Energy and Commerce. The telephone industry as a whole paid for 195 trips by congressmen, senators, and congressional staffers from the House and Senate commerce committees between 2000 and March 2004, valued at $276,071, according to documents.

The congressional office of former Rep. Billy Tauzin (R-La.), former chairman of the House Committee on Energy and Commerce, racked up the most miles at industry expense by a large margin. Tauzin or members of his staff went on 23 trips at a cost of $36,568. Tauzin personally took trips home to New Orleans and to a number of California locales, including San Diego and Palm Springs, at a cost of $6,894 to the industry. Tauzin's wife Cecile accompanied him on the California outings, while son Michael went along on one of the trips to New Orleans.

Revolving Doors

At least 58 current and former FCC, congressional, and executive branch officials have worked for the phone industry. At least 23 have worked for the Bells or their trade association, the United States Telecom Association. Many others have been hired as outside consultants and contract lobbyists on the industry's behalf.

The companies that enlisted the help of these commission veterans have reaped a substantial return on investment. In particular, the Bell companies have scored a series of victories at the FCC since former FCC Chairmen Reed Hundt, who openly favored new competitors at their expense, left the FCC. Hundt's successor, William Kennard, another Clinton appointee, hewed closer to the Bells, while former Chairman Michael Powell, a Bush appointee, went even further by favoring the Bells on most key decisions. The industry's political largesse meant that the Bells had allies among Republicans as well as Democrats on the powerful House and Senate panels poised to undertake a wholesale realignment of the telecommunications laws. A federal judge threw out rules opposed by the Bells, and the administration chose not to appeal—a decision that left Bell competitors to face price hikes. Attempts to pass a major rewrite of telecommunications laws in 2006 were unsuccessful.

PHONE CASE: THE REVOLVING DOOR FROM K STREET TO THE CAPITOL

Some call it cashing in. Some call it an opportunity they can't pass up. Others call it an offer they can't refuse. The tight-knit group of congressional aides who crafted the pro-business Telecommunications Act of 1996 and then went to work as K Street lobbyists call it making a living. "Some people had no choice but to go through the revolving door because you have to make a living on what you do and what you know," said Christopher McLean, who held senior roles in the Clinton Administration and then left government for a Washington, DC-based lobbying firm following President George W. Bush's 2000 election victory. Whatever their motives, the aides have taken the common route from political staff member to champion of private interests.

Through press accounts, primary sources, and detailed interviews, The Center for Public Integrity has identified a short list of the 15 staffers most involved in the years-long process of drafting the 1996 act. Thirteen of them, it turns out, became lobbyists, and another is a lawmaker who gets generous financial support for his campaigns from telephone companies. Another worked for a telecom lobbyist before helping to draft the law. Their clients include Nextel Communications, AT&T, AT&T Wireless, Winstar, Gateway Inc., COMSAT, Network Affiliated Stations Alliance, XO Communications, Covad Communications, Allegiance

Telecom Inc., the National Cable and Telecommunications Association, Qwest Communications, Verizon, SBC Communications, NextWave Telecom Inc., News Corp., BellSouth Corp., Motorola, and MCI Inc. (now part of Verizon Communications Inc.).

Those staffers have defended the integrity of their work on the 1996 act, but some telecom observers have their doubts. Adam Thierer, for example, formerly an analyst at the libertarian Cato Institute, called it a fundamentally flawed law loaded with contradictory giveaways and payoffs to various industry players.

The Center identified at least 199 current or former FCC officials and congressional staff who have worked as registered lobbyists, many leaving the FCC to work for the communications industry and later returning to government. "I think there's no doubt that there's what some people call an almost incestuous relationship," said Thierer of Cato. "The real question at the end of the day for someone like me is, What are they fighting over? They're fighting over control of the government's ability to influence private decision making."

Rep. Chip Pickering (R-Miss.), who worked on the act as a staffer to Senator Trent Lott (R-Miss.), now holds the most prominent position of the group, having run for office immediately after leaving Lott's office. Pickering may not have gone to work for the industry, but the phone companies nevertheless found a way to boost his career: Along with their employees, they donated $264,000 to his campaigns from 1998 to September 2004. That's just a small slice of the more than 35,000 contributions—totaling more than $65.5 million—the industry made to House, Senate, and presidential candidates. Records show that between 2000 and March 2004, traditional telephone interests paid for 10 trips for Pickering or members of his staff, at a cost of more than $11,000. Among the trips: a 2-day outing to New Orleans during Mardi Gras that Pickering took at the expense of the trade group representing cell phone providers. Pickering's office did not respond to the Center's request for an interview.

John Windhausen Jr., a former Senate Committee on Commerce, Science and Transportation staffer reporting to Senator Ernest "Fritz" Hollings (D-S.C.), went on to head the Association for Local Telecommunication Services, a trade group for local phone-service providers that compete against the major phone companies. "It wasn't going to get any better than that. After the '96 act I'd accomplished everything a staffer could accomplish," Windhausen told the center.

Of course, congressional and FCC staffers are not alone in moving from public servants to telephone industry lobbyists. They are joined by FCC commissioners, administration officials, and even members of Congress. Indeed, at least two congressmen and one senator who played key roles on the 1996 bill became lobbyists whose clients included one of the former "Baby Bell" companies that now appear to be the ultimate winners from the law: former Reps. Jack Fields (R-Texas) and Thomas Bliley (R-Va.) and former Sen. Larry Pressler (R-S.D.). At the time the 1996 act passed, Pressler was chairman of the Senate Committee on Commerce, Science and Transportation, but lost a tough re-election fight later that year. Bliley was chairman of the House Committee on Energy and Commerce. The only other lawmaker centrally involved in crafting the 1996 law who has left office is Senator James Exon (D-Neb.), the veteran politician who retired after his term ended in January 1997.

PART III: THE CABLE GUYS

Overview

The cable television industry has steadily increased the amount it spends to sway opinion in Congress and at the FCC since 1998. The industry spent $15.5 million on lobbying in 1998 compared with $20.2 million in 2004, a jump of 30%, according to records. All told, the industry spent $116 million on lobbying from 1998 to 2004. The cable television industry also spent $25.7 million on political campaigns from 1998 to 2004 and sponsored 127 trips for members, family, and staff of the two congressional committees that oversaw the industry from 2000 to March 2004 at a cost of $225,670. Center researchers also identified 16 former senior government officials (FCC, House, and Senate commerce committees) who went to work for the cable industry since 1996.

Lobbying

The top spender on lobbying among cable systems is the National Cable and Telecommunications Association (NCTA), the cable industry's Washington representative, at $36.8 million. Second on the list is Time-Warner Inc., at $29.2 million. The total includes spending from non-cable divisions of the media conglomerate. Time-Warner is the nation's second largest cable systems operator, with more than 14 million subscribers. Its holdings also include several cable television networks, including news channel CNN. Third is The DirecTV Group Inc., at $14.7 million. The nation's largest cable television provider, Comcast Inc., follows with $12.4 million.

The most important issues facing the cable industry deal mostly with staving off federal regulation. The industry successfully kept its broadband Internet service free from similar "common carrier" rules that once affected telephone companies. The industry is also fighting a proposal that would require them to carry multiple digital broadcast signals on their systems from a single broadcaster. And it is in the midst of a massive lobbying effort to snuff out attempts to require that channels be made available on an "a la carte" basis rather than in tiers (discussed in detail later).

Although cable television is still by far the most popular means of receiving video programming, direct broadcast satellite has become a legitimate competitor.

The DirecTV group is now the second-largest provider of pay-television service with 11.6 million subscribers. EchoStar Communication Corp. ranks fourth behind Time-Warner with 8.8 million customers. These two companies have also become big Washington influence players.

Contributions

Time-Warner is the top contributor in the sector at $9 million. Comcast is a distant second with $3.83 million and the NCTA is third with $3.77 million. The partisan preference of the cable television industry skews Democratic. Contributions were split 52.5% for Democratic candidates and party organizations and 47.2% for Republicans. The top recipient of cable contributions was Bush at $399,110. Kerry was second at $354,629, whereas now retired Senate Minority Leader Tom Daschle, (D-S.D.) was third at $266,800.

Junkets

The cable industry is also a frequent sponsor of trips for members, family, and staff of the two congressional committees that regulate the industry—the Senate Committee on Science, Commerce and Transportation and the House Committee on Energy and Commerce.

The NCTA, which puts particular emphasis on its annual trade show, is far and away the cable industry's largest underwriter of travel for members of those committees; in fact, it is the leader in the communications industry as a whole. The NCTA spent $191,727 on congressional travel during the study period. (The NCTA also spent $192,609 on 102 trips over an eight-year period flying FCC commissioners and staff to various events, according to a 2003 center study. The practice all but stopped due to congressional pressure following the center's report.) Second for spending on congressional travel was Time-Warner, with 16 trips worth $22,495. Comcast was a distant third, with $2,810 worth of trips.

The most well-traveled congressional office—with 21 trips valued at $38,037—was that of former Rep. Tauzin (R-La.), who was chairman of the House Commerce Committee. Second, with 12 trips valued at $15,514, was the office of Senator Conrad Burns (R-Mont.), a member of the Subcommittee on Communications of the Senate Commerce Committee.

Revolving Doors

Among former government officials now working for the cable industry, perhaps the most well known is Victoria "Torie" Clarke. Clarke was the Pentagon spokeswoman during the most recent Iraq war and is widely credited with creating the "embedding" system of attaching reporters to troops in the field. Clarke, who had previously served as chief spokeswoman for the NCTA, took a job with Comcast as senior adviser for communications and government affairs.

One of the top cable lobbyists in Washington is Daniel Brenner, senior vice president for law and regulatory policy with the NCTA. Brenner was senior legal adviser to former FCC Chairman Mark Fowler from 1981 to 1986.

Among Time-Warner's heavy-hitting lobbyists is Susan Brophy, who was director of congressional relations for the Clinton–Gore transition team, and then deputy director of legislative affairs at the White House. She joined Time-Warner as senior vice president of domestic policy in 2001 and now serves as vice president for global public policy. As for satellite broadcasters, DirecTV hired Susan Eid as vice president of government relations. Eid was senior legal adviser to FCC Chairman Michael Powell before joining DirecTV.

CABLE CASE: CHOOSING CHANNELS IS ANTI-MINORITY?

A La Carte Cable Programming

When a conservative member of Congress floated the idea of allowing consumers to pick which channels they want to pay for rather than having to buy a "bundle" of channels they may never watch, it seemed like a pretty good idea. Rather than pay a flat fee for dozens of channels, consumers could choose a handful of channels "a la carte," or from a menu—and possibly pay far less for their service.

Among the hundreds of initiatives tracked by the cable television industry and its army of lobbyists, a la carte is among the most feared. A re-regulation of the industry could cost cable companies control over programming and jeopardize an economic model that has helped the industry maintain huge profits over the years. The industry has argued that a la carte service would increase costs to consumers rather than lower their cable bills. In addition to warning of higher prices, the industry's chief argument for bundling channels is that cable networks that cater to minorities will wither and die if people are allowed to choose their own programming.

The cable industry, largely unregulated since 1996, has had a problematic track record with consumers. Cable prices have risen 40% from 1997 to 2002 and subscribers in 98% of markets have only one cable provider, according to a study by the Government Accountability Office. Surveys by J.D. Power show a low level of public satisfaction with cable service providers.

But rather than spark an uprising of consumer support, the proposal to offer channels on an a la carte basis was blasted by elected officials at every level of government. Joining the chorus were nationally known and respected civil rights organizations and a large collection of women's groups, all of whom appeared to be singing from the same page.

Grass Roots Support or Industry Influenced Campaign?

A review of hundreds of filings with the FCC, lobbying reports and other documents reveals that the "grass roots" opposition to a la carte is actually a highly sophisticated lobbying campaign where seemingly disinterested third parties—like nonprofits and legislators—are spreading the anti-a la carte message using minority programming as the key issue. In fact, rather than being disinterested, these third parties have much to gain.

The center identified hundreds of thousands of dollars in donations and other benefits showered by cable companies on some of these nonprofits. The Center also found one instance in which free airtime was made available to a mayors' group and identified nearly $60,000 in contributions to one key pro-cable congressman.

One measure of the campaign's effectiveness is an August 13, 2004, comment filed by the NCTA, the key lobbying arm of the cable television industry, with the FCC, pointing out the overwhelming number of letters that had been filed in opposition to a la carte. Among those letters pointed out by the NCTA, some 64 were from nonprofits and politicians whose comments were derived from one of five basic form letters, one of which was written by four cable executives, another circulated by an industry lobbyist.

It is not unusual for a special interest group to orchestrate mass-mailing campaigns, on either side of the political spectrum. But what's not so common is for nationally recognized civil rights leaders, members of Congress, city council members and state legislators to sign their names to an industry-generated letter. Among the signers of one of the five letters were two members of Congress, six mayors, seven members of the New York City Council, six members of the New York state legislature, one Florida state legislator, one Arizona state legislator, one Illinois state legislator, and the comptroller of the state of Connecticut.

"The difference with this campaign, the ethnic community is at the vanguard," said Brian Woolfolk, who is working pro bono for small cable networks that are in favor of a la carte pricing. "Strategically, it completely disarms the Democratic caucus. The Democratic caucus is going to be sensitive to civil rights organizations and civil rights issues." Among the minority nonprofits who signed form letters in opposition to a la carte pricing were the National Urban League, the Rainbow/PUSH Coalition and the National Congress of Black Women Inc.

The cable a la carte issue actually dates back to the 1980s, when cable companies offered it as a pricing option. Although a la carte service never caught on, demand for the option resurfaces periodically depending on the level of public dissatisfaction with service and cost. By March 31, 1999, provisions of the Telecommunications Act of 1996 had phased out cable price regulation and prices began to spiral upward. But the a la carte issue got red hot exposure when conservative groups started pushing the issue.

The Janet Jackson Super Bowl stunt emboldened social conservatives like Rep. Nathan Deal, a Republican from Georgia, to push for a la carte as a means to keep risqué programming out of households that don't want it. Joining the cable a la carte debate was Consumers Union and its senior director of policy Gene Kimmelman, albeit grudgingly. "A la carte emerged in the consumer agenda only after efforts to prevent concentration of ownership and price gouging all failed," he said. "It was kind of a fall-back position. It was never viewed as the totally optimal approach." Consumers Union is the respected nonprofit publisher of Consumer Reports magazine. The organization accepts no corporate contributions. Kimmelman and others expected resistance from the industry. But they were certainly caught off guard by the allies that cable mustered to its side.

The Congressional Black Caucus, which was founded in 1969, is a group of 39 lawmakers who list among their legislative priorities the creation of a living wage, guaranteed health insurance for all Americans, and strengthening and better enforcing civil rights laws. The caucus has long been concerned with the lack of diversity in cable programming, ownership and in the senior ranks of system operators. In early 2003, the caucus, led by Rep. Bobby Scott (D-Va) signed a letter addressed to the cable industry threatening severe regulatory consequences if cable's prospects for minorities did not improve. A portion of the letter was published in a trade magazine, but it was never actually mailed, according to Woolfolk, who actually penned the letter. A series of meetings between cable television representatives and the caucus followed, and the diversity issue appeared to fade away.

Not long after the meetings, the a la carte movement began to get some traction thanks to the Janet Jackson scandal and general public outrage regarding indecent broadcasting. Deal authored an amendment to a bill that would require cable and satellite companies to give customers the option of buying channels one at a time, rather than as a package.

Alfred Liggins, chairman of TV One, a new African-American-themed cable television network co-owned by broadcaster Radio One and cable giant Comcast Corp., wrote an April 12, 2004 op-ed piece in the *Washington Times* arguing against the idea. Liggins, whose network benefits from being bundled with other channels, argued that minority networks would be squeezed out if consumers were allowed to choose their own channels.

On May 12, 2004, the Congressional Black Caucus went from being a key cable industry threat to an ally. The caucus sent a letter to Reps. Joe Barton and John Dingell, then chairman and ranking member of the House Committee on Energy and Commerce respectively, arguing against the a la carte idea. The letter, which quotes the Liggins piece in the *Washington Times*, shifted the focus of the a la carte debate from lowering consumer costs to preserving minority programming.

Over the past few years, the nonprofit Congressional Black Caucus Foundation—a separate and distinct entity from the caucus that nevertheless boasts 11 caucus members on its board of directors and two caucus members among its top officers—has drawn considerable support from the cable industry.

In its 2002 annual report, the foundation, which ranks donors within ranges, lists Time-Warner (then AOL Time-Warner) as contributing between $50,000 and $99,999. In the 2003 report, the company jumped a tier, contributing between $100,000 and $249,999. Comcast also upped its contribution, from between $30,000 and $49,999 in 2002 to between $50,000 and $99,999 in 2003. Viacom companies are listed as contributing between $20,000 and $44,998 in 2002, but in 2003 only Viacom was listed as contributing between $5,000 and $14,999. The NCTA donated between $5,000 and $14,999 in 2002 and 2003. In addition, Comcast, the NCTA, Time-Warner and Viacom are all regular sponsors of the caucus' annual legislative conference. In all, over the 2 years, those companies contributed as much as $589,991.

Calls to the director of communications for the Congressional Black Caucus were not returned. The same day the caucus sent its letter—May 12—another missive was directed at the Commerce Committee. Four cable executives—Debra Lee (BET Holdings Inc.); Jeff Valdez (Sí TV, a Hispanic-themed network whose investors include Time-Warner and EchoStar); Johnathan Rodgers (TV One), and Kent Rice (the International Channel, which was acquired by Comcast)—wrote the committee a letter objecting to the a la carte concept. The executives' letter was posted on NCTA's Web site; the trade group has cited it in its filings with the FCC. Portions of the letter were identical to the op-ed Liggins penned for the *Washington Times*.

A week later, Congress opted to bail out of the a la carte debate and let the FCC deal with it. Leadership of the House committee that regulates cable wrote a letter to FCC Chairman Michael Powell asking the FCC to study the issue. Senator John McCain, former chairman of the Senate Committee on Commerce, Science and Transportation, also wrote a letter, arguing forcefully for an a la carte option. On May 25, 2004, the FCC officially opened the issue for comment, in anticipation of filing a report in the fall. And the floodgates were opened.

On July 15, 2004, the NCTA released a study funded by the association that said a la carte would hike rates. The study, conducted by Booz Allen Hamilton Inc., determined that if cable operators had to offer all channels a la carte while still offering tiered programming, costs would rise between 7% and 15% just to outfit cable systems with the a la carte technology. Booz Allen argued that consumers would end up paying more for fewer channels, and an a la carte system would make it difficult for new networks (like minority-themed channels) to get into the business.

On the same day, the nation's "oldest, largest and most diverse civil and human rights coalition," the Leadership Conference on Civil Rights (LCCR), also filed comments with the FCC opposing a la carte on grounds that it would hurt diversity in cable programming. The organization quoted Liggins in its filing. The LCCR counts among its victories passage of the Civil Rights Act of 1964, the Voting Rights Act of 1965, and the Fair Housing Act of 1968. More recently, the group has worked to eliminate the so-called "digital divide" in America and worked to provide the advantages of Internet access and other technology to disadvantaged schoolchildren. In the a la carte issue, the civil rights organization sided with the cable companies.

On June 23, 2004, the LCCR, in conjunction with TV One and others, sponsored a lunch briefing opposing the a la carte effort. Titled "Cable A La Carte: The Beginning of the End for Media Diversity on Cable TV?" the event was organized by its then-lobbyist, Leslie Harris of Leslie Harris and Associates. Among Harris's other clients is Time-Warner Inc. Leading the charge for LCCR against a la carte was the group's deputy director, Nancy Zirkin, who targeted Kimmelman and Consumers Union specifically because he was supporting an issue that was also backed by conservative groups like the Parents Television Council and Concerned Women of America.

The AOL Time-Warner Foundation reported in its calendar 2001 tax filing with the IRS that it pledged $100,000 to the LCCR and another $200,000 to the Leadership Conference Education Fund. For 2002, the foundation reported a payment of $100,000 to the LCCR. Among contributions to other nonprofits who signed letters opposing cable a la carte, the AOL Foundation gave $300,000 to City Year, $100,000 in 2000, 2001 and 2002; and the Comcast Foundation contributed $50,000 to the National Urban League, $78,600 to the Women in Cable & Telecommunications, $70,000 to City Year Philadelphia, and $50,000 to City Year Detroit. Each organization signed form letters supporting the cable industry.

The FCC Letter Campaign

Then the letter campaign began. First to be heard from were the civil rights groups. The first was from the National Congress of Black Women Inc., signed by national chairwoman the late C. DeLores Tucker on June 28, 2004. Tucker's letter was nearly a word-for-word copy of the one written by the cable executives. She also penned an op-ed piece published in the *Boston Globe* and *Chicago Sun-Times* arguing that the proposed regulation would kill cable programming diversity, one outcome of the Civil Rights Movement. Tucker had earlier led a high-profile crusade against misogynist lyrics in rap music. Ironically, Tucker's chief target at that time was Time-Warner's music division. Other groups who signed the same letter to the FCC as Tucker include the National Urban League, the Rainbow/PUSH Coalition, and the Brotherhood Crusade.

Also submitting the same basic letter were three members of the board of directors of the National Conference of Black Mayors—Harvey Johnson Jr., mayor of Jackson, Mississippi, and president of the conference; Irene Brodie, mayor of the village of Robbins, Illinois and assistant secretary for the group; and Roosevelt F. Dorn, mayor of Inglewood, California who is second vice president of the mayoral conference. The letters were dated July 12, 2004. On April 27, 2004, the group hosted its annual conference in Philadelphia, the hometown of Comcast. The conference's "premium" sponsors were Comcast and TV One, employer of Alfred Liggins. TV One president and CEO Johnathan Rodgers spoke at the event, which also included an hour's worth of video interviews with mayors about major issues of urban cities.

Another mayoral organization chimed in on the debate—the National Conference of Democratic Mayors (NCDM). The NCDM lists the NCTA as a sponsor. Among cable television's best friends is Democratic Rep. Edolphus Towns, a member of the Congressional Black Caucus who signed the May 12 anti-a la carte letter. Towns also submitted another letter, this one to the FCC, on July 8. The letter was nearly identical to five others—one by Florida state Rep. Bob Henriquez, three by members of the New York state legislature, and one by a New York City Council member.

But unlike those elected representatives, Towns is in a position to do more than comment on the issue. The one portion of his letter that does appear to be original refers to his role as a senior member of the House Committee on Energy and Commerce, and his plea to reject the a la carte system in order to maintain diverse programming. Any legislation that would re-regulate the cable industry would originate from that committee. Towns submitted his comment despite the fact that Chairman Barton, ranking member Dingell, and three other representatives had asked the FCC to prepare the report on the feasibility of a la carte. From 1998 to September 2004, Towns accepted $58,897 from the employees, their immediate family members, and the political action committees of the nation's largest cable television companies, sixth-highest among House Democrats over that period.

Towns spokesman Andrew Delia submitted a statement to the center in response to a set of detailed questions. "Congressman Towns believed and continues to believe that on the whole, an a la carte system would hurt niche and diversity programming. While there are legitimate concerns about diversity in the media, a la carte is not the solution to such a problem," the statement reads. Towns contended that some channels would disappear if they were located on fewer systems because they would be unable to generate advertising revenue. Programmers would have to raise rates to make up the difference and niche channels would suffer. The result would be higher costs for consumers for fewer channels.

The response from the congressman's office did not answer the question of who wrote the letter he submitted to the FCC or whether the cable industry's support of his campaigns or the Congressional Black Caucus Foundation—Towns is a member of its board of directors—played a role in his position.

The cable industry found support in other quarters. Rep. Raul M. Grijalva of Arizona was one of 12 elected officials to sign another version of one of the industry's form letters. Grijalva is a member of the Congressional Hispanic Caucus. Among others who submitted the same letter were Hispanic officeholders in the New York state legislature and New York City Council in addition to a number of nonprofits.

In case the FCC was still unconvinced, the cable lobby had one more "grass roots" coalition join the battle—and this would be the largest outpouring yet. In late July and early August, a truly impressive cross-section of women's groups began sending letters to the FCC, asking it not to support cable a la carte pricing. Once again, the letters sounded familiar. In fact, 30 of them were virtually identical. Among the signers were the Sexuality Information and Education Council of the United States, the Global Fund for Women, the Feminist Majority, American Women in Radio and Television Inc., and the National Council of Women's Organizations. In this case, it wasn't hard to find the source. Leslie Harris, the Time-Warner lobbyist whose firm also represents the LCCR, orchestrated the campaign. Her firm advertised its role on its website.

After being retained by a "major media company" Leslie Harris and Associates coordinated a meeting where minority and women's programmers presented the case against a la carte programming, and then helped more than 30 women's organizations—most notably, Oxygen Media—write and file comments with the FCC. Harris filed a lobbyist registration statement listing Time-Warner as a client on July 23. Harris, through a spokeswoman, declined to comment for the record. Oxygen Media is a 24-hour cable television network aimed at women. Founded in 1998, the network is independently owned and now available in more than 52 million cable households, according to its website. Its chairwoman and CEO, Geraldine Laybourne, is also on the board of directors of NCTA. Time-Warner is an investor in the network.

Eleanor Smeal of the Feminist Majority submitted one of the form letters. It stated in part, that the a la carte proposal could have a "significant negative impact on the so-called niche networks, such as Oxygen, that cater to particular interests or demographics such as women, minorities, gays and lesbians and non-English language speakers."

In a detailed e-mail to NCTA senior director of communications Brian Dietz, the center asked whether it was disingenuous to point to a groundswell of opposition to a la carte when so many of the letters were industry generated. Dietz replied that "As a matter of policy, NCTA does not comment on our lobbying or PAC activity." But the association included a generic statement: "Many of the individuals and organizations that have voiced opposition to a la carte are concerned that such regulation would further reduce diversity in media, which directly affects the constituents they represent. NCTA welcomes this support and will continue to work with organizations that are concerned about these issues," it reads in part. So the question is, are the commissioners aware that the anti-a la carte comments they are being inundated with are not as spontaneous as they seem? And does it really matter?

The center sent a list of detailed questions to former FCC Chairman Michael Powell through a spokesman. Powell did not respond, but the center was provided with a written statement from then-FCC Media Bureau Chief Kenneth Ferree. "In the A La Carte proceeding, as with all FCC proceedings, we value, and read, comments from any and all persons and parties on all sides of the issue. Our job is to then evaluate the pros and cons, and strengths and weaknesses, of all the arguments, positions, and studies presented to us, and then make our best public policy judgments based on the entire record," it reads. "It is interesting to note that the practice of a party on one side of an issue soliciting letters of support is not uncommon. For example, most recently, a very high percentage of letters in the media ownership rule proceeding were form letters generated by self-interested organizations. As to the A La Carte proceeding, we are in the process of evaluating the record at the present time and do not comment on any individual filings."

Currently, the anti-a la carte comments outnumbered pro a la carte comments by about 6 to 1. Among the pro camp are Consumers Union, the Consumer Federation of America, the New Jersey Ratepayer Advocate, the Urban Broadcasting Company, and the city of Seattle. There is no indication of coordination among the comments filed supporting a la carte.

Cable Industry Influence Behind the Debate on A La Carte Programming

From 1998 to 2004, the nation's cable television companies spent $25.7 million in campaign contributions to national party organizations, members of Congress and presidential races. The money came from political action committees, soft money donations (before they were banned by the Bipartisan Campaign Reform Act of 2002), and individual contributions from company executives, employees, and their families. In addition to the donations, cable television companies can provide something worth more than greenbacks—airtime. No politician, no matter how skilled or principled, can turn his or her back on reaching large blocs of the voting public through public affairs programming.

Corporations can also play the sponsorship game. With millions to give away, the Time-Warner Foundation funds good works all over the country, including the activities of civil rights groups. As detailed at the beginning of this section on cable, the industry's trade group, the NCTA, is among the most deep-pocketed and influential lobbying organizations in Washington, as well as sponsoring junkets for Congress, FCC, and staff. The NCTA's top lobbyist, Daniel Brenner, was senior legal adviser to former FCC Chairman Mark Fowler in the 1980s. While campaign contributions get the most coverage in Washington politics, the real money is being spent on lobbying.

The cable television industry has been steadily increasing the amount it spends to sway opinion in Congress and the FCC, from $15.5 million spent on lobbying in 1998 compared with $20.2 million in 2004, a jump of 30%. Although the dollar amount is impressive, it is the people who are doing the lobbying that are the real key. The cable, broadcasting, and telecommunications lobby in Washington is stacked with former FCC commissioners, bureau chiefs, top-level aides, and former congressional staffers who have helped write communications laws.

As to the a la carte issue specifically, the NCTA is using Brenner as well as Jill Luckett, vice president of program network policy, formerly a special adviser to former FCC Commissioner Rachelle Chong, and before that, legislative director for former Senator Bob Packwood of Oregon. Leslie Harris and Associates, now defunct, employed Jon Bernstein who, before becoming a lobbyist, was an attorney adviser with the FCC. Prior to that, he was a lobbyist for the National Education Association. Working for Viacom on the issue is Wiley Rein & Fielding. Richard E. Wiley heads the firm's 70-attorney communications practice, which it bills as the largest in the nation. Wiley was chairman of the FCC from 1970 to 1977 and was a chief advocate of deregulation of the communications industry. Lawrence W. Secrest, III, a partner with the firm overseeing the media practice and Wiley's former assistant at the FCC, is also on the a la carte beat.

That's a lot of high-priced talent to be spending on what seems like a relatively small issue that is barely on the regulatory radar screen. The FCC was merely collecting comment for a report (a report, not surprisingly, that gave a negative view of a la carte programming.) There is no realistic expectation that the issue will make it into any legislation. Kimmelman told the Center he believes the lobbying campaign has more to do with the issue's potential than what it means right now. He believes that "there was a potential out-of-control grass-roots uprising against the way large cable and broadcast companies control programming and how it's distributed to the public," he said. "I think they recognized this was so attractive to people from all political perspectives and so volatile in terms of spiraling cable rates and concerns about smarmy programming—they needed to deep six and absolutely wipe the debate off the ledgers before policymakers got into it."

CONCLUSION

For a relatively brief period of time in late 2003, the public got a glimpse of the arcane and largely invisible sausage machine that grinds out the nation's media and telecommunications policy. The debate over media consolidation and controversy involving television owner Sinclair led to a greater public understanding of what the broadcast industry really is—an industry—an industry with powerful friends, tremendous political clout and the ability to spend tens of millions of dollars to get its way in Washington.

Since 1998, lobbying expenditures by the broadcast industry have risen 80%. Activity peaked in 2003, the same year the FCC proposed significant relaxation of media ownership rules. The FCC voted to allow corporations to own more media outlets than ever before and reach a greater percentage of the national audience.

Former FCC Chairman Michael Powell, who was appointed by Bush, was a supporter of deregulation. That was made clear during Powell's first public appearance as chairman-designate. At a meeting of the Association of Local Television Stations on January 22, 2001, the commissioner agreed with fellow Republican FCC Commissioner Harold Furchtgott-Roth that the agency had in effect become a needless third federal antitrust regulator, given its restrictions over media ownership.

Gradually, the impending deregulation found its way into the national discourse. By June 2003, the FCC had received more than 700,000 public comments about the issue, with 99% opposed to any further deregulation. They constituted the largest public response in the agency's history. Nevertheless, on June 2, 2003, Powell led the other Republican commissioners in voting to increase the national broadcast television audience cap to 45% (up from 35%) and eliminate the so-called duopoly rule (a prohibition against owning two TV stations in the same market) and the media cross ownership rules.

Perhaps convinced by the public outcry, however, the Senate Commerce Committee quickly called on the FCC to defend the new rules and, at the same time, crafted a bill that would override the attempt to set the national audience cap at 45%. Ultimately, the White House stepped in and the 45% cap was reduced to 39%. Some of the nation's largest broadcasters, including Fox and Viacom, are all bumping against or exceeding that limit. Since then, a federal appeals court sent the new rules, other than the audience cap, back to the FCC for re-justification, forcing the agency to re-submit the proposed changes with new supporting research. Eventually, the agency will do just that, and the ultimate decision as to whether the rules will be loosened further will rest with the federal government.

It is disturbing to many to think of media companies as politically active. But it would be naïve not to. Big broadcasters are required to consider both the public interest and the interest of their shareholders, but too often the shareholders' interest wins at the expense of the public's.

All good journalists follow the money. And in the media consolidation debate, the money has been flying out of the bank for lobbying, contributions, and junkets, with pervasive involvement from revolving door officials and lobbyists. In addition to the impact on national audience caps, the decisions also affected so-called cross-ownership—how many different kinds of media outlets, such as newspapers and television stations—a company may own in one market.

Belo Corp., owner of television stations and newspapers, had consistently spent $200,000 a year from 1998 to 2002. In 2003, however, the company spent 50% more—$300,000—and even more in 2004, spending $330,000. Clear Channel Communications Inc., the largest radio station owner in the country, went from an average yearly lobbying total of $68,675 for 1998–2002 to a whopping $1.9 million in 2003. Both the National Association of Broadcasters and the Washington Post Co., which also owns television stations, saw significant increases.

The most obvious sign of this shift comes not from hard-core Republicans like David Smith and his three brothers, who control Sinclair, but from the 84-year-old self-described "liberal Democrat" who controls Viacom Inc., owner of CBS, and one of the largest media conglomerates in the world. Sumner Redstone tacitly endorsed George Bush for re-election in late September, at a meeting of CEOs in Hong Kong. The election of a Republican administration, Redstone told his audience, "is a better deal" in Viacom's view, "because the Republican administration has stood for many things we believe in, deregulation and so on."

Newsweek reported that Redstone's remarks were viewed by many as a breach of an understood code of silence regarding political endorsements that most media conglomerates respect. Several other executives from the top broadcasters were asked to comment, and most repudiated any notion of political favor, while a public interest lobbyist pointed to the remarks as evidence of what many have known for a long time—a comfortable relationship between the industry and its government regulators.

13

CONSIDERING GLOBAL MEDIA

Sociological Contributions

Denise D. Bielby

Molly Moloney

In Spring and Summer 2003, the Federal Communications Commission (FCC) was poised to implement one of the most significant overhauls of media ownership policy in a generation. The changes—since struck down by the courts—relaxed or eliminated a wide-ranging set of media ownership rules that are profoundly altering many longstanding, distinguishing features of the media landscape. They would have significantly expanded the number of local TV stations networks could own, allowing for more cross-ownership of newspapers and TV stations, and generally facilitating the growth of major media conglomerates. These changes were not met without resistance—indeed there was unprecedented vocal opposition, focusing on the dangers posed: the homogenization of entertainment programming, the supplanting of local news coverage with national broadcasts, and a reduction in the economic advantage of local stations to offer independent programming (Labaton, 2003). These questions continue to be pressing, as Congress revisited the sweeping 1996 Telecommunications Act in 2006, and, unable to reach resolution, plans to return to the matter in 2007 (Untangling the Telecom Act, 2007).

Concern about the homogenization of cultural products, about the dominance of outside programming over that which is locally produced, and the diminished capacities for smaller and independent producers to create and distribute are amplified when transferred from the domestic U.S. context to the broader arena of international and global media systems. Also amplified is the complexity of the task of making sense of these issues. In this chapter, we consider what a sociological vantage point can contribute to scholarly analysis and debate about global media concentration, using the case of the international export market for television. This approach involves attending to middle-range, empirical research and combining cultural, institutional, and organizational analyses of the culture worlds of global media.

MEDIA CONCENTRATION IN THE UNITED STATES: THE DEMISE OF "FIN-SYN"

Our interest in this issue of media concentration and its cultural effects emerges from research on the entertainment industry in the United States that examines the consequences of media consolidation on one group of media creative producers: television writers (W. Bielby & Bielby, 1999, 2003). Somewhat obscured in the 2003 fray was the significance of the earlier demise of the Financial Interest and Syndication Rules, or "fin-syn" rules. The fin-syn rules, implemented in 1971 and fully in effect until 1990, constrained the then-three networks (ABC, CBS, and NBC) from producing all but a small number of the programs they broadcast in prime time and barred them from participating in the syndication of prime-time series (a source of large profits as there are few production costs).

The rationale for the design of the fin-syn rules in the early 1970s was explicitly to intervene in the market to promote diversity and competition in the supply of prime-time entertainment programming and to forestall the kind of vertical integration that dominated the film industry during the studio era. Vertical integration—the ownership by one firm of multiple links along the production supply and distribution chain—is of concern owing to worries about the power this gives distributors to effectively shut-out independent producers by closing off access to distribution channels or significantly favoring only their own productions. A study of the effects of the demise of the fin-syn rules shows two outcomes of consequence to media concentration in the television field: the percentage of network-owned television programs soared, while independent shows all but disappeared; and television writers were increasingly working for the networks' studios and not independent production companies.

Although it took a while for effects of the elimination of the fin-syn rules to be fully felt, by 1995 a significant shift occurred. Prior to 1990, the networks were barred from a financial stake in the series they produced, but by 1994, modified fin-syn rules allowed the networks to have a financial stake in up to 40% of their prime-time entertainment schedules, and after that the limits were lifted entirely. Between 1996 and 2002, the networks increased their participation in program supply substantially. Our analysis revealed that by 2002, ABC, CBS, NBC, and Fox collectively had an ownership stake in more than 70% of the pilots developed from prime time, and the networks owned or co-owned more than 75% of the series scheduled to debut on the 2002–2003 prime-time schedule. These trends continued: the pilots for the 2005–2006 season were overwhelmingly network-studio productions. With the merger of NBC and Universal Pictures, all six major networks had sister-production companies/studios within their corporate families and only one major studio (Sony) remained without a network. One of the few remaining independent production companies, Carsey-Warner, produced only two series that year, had only one show returning the following year, and announced that it was essentially shutting down its production arm (Collins, 2005; Littleton & Andreeva, 2005).

These changes in ownership had profound implications for the employment of television writers—who are now increasingly employed only by the largest firms in the industry. We found a significant increase in concentration of writers' employment since the demise of fin-syn. Whereas in the early 1980s less than 50% of television writers were employed by the networks or major studios, by 2000 close to 75% were. In short—consistent with trends across myriad media, from film, to music, to book publishing—compared with the late 1980s, in 2006 a handful of large media conglomerates account for a greater share of television narratives.

Although this research did not extend to the measurement of the effects of concentration on program content, research by sociologists on other culture industries, such as the recording industry, found that industrial concentration of the sort observed here produces homogeneity of culture products (Dowd, 2004; Peterson & Berger, 1975). In short, market structure and organizational strategy shape the diversity and form of popular culture products. Unless and until the technological developments that are lowering barriers to entry and revolutionizing the production and distribution of recorded music (e.g., through the Internet) take hold in the television industry, for the near term, the "creative community" of program suppliers to the television industry remains locked into a system that is increasingly controlled by a dwindling number of companies and that has traditionally valued creativity only within the parameters of well-established genres and narrow formats. With an even closer eye to the bottom line because their own profits are now at stake, these companies are ever-more risk averse to creative ideas that may challenge audience tastes by breaking new ground or defying prevailing conventions. So far, the most visible impact of domestic deregulation has been a reduction in the number of organizational settings in which those who create television series are employed, and an increase in corporate control over the circumstances under which they practice their craft, as evinced by our two studies (W. Bielby & Bielby 1999, 2003).

Although the examples given here arise out of the domestic U.S. context, increased consolidation and conglomeration within media industries looms just as large, if not more so, in the broader global context. Studying equivalent shifts at the global level, however, is profoundly more complicated. A significant challenge entails defining the kind of evidence to consider for evaluation, because what one identifies as relevant depends in large measure on disciplinary orientation. In the following sections, we discuss selected aspects of the difficult terrain one traverses in pursuing questions pertaining to concentration, conglomeration, convergence, and creativity at the international level, drawing from our ongoing work on the global market for exported television.

After briefly reviewing key currents in some of the literature of the field, we address what sociology can contribute to the debate about international media conglomerates. We consider evidence from the global market that illustrates the complicated nature of empirical verification. We discuss an analytic strategy that engages meso-level investigation into the organization, operation, and cultural logics of global television production, distribution, and marketing; this level of analysis may be a key area in which sociology can contribute to the necessarily multidisciplinary task of understanding global media today.

ENCOUNTERING THE GLOBAL MARKET-MEDIA FLOWS

One of the most recognizable features of the global television industry occurs to scholars and nonscholars alike while traveling abroad and stumbling upon American programs airing in other countries or learning through news reports of the popularity of a series in other parts of the globe. *Friends* is an enormous success in Italy, England, and many other countries, as is the soap opera *The Bold and the Beautiful. Dallas*, the prime-time serial of the 1980s, was a widespread hit throughout Europe and parts of the Middle East, although it was a failure in Japan. As Tables 13.1 and 13.2 show, industry trade publications regularly feature lists of the top ten American shows in selected countries, and although such lists do not reveal how those shows rank relative to imports from other countries or to local fare, they do reveal what has been sold to particular locales.

The now-classic UNESCO study by Nordenstreng and Varis (1974) provided one of the earliest attempts to systematically document exported television programming, revealing what they termed a *one-way flow* of television content from the United States and a few western European countries to the rest of the world. They demonstrated that by the early 1970s, the United States, Britain, France, and the Federal Republic of Germany were the overwhelming leaders in television production and distribution worldwide. According to that research, the United States led this group with an estimated 150,000 hours of programming per

TABLE 13.1. *Top U.S. Shows in Selected Countries*

Rank	France, December 2005	Spain, February 2005	United Kingdom January 2004	Germany, December 2003	Germany, February 2003
1	*Walker, Texas Ranger*	*CSI: Miami*	*The Simpsons*	*Quincy*	*The Dating Game*
2	*CSI*	*CSI*	*CSI*	*World of Wonder*	*World of Wonder*
3	*Crossing Jordan*	*Everwood*	*Fresh Prince of Bel-Air*	*Friends*	*Quincy*
4	*NCIS*	*Walker, Texas Ranger*	*The Osbournes*	*Scrubs*	*Sex and the City*
5	*Alias*	*Fresh Prince of Bel-Air*	*CSI: Miami*	*ER*	*ER*
6	*Charmed*	*She Spies*	*Will & Grace*	*24*	*Charmed*
7	*My Wife & Kids*	*Malcolm in the Middle*	*Buffy the Vampire Slayer*	*Autopsy*	*The A-Team*
8	*Friends*	*Will & Grace*	*Law & Order: SVU*	*CSI*	*Buffy the Vampire Slayer*
9	*Relic Hunter*	*Relic Hunter*	*Murder, She Wrote*	*Relic Hunter*	*The X-Files*
10	*Dragnet*	*MacGyver*	*Scrubs*	*Medical Detectives*	*Airwolf*

Source: *Hollywood Reporter*, March 2, 2005, p. 10; February 2, 2005, p. 14; January 7, 2004, p. 10; December 3, 2003, p. 14; February 5, 2003, p. 8.

year, followed by the United Kingdom and France, estimated at 20,000 hours each, with West Germany exporting approximately 6,000 hours.

To some extent, these statistics merely reflect the number of hours of air-time that needed to be filled in newly emerging national television systems. With the costs of producing local content vastly outstripping the costs of importing programs from abroad, numerous nations relied heavily on foreign, often predominately U.S., programming. For the most part, the leaders in exported programming sent their programs to countries whose broadcast systems were sufficiently developed to have schedules to fill, or who shared the same language.

Although these nations were the largest exporters, several other countries were major producers of programming for international distribution within regional markets. For example, programs from Mexico were distributed throughout Latin America and in areas of the United States where Spanish is widely spoken, and Lebanon and the United Arab Republic were major producers for the Middle East. Analysis of flows among non-Western countries published a decade later showed that among those flows was a trend toward greater intraregional exchange, especially among Arab countries and also within Latin America (Varis, 1985). In addition, countries with large internal markets, such as Brazil's and India's television industries, now supply not only their own market but also send exports internationally. Sales of Brazilian telenovelas, for example, are very successful in eastern Europe, Russia, and increasingly, the Middle East. Television exports comprise a growing share of the revenues of television production companies across Latin America, with the United States (with its growing Spanish-speaking population) increasingly importing these shows. Top Mexican broadcaster Televisa, for instance, gets about one fourth of its annual revenue (around $160 million) from exports, with about half of this from licensing to U.S. network

TABLE 13.2. *Top U.S. Shows in Australia, Selected Years*

Ranking	October 2001	October 2003	July 2004	March 2004 (All Series)
1	*This is Your Life*	*CSI: Miami*	*Law & Order: SVU*	*CSI*
2	*Friends*	*ER*	*CSI*	*CSI: Miami*
3	*Boston Public*	*CSI*	*ER*	*Friends*
4	*CSI*	*Law & Order: SVU*	*Without a Trace*	*National Nine News Sunday Ed.*
5	*Malcolm in the Middle*	*Without a Trace*	*Law & Order: CI*	*Malcolm in the Middle*
6	*Providence*	*Becker*	*CSI: Miami*	*Without a Trace*
7	*Law & Order*	*Friends*	*Law & Order*	*Who Wants to Be a Millionaire*
8	*Everybody Loves Raymond*	*Alias*	*Friends*	*Australian Grand Prix*
9	*Law & Order: SVU*	*Everybody Loves Raymond*	*Crossing Jordan*	*60 Minutes (Australian)*
10	*Becker*	*Frasier*	*Frasier*	*Law & Order: SVU*

Source. Hollywood Reporter, November 7, 2001, p.10; October 1, 2003, p. 8; September 1, 2004, p. 10; March 16-22, 2004, p. 62.
Note. "All series" refers to the top 10 shows regardless of country of origin.

Univision and the other half from sales to eastern Europe, Asia, the Middle East, and Latin America (Higham, 2003). Japan is a major supplier throughout Asia and South Asia, especially of animation, and its global reach includes the West (Iwabuchi, 2002). In short, since the pioneering work by Nordenstreng and Varis, vital regional markets based on geolinguistic, cultural, and geographic similarities have developed (Sinclair, Jacka, & Cunningham, 1996; Straubhaar, 1991)

Still, however, the United States remains the dominant television exporter; one study estimates that 85% of all children's programming, 81% of television movies, and close to 75% of dramatic television programs sold in the global television market are of U.S. origin (DCMS, 1999a cited in Freedman, 2003), whereas another estimates that more than 60% of the global trade in television, about $4 billion, is controlled by U.S. companies (Colitt, Fifield, Johnson, Marcelo, & Larsen, 2003).

Statistics documenting the degree of U.S. domination of television markets abroad are often provocative and compelling; they can be a helpful starting point for trying to understand key questions about the power of U.S. media culture around the world and about the landscape of diverse media markets. Indeed, some of these statistics helped to launch a debate about media domination that has now raged for three decades. However, too often discussions of global media ownership *end* rather than start here. This is a problem because these broad statistics about the distribution of shows, although interesting and obvious, also lock us into assumptions and narrow our explanations and understanding of the global market. Important questions are missed when one focuses only on the broad picture of television trade flows. For example, how are television programs transformed by touching down in diverse international contexts of reception and interpretation? The problems are not isolated to the level of reception and audiences, however. It also obscures understanding of the *particulars* of production and distribution processes. Why are specific programs, for instance, marketed, adopted, or rejected by distributors? To understand the dangers that U.S. domination of global television markets and the increasing consolidation and integration of global media mega-conglomerates may pose requires a much more fine-tuned understanding of the organization and cultural logics of these markets than can be gleaned from such crude evidence of program flows alone.

THE MEDIA HEGEMONY DEBATES

Discussion of the export of television programs needs to be placed in the context of broader debates about global media. Among critical scholars of media communication and culture, much important work has been done in recent years, and it is a burgeoning field. At the same time, however, sometimes it seems as if it has reached an impasse. There is a too-easy way to caricature contrasting sides and approaches to critical analysis of the globalization of media. This caricature pits macrostructural analyses focused on issues of economics and power writ large against more micro-oriented analyses of media texts and audiences within localized contexts, with each camp bemoaning the oversimplistic outlooks of the other.

On the one hand, there are those who focus on political-economic and structural issues. Their generally macro-oriented, top–down analyses note with alarm the increasing power and scope of a small number of media corporations around the world. Conglomeration and consolidation within media-culture industries has meant that a smaller and smaller number of larger and larger (global but largely U.S.-based) corporations have come to dominate not only American media, but media around the world. Work in this tradition points to the uneven, unequal, and generally unidirectional flows of media products—from North to South, from the United States to the rest of the world. They highlight threats posed by these global corporate media outfits and this uneven flow: threats of displacing native or national cultures; eroding the foundations of or potential for democracy and the public sphere; spreading a lowest common denominator-debased and homogenized culture; and, imposing American, capitalist interests on all others; a form of media imperialism.

Critics of such work point to myriad problems in these structural analyses with a variety of charges, often painting a caricature of this kind of work: They are economic reductionists who do not grapple with the specifically *cultural* aspects of culture industries; they deny any agency to audiences, or are unable to acknowledge the capacity for audiences to engage with, interpret, reject, or resist media-culture programs and texts; they ignore context and the variations of the specific, local experiences of global media industries by remaining too focused on the big picture; they treat as an automatic, unidirectional process of domination and/or cultural imperialism, what may actually be the result of complicated and sometimes undetermined negotiations and struggles; they ignore growth in national/regional television production and the flow of television from places other than the United States; they assume rather than prove the deleterious effects of global media consolidation; they are prone to conspiracy theory rather than nuanced analysis.

On the other hand, these critics of the macrostructural, top–down approaches to media globalization argue instead for analysis grounded in more microlevel concerns that focus around meaning and the interpretation of texts, contexts, and audiences from a more bottom–up perspective. This work comes from a variety of interpretive traditions, often focusing on issues of meaning and pleasure, of ways in which local audiences may retool, reject, or resist global media products.

These textual- and/or audience-centric analyses have also been the subject of searing criticisms and caricatured presentations: They ignore real issues of power and politics; in trying to eschew economic reductionism, they end up unable to account for any political-economic dimensions; their celebrations of audience agency naively assume some sort of level playing field, ignoring real power differentials between mega-media corporations and relatively disenfranchised individuals; and that unwittingly or not such work results in a defense of the status quo and uncritical market explanations of global media culture.

The problem is not that these caricatures of the two positions have no basis in reality. Certainly, we could find examples of work that truly are overly simplistic, particularly in some older work in the field; there is a reason why such criticisms recur. However, in recent years work on both "sides" has become much more nuanced and more attentive to the problems highlighted by their critics, at least on the theoretical level. Thus, revisiting once again critiques of particular texts does not get us very far. Textual and audience analyses are less likely to entirely ignore issues of power, politics, and economics. Political-economic analyses attempt to engage with more complicated models of active audiences and conflicting texts. Additionally, perhaps more significant are calls to get beyond the divide, to stop treating these approaches as mutually exclusive, and to truly integrate macro- and micro-; top–down and bottom–up; and political-economic, textual, and audience-centric work (for two strong examples of this, see Miller, Govil, McMurria, & Maxwell, 2001; Parks & Kumar, 2003).

Yet, although work has grown increasingly theoretically sophisticated, there seems to still be a bit of an impasse at translating these theoretical calls for integrative approaches to global media into concrete, empirical analyses that are able to move this work forward. Such an endeavor will require the difficult task of truly interdisciplinary engagement as well as drawing on the unique strengths of various disciplines to fit the pieces of the global media puzzle together. The discipline of sociology—although it has too often been missing from critical engagement with questions of global media in the past—may have something to offer to such a project. One aspect of this contribution may be an engagement with meso-level analysis and middle-range research.

THE IMPORTANCE OF MIDDLE-RANGE SOCIOLOGICAL CONSIDERATIONS

Our interest in consolidation and conglomeration directs us to so-called "middle-range" approaches that bridge micro and macro approaches and target meso-level conceptualization, evidence, and analysis. Meso-level analysis attends to the concrete operations and cultural logics of global television markets: the features and operation of the marketplace rather than the market per se; the cyclicality of the forms and practices of the business of the industry rather than a presumed teleological development; industry participants who are located at all levels of organizational settings, rather than just industry leaders or other highly visible decision makers; and the possibility that television series are malleable cultural products rather than immutable texts with inherent meanings.

So, how does one empirically engage in a meso-level of analysis? In early sociological work on the international television market conducted nearly two decades ago, sociologists Muriel and Joel Cantor (1986) noted that television import/export is a chaotic business in a chaotic market, that this chaos is one of its defining features. Observing this unpredictable marketing environment, they described it as one "with no one really in charge and no one knowing, for example, why certain programs sell one year and do not move off the shelves the next" (p. 514). In many respects, the statement that chaos defines the global television market is even more true today. Getting a hold of, interpreting, and explaining this chaos is one of the great tasks, then, of an empirical, meso-level analysis of the global television marketplace. There are numerous ways in which one can study the tumult of the marketplace. Relevant sites to study include trade fairs, industry seminars, and other venues, and the discourses that comprise television's "culture world" (D. Bielby & Harrington, 2004). In this chapter, we focus on selected meso-level aspects of the culture world of international television to illustrate its importance as a conceptual approach for breaking through the research challenge posed by the global media question.

The bottom line is that uncertainty rules because ultimately the international audience accepts or rejects on a show-by-show basis, just as it does in the domestic market. Although some would suggest that television programmers simply respond directly to these audiences' choices, giving them what they want (e.g., Tracey & Redal, 1995), these "choices" must be put in the broader structural context of uneven development and global inequalities (Rajagopal, 2002) as well as the ways in which the increasingly commercialized media world shapes and limits audiences choices to programs that are amenable to the concerns of advertisers and media conglomerates (Herman & McChesney, 1997; McChesney, 2004). However, although their choices are substantially constrained, it is also not true that audiences simply accept whatever is placed before them. Regardless of the predictive validity of proven indicators such as audience ratings and appeal among key demographics, the fact remains that local tastes, and programming factors such as placement on the program schedule, can trump all in determining whether a show becomes a hit, as is shown here.

TARGETING THE "CULTURE WORLD" OF THE GLOBAL TELEVISION MARKETPLACE: MECHANISMS AND CONVENTIONS

Our focus on the culture world of the global television market foregrounds the social organization and institutional arrangements that underlie its logic. We borrow the concept of *culture world* from cultural sociologist Diana Crane (1992), who devised it to take into account the myriad contributors to the production of popular culture and the arts, including gatekeepers, culture creators, and audiences. Crane highlighted some macrolevel components—the organization of the economy for instance, or whether an audience is urban. But, more importantly for our purposes here, she acknowledged *mechanisms*, which are accounts of what brings about change and reveal how institutional components interrelate (Sorensen, 1998). So in her scheme, the fact that an alarmingly small number of huge media corporations—including News Corporation, Disney, Time-Warner, and Viacom—now dominate the majority of media industry markets from film to publishing to music recording to television is a very useful *starting* point for investigating global television ownership and control and its consequences (Herman & McChesney, 1997). But research on global television must move beyond this important starting point.

Understanding the institutional logic of the mergers and acquisitions underlying this global consolidation as business practice could be central here, and would permit us to ask informed questions that speak to anthropologist Mary Douglas' (1986) important quest to penetrate "how institutions think." It would allow us to tie trends in bureaucratic hierarchy, such as the institutionalization of international divisions within major studios, with trends in the domestic market, such as the promotion of successful domestic industry leaders to head those international divisions. It would also challenge us to take into account recent trends in the disaggregation of global conglomerates when business practice dictates—from the fabled failed synergies of AOL-Time Warner (now back to "Time-Warner" in name) to media giant Viacom's recent division into two companies, splitting its broadcast and cable assets (Hofmeister, 2005; see also Orwall & Peers, 2002). Again, however, it is important to not just note these trends, but to analyze the mechanisms underlying them, a few of which are outlined here.

Mechanisms, we observe, exist in response to managing uncertainty in the marketplace. Within the business literature, there are three broad areas of uncertainty in business transactions; these pertain to the search for business partners ("search/contact"), agreement among the parties to rely on collective standards and norms governing membership in the industry and associated groupings ("agreement/contract"), and the extent to which contractual agreements are formalized to pre-empt opportunism and to structure business transactions (Whitley, 2001). These would translate sociologically into what we have termed co-orientation by culture world participants around production of programs (creating new ones and transforming/adapting existing ones), and the transformation/adaptation of the features or elements of the cultural products themselves in order to transact sales and to improve the penetrability of markets. Applying this idea to the culture world of television would include attention to organizations, participants, market features, location, and properties of the cultural products (which pose a management issue because they are so culturally/symbolically laden and their success or failure so difficult to predict). More importantly, it would call for concerted attention to the locally specific ways in which such mechanisms are culturally understood, practiced, and enacted (such as in China, where socially based relational rules play a pivotal role in economic transactions [Chung & Hamilton, 2001]).

For our purposes, Becker's (1982) germinal study of art worlds offers us a useful illustration of the utility of mechanisms in a culture world analysis of media concentration. His key mechanism is the notion of artistic *conventions*—shared understandings among art world participants. As mechanisms, artistic conventions align participants with disparate but interdependent skills, specialties, and backgrounds so they can orient to the task at hand and so that the creation of an artistic product can be accomplished. How this co-orientation works (or doesn't work) in the culture world of global television can reveal a lot about the consequences of media concentration and conglomeration, on the ground.

For example, China may have been penetrated by News Corp.'s Star satellite distribution system, and it now seeks Western programming to fill its expanded broadcast system. But doing business in China without *quanxi*—business contacts that originated in lifelong school friends or by weathering the Cultural Revolution together—dooms to failure Hollywood's conventional practice of combining bonding and business over lunch (Brennan, 2003). Earlier plans of Rupert Murdoch's Sky Satellite services to establish a pan-Asian and/or pan-Chinese network were forced into adaptation and eventually abandoned in favor of establishing multiple smaller, more culturally specific channels. The plans were scuttled by key miscalculations regarding varying internal taste markets, language differences, logistical difficulties of competition with local broadcasters, and direct government intervention against these plans (Curtin, 2005); the guiding conventions that worked elsewhere for the company did not apply here. In this instance, the company's organizational mechanisms (business strategy for growth, organizational expansion, development) were disrupted fundamentally by China's culturally distinct expectation that markets be subordinated to political institutions and ideologies. Thus, conventions (shared understandings) about market/corporate growth were not shared and remained unaligned. In short, the cultural assumptions underlying the requisite co-orientation were absent, miscommunicated, out of reach, unknown, or unattainable, and the business plan could not proceed as Murdoch intended, because it ran counter to China's practice of subordinating the business of its economy to government.

MANAGING RISKS, SPANNING BOUNDARIES

Another stream of research we draw on for understanding mechanisms shaping the culture world of global television exports comes from sociologist Paul Hirsch (1972, 1980), who extensively studied the mechanisms that drive culture industries, from fashion to publishing to television. He argued that the processing of "fads and fashions" is the characteristic mechanism that drives culture industries and the often loosely coupled markets in which they are based. To deal with the inherent uncertainty of how the next fashion will originate or the uncertainty associated with whether it will become a fad, these industries are characterized by an abundance of product. So, just about any series ever produced is a contender for the global market—consider the fact that the vintage sitcom *I Love Lucy* and the western *Bonanza* still air globally. But shows that are big hits domestically can fail to satisfy international audiences, and even series that are cancelled after just a few weeks in the United States can sometimes become successes in export abroad. For example, in Venezuela, although the hit U.S. sitcom *The Nanny* was a ratings failure, the controversial and short-lived U.S. drama series *Nothing Sacred* was a big success (*TV International Sourcebook*, 1998).

How various participants in this culture industry attempt to manage the risks associated with these uncertainties is important to analyze. One Sony executive, for example, attributed risk-reduction concerns as defining the structure of relationships between his company and programmers/networks in Latin America. Relationships typically first begin by simply selling dubbed shows from their vast catalogue of American products; this may or may not achieve success. Then, his company often enters into co-productions of reformatted American series, rewritten for the local context. And only after this do they tend to develop co-productions of truly local fare, not based on a pre-existing import (these latter two practices are discussed in greater detail later). "'Basically we use sitcoms or drama remakes as a lower-risk way of having a channel get into business with us for a first time,' says [Sony exec] Fitzgerald. 'It sort of lowers the apprehension bar'" (Johnson, 2005).

Hirsch also highlighted the importance of attending to pivotal figures who work within these industries, particularly industry participants who act as boundary spanners in order to target innovation and move products along. One researcher who has begun to apply this concept to the television culture world is Havens (2003), who studies global television trade fairs and the role that television buyers play as intermediaries between exporters/producers and potential audiences. He finds that

> because buyers function as surrogate consumers in international television, distributors can focus their promotional efforts on courting their favor, rather than trying to create programming that appeals to viewers around the world with far-flung tastes. This practice rationalizes the process of international television trade and makes manageable the otherwise insurmountable task of trying to understand the cultural affinities and dislocations between specific national and sub-national groups and specific television series or films. (p. 29)

Havens' small study, although provisional, is an example of a type of research we need more of: critical analysis of the micro-business practices that shape international flows of television shows and the organizational contexts in which they operate.

Pertinent to media concentration, Hirsch's insights about boundary spanners invite considering that co-productions—strategic business ventures between firms from different countries to jointly produce television series that are specifically intended to penetrate geo-linguistic regions and share the high costs of production—are often exceedingly difficult to get off the ground. Co-productions initially emerged as a strategy to get around limits on the airing of imported programming (Segrave, 1998), but these international ventures often result in frictions due to cultural differences in production conventions. Germany may require German actors in key roles and prefer to avoid certain plot devices, neither of which may be selling points to a co-production partner; Canada may require production in that country, which may be financially impossible for an interested party; and so on (D. Bielby & Harrington, 2002).

Although co-productions can be a source of intercultural friction, they are often adopted precisely as a strategy for avoiding these problems. Hubka (2002), in studying the global market for children's animated television, for example, demonstrated that co-productions in that genre are often motivated not just by economic concerns (e.g., risk reduction, market access, raising capital, avoiding quotas, although these clearly remain important) but also for creative reasons: to enable the production of material that can translate culturally into multiple markets. Boundary spanners are often embedded in satellite offices of production or production-related firms (such as the United Kingdom increasingly has in the United States, or the United States in the United Kingdom, and Australia tried here in the 1990s), or they may be employed in pivotal roles by talent agencies (e.g., the pivotal role that agent Mark Itkin played as worldwide head of syndication, cable, and nonfiction programming at the London office of the William Morris Agency in bringing European television formats to the United States [Winslow, 2004]). Although remote offices may, on their face, be viewed as evidence of global expansion, bringing a concept, format, or series "pitch," whether as a format sale or co-production, to a successful resolution is far from assured.

THE TELEVISION COMPLEX: CONTENT, GENRE, FORMAT

A last essential point to draw here is that such a loosely coupled and globally dispersed market invites a focused *empirical* examination of just how permeable cultural borders are thirty years after the work of Nordenstreng and Varis on television flows. The work of cultural sociologist Wendy Griswold (2000) invites rich empirical exploration of the role textual elements such as genre play in meso-level approaches to media consolidation. She found in her study of the Nigerian novel that the globally dispersed authors, editors, booksellers, and readers who shape that genre operate in a milieu—which she referred to as a *fiction complex*—characterized by a loose, globally dispersed set of interacting participants whose activities pass with ease through cultural borders as they orient to the production and consumption of these novels. Griswold's work calls attention to the conditions under which participants in such milieux preserve, constrain, or modify defining properties of cultural imports, and it underscores how seemingly static cultural properties such as genre and aesthetic elements become important "culture matters" that need to be strategically managed before cultural products cross borders. These Nigerian novelists relied on local newspaper printers to publish their work under the name of a fictitious publisher when no such organizational entity exists, adapted the narrative structure of Western romance novels to fit oral traditions, and made males the protagonist at least as often as females (Griswold, 2000, 2004; Griswold & Bastian, 1987).

In the very different context of the global market in television, we still find the salience of issues of genre and aesthetics in organizing the behavior of culture world participants (D. Bielby & Harrington, 2004). An examination of the relevance of genre for framing the sale of television programs revealed the *fluidity* of genre both as a system of categorization and as an organizer of understandings. Series that air as a continuing-format daytime serials in the United States ("soap operas") are marketed for export as noncontinuing dramas, with genre type unspecified, and prime-time, action-adventure programs are bundled into programming "packages" with daytime talk shows. Whereas genre is often simply treated as a static label "assigned" at the level of domestic creation, to remain attached to a television program/concept/format throughout the import–export process, we found that genre categorizations and understandings are strategically modified at the level of global distribution and thus are dynamic, negotiated throughout the process of international syndication (D. Bielby & Harrington, 2004). Griswold's work invites rich empirical exploration of the contribution of literary properties and other aesthetic elements such as genre to meso-level analysis of media consolidation, including the relevance of program formats. The properties of program formats and their marketing on aesthetic considerations are important ways in which buyers and sellers in the global market co-orient to one another.

What, then, are some of the organizing mechanisms of the global television complex? In the culture world of television, "content is king"—that is, the programs themselves are the pivotal cultural product that drives this industry and its markets. Economists may consider a television series to be merely "an asset consisting of a bundle of broadcast rights" (Litman, Shirkhande, & Akn, 2000; Owen & Wildman, 1992), but it is clearly much more than that. Television is a product that embodies cultural substance reflecting interests and values. It originates in the creative process of writers, and it is evaluated by critics and audiences who apply aesthetic criteria that ultimately determine the fate of individual series (Bielby, Moloney, & Ngo, 2005). Although those in the business are motivated by profit, a central feature of this culture world is the way collaboration among individuals with disparate understandings about the cultural product shape its production, distribution, and reception globally. As in Griswold's fiction complex, the cultural product of television is a text, in this case the television program itself, but it is more than just written text—it is a visual and aural medium that must be described, presented, and adapted for use. Thus, the organizational or institutional mechanisms that consider, deal with, or translate the elements of the shows themselves are crucial to the operation of this culture world. In D. Bielby and Harrington (2004), Bielby goes into some detail about how participants in the international marketplace organize their understandings of the products they buy and sell and the business and cultural contexts in which they operate, and we offer a brief illustration here. In studying how *Pop Stars* became a global phenomenon, it became apparent that the way in which this format entered the United States was an anomaly. Typically, the major Hollywood studios monopolize the business of licensing formats and acquiring broadcast rights from abroad for production here. However, in this particular case a small, independent production company (Stone Stanley Productions, which is now defunct) landed the rights to the format, who only then entered into a production agreement with a major studio/network, the WB.

What made this deal possible when industry practice favors large, established studios? On first blush, the format from abroad wasn't all that compelling and needed further development. To avoid risking a flat out rejection, the rights holder first tapped the small, independent company, knowing their reputation for attention to the kind of detail that could successfully transform the format and increase the likelihood of attracting a major U.S. studio. But even so, the producers of the independent company had to be convinced that such an investment was worth their while. In this instance, it was "the buzz" about the format's "emotional authenticity" that provided a convincing counter to the business uncertainty stemming from the inability to predict success. In short, industry participants relied on indicators that went beyond standardized "hard" measures such as proven ratings and projected revenues in their decision-making process to include shared understandings about a show's entertainment payoff (D. Bielby & Harrington, 2004). Such co-orientation to shared understandings about cultural products is but one example of the kind of mechanisms that render this culture world operational.

Cultural sociologists underscore the importance of empirically grounded analysis of institutional logics (the cultural determinants of organizational decisions) and production logics (the social contexts and historical contingencies that shape markets and mediate the effect of concentration and competition on product homogeneity). Study of institutional logics, for example, entails analysis of which organizational issues and problems stakeholders attend to in order to survive, or what answers and solutions are available or deemed appropriate (Thornton, 2004). Analysts of production logics study how new technologies transform markets, noting that successful strategies toward technology, not the technologies themselves per se, are what prompt market changes (Dowd, 2004).

Researchers using these concepts have been attempting to introduce cultural analysis into traditional organizational and institutional analyses. We are saying that studying institutional and production logics is very important, but to better understand the mechanisms by which they operate, it is essential to consider the flip side—that it is important to bring evidence of organizational, institutional, or economic issues into *cultural* explanations. Studying trade flows is an important beginning in the project of grappling with pressing global media issues and concerns; but to move beyond the current impasse in global-media scholarship, it is necessary to regard such information as only a beginning.

What insights can cultural sociology bring to this research challenge? To return to the evidence of the top-10 U.S. television shows introduced earlier, although such "snapshot" lists or similar information such as distribution flows are inherently interesting and somewhat informative, at a fundamental level they are inadequately socially contextualized—that is, they tell only part of the story one needs to know in order to understand the effects of global media at the local level.

GETTING BEYOND/DIGGING BENEATH TRADE FLOWS

Regulation and Policy

First, lists and flows of series placement and ratings do not reflect culturally specific regulatory practices of receiving countries. Despite globalization, regulations and policies at the nation-state level can still have a profound influence on what television programming audiences have access to. In addition to the types of laws and regulation discussed in the beginning of this chapter (with our U.S. example of the fin-syn rules)—restrictions aimed at inhibiting or facilitating varying degrees of horizontal and vertical integration through ownership caps—policy and regulation shape the specific, national contexts of global media penetration in many other ways as well. Government policy and regulation shaping the global dimensions of the television world range from quotas for domestically produced programs or limits on imported programming, to government censorship of content, but also include subsidies or tax breaks that favor local/national producers, regulations for allocation of broadcast frequencies, and government policies leading to the privatization or commercialization of previously public service-oriented television networks.

Most obviously affecting global television exports and imports are quotas specifying certain amounts of domestic programming. Often justified in the name of preserving cultural heritage or identity, the specifics of these quotas vary greatly from country to country. In Australia, 55% of terrestrial television programming broadcast between 6 A.M. and midnight must be made by Australians

(Colitt et al., 2003). In the European Union (EU), 60% of prime-time programming must be of European programming, with 10% coming from independent European production companies; some countries within the EU have additional restrictions, for example in France 50% of fictional shows must be French. Other countries regulate not just broadcast content but cable content as well; for example in Korea, 70% of cable TV programming must be Korean (except for science, technology, and sports channels for which the quota is 50%; Feigenbaum, 2004). In Vietnam, only 20% of TV programming can be imported, but in Malaysia the limit is 30% (Chadha & Kavoori, 2000). Other restrictions are based on foreign ownership of media companies. However, under the banner of "liberalization" and "deregulation" such laws have increasingly been eroded; by 2001, for example, there were no restrictions on foreign ownership of either terrestrial/broadcast or cable television in such countries as Belgium, the Czech Republic, Denmark, Finland, Ireland, Italy, the Netherlands, New Zealand, Norway, Portugal, Sweden, or the United States (OECD, 2001).

Another approach governments have taken to increase local/national production, or to shore it up against perceived threats from foreign/global media, is to encourage local production through subsidies and tax incentives (Cavallin, 1998). Some analysts predict that subsidies and other positive supports for local productions will have to increasingly supplant quota/limit strategies because rules requiring certain amounts of national versus foreign television programming are increasingly difficult to police due to the border-spanning properties of newer technologies such as the Internet and satellite television (Feigenbaum, 2004). While Feigenbaum was discussing the future of media policy in Europe, many Asian countries (including Singapore, Malaysia, Vietnam, Brunei, and China) have banned private ownership of satellite receivers because they are seen as bringing in socially detrimental, unregulated programming from the West (Chadha & Kavoori, 2000; Curtin, 2005).

Although quotas and limits regarding foreign/domestic programming are obviously directly aimed at the issue of global media and concerns about its deleterious effects on national or cultural identity and culture, other forms of government intervention in media systems can importantly impact global television programming options as well. Exporters of television programming may have difficulties navigating government censorship of program content. Some countries forbid certain kinds of violence. Mexico, for instance, routinely denies import of any episodes of a series that depict kidnappings. In contrast, China regulates depictions of violence based on frequency, not kind; violence cannot exceed a certain number of occurrences within a given interval, and it "has to be justifiable," a Chinese buyer told us. Moreover, censorship of violence occurs at two levels in that country, one prior to or at the time of purchase, and one after acquisition, which can prevent a purchase from ever airing and the selling country from ever receiving payment for its sale (D. Bielby & Harrington, 2002).

Until nearly 10 years ago, India by law would not allow romantic kissing onscreen, which hindered the importation of any show with such depictions among lovers. When one of India's most popular cinematic actors, Kabir Bedi, appeared in the U.S. soap opera *The Bold and the Beautiful* in the mid-1990s as the lover of one of its leading female characters, the domestic U.S. production company was faced with a challenge. Even though the production company (Bell Serial Productions, Inc.) is quite skilled at exporting the show around the globe—it now airs in more than 130 foreign markets (http://www.c21media.net/resources/detail.asp?area=78&article=32427)—and is well versed in adapting its programs to local concerns, Indian custom and law derailed the series' import and hampered entrée into the enormous Indian audience (although it eventually aired on satellite channel Star-TV). Thus, larger policy structures strongly influence just what may be shown, what people are exposed to—and thus what kind or degree of media domination may be possible.

Scheduling and Programming

Second, country-level placement information, such as import lists or distribution flows, does not represent the likelihood of audience exposure to a show. In that regard, concrete and local programming practices may be of greater importance to actual viewer exposure, as some have suggested. Australian scholars who studied the U.S. failure of the Australian soap opera *Neighbours*—a huge success in Australia and the United Kingdom—identified aspects as mundane as placement on the network schedule and the brevity of the series' run as crucial to the series' efforts in reaching its targeted demographic—the result was that *Neighbours* was almost impossible for potential viewers in the United States to find, much less have any significant cultural impact. However, it would not be accurate to attribute the show's failure to take off in the United States solely to scheduling problems. Others argue that *Neighbours*' "nonexceptional realism"—a staple in many Australian and British soap operas, significantly separating them from American soaps—may have also been off-putting to American audiences (Crofts, 1995), demonstrating the importance again of cultural content and audiences' expectations.

Simply focusing on the number of hours of television shows imported may obscure the fact that these shows can end up in odd time slots or less popular or accessible channels. So, the American shows in Europe that do air in prime-time hours have, over the past decade, tended to be on smaller networks (Higham, 2003). In Italy, for example, U.S. imports tend to end up on Berlusconi's Mediaset channels, which have smaller viewerships. In a study of Latin American programming, the concept of audience hours—defined as the number of hours of imported programming divided by size of audience for each program—was employed to reveal the likelihood of exposure to a particular show (Antola & Rogers, 1984). In 1 year of the study in Mexico, for example, the 50% of programming hours filled by imported programming was found to translate into just 33% of the total possible audience hours of viewing, and these were almost all concentrated on just two channels. In Europe, the amount of American programming on European channels has been rising in recent years, due to the proliferation of new channels. However, within this time period there has also been a decrease in the tendency to schedule U.S. shows in prime-time hours, relegating the shows instead to off-peak time slots, reserving the prime-time hours for locally produced fare (Higham, 2003; although one U.S. industry executive reported a recent renewed interest in scheduling American shows in prime-time slots in Europe; Guider, 2005). Similarly, prime-time programming in Brazil has long been dominated by its national genres of telenovelas, light comedy/variety, and music programming, pushing U.S. imports into off-peak hours (Straubhaar, 1991).

This same issue must be examined when considering the potential impact of "counterflows" of television programming, from the third to the first world (such as is increasingly occurring on U.S. cable systems that elect to carry networks from other countries in order to reach audiences of immigrant populations within subscriber markets (see, for e.g., Gomes, 2002). Sometimes lost in the excitement over apparent successes of Latin American programming (especially telenovelas) in Europe is the fact that such programming too predominately ends up in off-peak hours, especially in northern and northwest Europe (where, after a brief surge of Latin American imports in the early 1990s in the wake of privatization and expansion of channels with hours to fill, the telenovela and other Latin American programming is all but disappearing from the dial; Biltereyst & Meers, 2000). Thus, evidence of a program "flowing" into a country cannot be equated with viewing by the entire population.

Local Preferences and the Cultural Discount

The programming of local/national content during prime-time hours, although sometimes required by national policies, also reflects the fact that, as researchers have documented in many diverse contexts and markets, audiences favor national over imported television programming, when available (Straubhaar, 1991; Tracey & Redall, 1995). These preferences are consistently reflected in audience ratings and domestic programming/scheduling decisions. What media economists refer to as the "cultural discount"—in which the country of origin of cultural products matters to audiences/consumers in a way that is not true of many commodities, depreciating the value of nonfamiliar/nondomestic programming—means that media culture may be traded quite differently on the global market than other "products" (Hoskins & Mirus, 1988).

Additionally, some have suggested that this may be particularly true for television programming, as compared with other media such as film in which American exports are more dominant; 18 of the top- 20 films in Europe in 1998, for example, were American (Freedman, 2003). Thus, television audiences prefer and seek out programming that is more relevant to them, culturally—first, programming originating from their country, or second, programming from other "culturally proximate" sources (countries with a shared history, language, culture, or region). For example, Straubhaar (1991) found in the Dominican Republic an audience preference for Dominican-produced shows, but in the absence of that, a preference for Mexican over U.S. imports, owing to shared language and culture. Chadha and Kavoori (2000) document that in the majority of Asian countries, more than 90% of the top 20 rated shows in each nation are locally produced. They argued that this is based not only on linguistic issues but also on cultural considerations (even among English-speaking audiences in India, for example, Indian-generated programming is preferred).

But in the assertion that audiences will turn to locally produced shows when available, the qualifier "when available" is crucial. For, one can never assume a level playing field or remove these issues from the broader global-economic structures that still favor the United States. Straubhaar's (1991) use of the concept of *asymmetrical interdependence* in the global television market may be a helpful framework for balancing these issues. This framework looks at the distinct dynamics of each nation's media industry that affect what is and is not aired in that nation—"key national issues include conflicts between domestic and transnational elites, interests of key national elites, entrepreneurial competition, the agenda and actions of key production personnel, and the effects of state intervention, particular as policymaker, provider of infrastructure, and advertisers" (p. 43) but it also always puts this in the context of "the limits placed on many nations' media systems by operating within subordinate positions in the world economy" so that unequal conditions "such as low income resources, lack of models for production, lack of support by weak governments, inappropriate models for production, and lack of trained personnel, keep a number of poorer countries from developing much local or national production, even if their audiences might prefer more national programs" (p. 45).

For example, because of the size of its domestic market, U.S. television companies are able to spend much more money per episode and still make a profit just within that domestic market, much less from what they earn additionally from foreign sales. As a result, they can sell their shows in foreign markets at a price often much lower than the costs of domestic production there, especially for comparable production values. Economies of scale can make it close to impossible for domestic producers to compete. This affects not only poor countries, but also wealthy countries whose markets are small, due to population size or language issues. In Canada, for example, the costs of producing an original domestic series are estimated to be 10-times the costs of importing a drama from the United States (Hoskins & Mirus, 1988).

There is mixed evidence regarding whether technological changes in the television industry may be changing these dynamics. On the one hand, an executive at Sony recently commented that "local programming around the world seems to have much more resonance with the audience than it did a decade ago. . . . The local programming itself is much better than it used to be . . . It's cheaper to produce good programming today than it was 10 or 15 years ago" (Lynton, cited in

Johnson, 2005), with the lowered costs of production attributable in part to technological changes in digital/video. On the other hand, with the expansion of digital transmission of television signals, both terrestrially and via cable/satellite, there are dramatically more hours to fill with programming, leading programmers to turn again to American shows to fill these schedules (Guider, 2005)—so that U.S. television exports in 2001 were nine times greater than in 1985 (U.S. Department of Commerce, 2002).

Translation and Adaptation

Third, lists of what series are imported do not reveal the transformations often necessary in the import process. Cultural factors that underlie a series' adaptation to local tastes include the adequacy of language and translation in the broadest sense. First, there are shows that are exported for the most part "as is," changed only by language (subtitling or dubbing) or technical standards. Even these most straightforward exchanges can encounter myriad difficulties or complications. Issues of language translation are more fraught than might initially be assumed, because they are crucial to establishing the cultural adequacy that is necessary for a cultural exchange to work (D. Bielby & Harrington, 2002). Translation is even sometimes deemed necessary when the language remains the same; South American shows, for instance, may be dubbed into Mexican-accented Spanish for distribution in the United States (Johnson, 2005). Some genres may be more amenable to dubbing and translation than others; animated programming, for instance, may be particularly suited to global exchange because of this (Hubka, 2002).

Likewise, technical translation varies culturally/nationally as well. One informant at a major U.S. media conglomerate revealed that Germany demands superior technical quality in its imports and is known to refuse to air them unless and until an improved version is provided. Shows exported essentially "as is" can fail miserably for myriad reasons. In racially integrated Brazil, for example, imports of the American situation comedy *All in the Family* (itself a "translation" of the British show, *Till Death Do Us Part*), which satirized the racial divide, made little sense (Segrave, 1998). Successful American comedies often fail abroad, again because of differences in cultural sensibilities—hits such as *Ally McBeal* and *Seinfeld* have been noted as spectacular failures in the export market.

Hence, even when linguistically translated, shows can fail to translate across borders in a cultural sense. To some degree, television producers may try to anticipate these potential pitfalls for global reception of their shows by not just modifying the shows for export after the fact, producing tweaked versions for international audiences (D. Bielby & Harrington, 2002; Hubka 2002), but also by shaping the shows' content or format at the very beginning of the production process: that is, creating export-friendly television. The extent of this practice, as well as its possible consequences, are somewhat in question, however.

Although once an afterthought, the export of television programming has increasingly become a consideration from the very beginning of the production process, some argue. This can take the form of creating programming with cultural references that are (or at least are designed to be) more accessible to audiences abroad or even specifically consulting with potential foreign buyers of television programming in the generative phases of television production. For example, in the recent sale of a U.S. show that debuted in Fall 2005, *Threshold*, representatives from Britain's Sky TV "had been intimately involved with the development... aware of its suitability for the UK market" and the head of 20th Century Fox's television distribution arm "noted that his division works hand in glove with the studio's development executives to produce programming with local appeal" with foreign broadcasters increasingly consulted in the development process (Brennan, 2005).

The implications of such moves, however, are not without controversy. One issue under question is whether a focus on creating exportable programming erodes the local and national character of television cultures. Freedman (2003), for example, argued that the current emphasis in British economic and cultural policy on expanding television exports may be resulting in a diminution of the aesthetic practices that have traditionally been the strengths and unique contributions of British television, particularly in its dramatic and documentary programming, in favor of an expansion of programming deemed more exportable on the global market, such as light entertainment and reality TV. This impacts not only what U.K. television producers offer up on the global market, but also what television domestic British audiences have access to, as resources are shifted away.

On the other hand, Paxman (2003), studying transformations in Mexican telenovela production and distribution, suggested that concerns over the erosion of the *Mexicanidad* of Mexican novelas with the aim of increasing exportability may be exaggerated. Some decisions may be shaped by export-related concerns; for example, including stars from other Latin American countries increases the success of the telenovelas in those countries. However, he argued that unlike production companies in Latin American countries with smaller markets, such as Colombia or Venezuela, which may have to focus heavily on exportability because their local markets alone cannot cover production costs, Mexico has a large enough market that exports, while important, don't have to drive content-related decision making.

Additionally, although the Mexican telenovelas are distributed widely internationally, the United States is by far the largest importer of the shows, and cultural connections between Mexican viewers and Mexican-American viewers is enough that the shows do not require "cultural translation" for successful export to that market. So, once again, an understanding of this phenomenon requires careful empirical analysis of the actual trends in specific national contexts.

Another approach taken by media companies to translate television programs for export is to simply re-shoot scripts in a new language or locale, with local actors and actresses. Although sometimes successful (e.g., currently, the U.S. version of Britain's hit-comedy *The Office* is doing well, for example), this practice can often sink a series, again due to difficulties of cultural translation. This occurred, for instance, when original scripts of *Charlie's Angels* were produced in Mexico in recent years. Although this idea was the brainchild of an experienced Latina television executive, in this instance the female characters did not translate well in this new context; they were deemed morally lacking by the Mexican audience and rejected. Similarly, simply shooting relatively unaltered scripts of the British hit sitcom *Coupling* did not last long in the in the United States as a result of differences in ironic sensibilities (Lloyd, 2003).

EXPORTING TELEVISION FORMATS

Increasingly, however, series may be exported as licensed formats—concepts that are bought and sold with specifications as to what elements stay constant and what should be adapted to local custom, and these series may fare better elsewhere. By adopting this strategy exporters attempt to avoid the previously discussed "cultural discount" in which audiences reject imported programming in favor of more culturally proximate local programming. Currently, across Latin America, a number of American sitcoms of decades past—from *The Jeffersons* to *The Nanny* to *Who's the Boss?*—have been updated and transformed and have received considerable ratings success. These co-productions of Sony Pictures and Latin American production companies take American concepts, and sometimes scripts, but often subject the shows to quite a bit of transformation. *Cancidos con Hijos,* Colombia's *Married With Children,* for example, "needed a lead character who was 'a little more warm, a little less defeatist [in] relationship with life, more of a fighter, because that's a little more of the Latin psychology'" according to one Sony executive (Johnson, 2005).

Of sociological interest are the ways in which different actors within this culture world—distributors, studio executives, domestic programmers, and so on—attempt to co-orient around shared understandings of the character, desires, needs, or cultural sensibilities of different audiences and cultures. Examples analyzed in detail elsewhere (D. Bielby & Harrington, 2004) include *Who Wants to Be a Millionaire*—a British format, and *Pop Stars/American Idol*—a New Zealand format that came by way of Australia and England. Each had to be carefully adapted to resonate locally in culturally specific ways. *Millionaire* is a hit in the Middle East because its host promotes a pan-Arabism in its questions and peppers his patter with quotes from the Muslim holy book (El-Magd, 2002). For *Pop Stars/American Idol* to generate interest in the United States its "documentary look" had to be replaced with stepped-up production values, more graphics, and a heightened look and feel (NATPE Educational Foundation, 2001).

The ways in which a common format may be transformed by being transported to a radically different social, political, or cultural context may be illustrated with some examples from recent Iraqi television. In recent months, popular shows have included *Iraq Star* (described as bearing "more than a passing resemblance to *American Idol*") and *Material and Labor*, described as "a homegrown Iraqi show inspired by *This Old House* and *Extreme Makeover: Home Edition*, but with a twist of *Apocalypse Now* (Wong, 2005). In a gruesome or at least poignant twist in the home-makeover genre, *Materials and Labor* entails the Iraqi satellite channel, Al Sharquiya, rebuilding homes destroyed by the war. The show's Iraqi writer declares the genre to be "the only good thing we've acquired from the American occupation," commenting that he considers the show's display of the ravages of the war as a critique of the American and Iraqi governments (Wong, 2005).

The dominance of the so-called reality television genre in these examples is no coincidence, for this genre has been particularly successful in the export market for television formats, as the genre is seen as especially suited for translation, adaptation, and cross-border trade. However, this too has raised concerns for some analysts, who suggest that with the increasing focus on exportability, this leads to greater resources being poured into the production only of the most exportable formats, such as reality TV and light entertainment, at the expense of other less easily translatable but perhaps richer or culturally distinct programming (Freedman, 2003). What formats and genres are successfully exportable also may not be stable over time. It has long been conventional wisdom within the television industry that comedy crosses borders poorly and the many failures of American situation comedies to find success abroad have tended to support this idea. Some industry analysts, however, suggest that this may be changing and that in Latin America, for instance, the sitcom format imported from the United States may be making inroads against the variety show/sketch comedies that have long dominated Latin American prime-time schedules, in part due to viewers' increasing comfort with sitcoms after exposure to them on satellite channels (Johnson, 2005).

The relevance of format as central to our sociological vantage point on the global industry cannot be underscored enough, for several reasons. In order to more fully understand the globalization of media, it is necessary to know—beyond considerations of entry and access—what form imported/exported programming takes on. Licensed formats such as the ones just discussed may so transform a series concept that it is, in effect, all but a local production, some would argue. Others, however, would characterize these as hybrid productions or as perfect examples of "glocalization," whereby global phenomena touch down in and are transformed by specific, local/national contexts and adaptations. Still others might argue that although they appear to be local productions they are still, at their core, foreign, external, and no less of a concern than other imported programming. Which of these three characterizations is most accurate is an *empirical* question.

Sociologically, we might not answer the question of whether or not the programs themselves are "truly authentic" or "foreign invaders," but rather examine the ways in which these categories themselves are constructed and deployed within the complex interactions in the television export market. Paxman (2004), for example, analyzes the transformation of the Mexican telenovela—initially a hybrid product, an example of "glocalization," the format was derivative of Cuban radio serials and American soap operas, but adapted for local use. But as the genre developed and evolved over time, Mexican telenovela creators began producing programs that may be described as distinctly national—and the *Mexicanidad* ("Mexican-ness") of the shows was a major marketing asset for their major producer in the 1970s and 1980s, Televisa. However, in recent years, with a renewed emphasis on exportation and with influences from South American telenovelas (which have a greater tradition of social realism) showing evidence, the shows may be more hybrid once again. The ways in which these shows are positioned and marketed as national or hybrid creatures and the ways this shapes production and reception contexts thus must be placed in specific, national, historical contexts, subject to empirical analysis.

A focus on format also invites consideration of the variety of co-productions, which originate and adapt in line with numerous factors pertaining to the global industry, and which requires extensive *empirical* documentation. This would include topics such as the vitality of national economies, firms' strategies for working within or circumventing import quotas, the degree to which sharing the risks of uncertainty are a factor in production, the emergence of new national markets or the closing down of established ones because local industries have developed and audience preferences for indigenous fare have asserted themselves, and so on. In this regard, we note the considerable news coverage, at least within industry publications, of evidence over the last decade of the decline in foreign markets for U.S. media products, and the recent resurgence in Europe for U.S. dramas (Fahri & Rosenfeld, 1998; Guider, 2005). Finally, an *empirical* focus on formats also invites consideration of their effect on creative workers in the industry, and in that regard, for example, how the licensing of unscripted formats appears to have seriously diminished the contribution of writers to the global market. Again, this is an *empirical* question worth seeking answers to, not a presumption.

CONCLUSION

We have argued that analysis of media concentration at the global level is considerably more complicated than at the domestic level. To get past seemingly irresolvable theoretical debates and reliance on evidence that is either inadequate or inappropriate for the conclusions drawn from it requires a middle-range conceptual approach. This would attend to the institutional and production logics of the global marketplace that emerge from a culture-world approach to global media. Furthermore, we have argued that because the products of this market are *cultural* ones, symbolically laden and imbued with contested meanings, we must move beyond traditional organizational and production insights; because participants in this global market must adapt to cultural practices, assumptions, mandates, and policies of other nations, we must attend in very fundamental ways to how culture shapes the assumptions of their institutional and production logics. Thus, although a vigorous profit orientation may be present, it is only one of any number of economic or policy logics that must be traversed for cultural products to cross borders. Similarly, a production logic must attend to the many ways in which the cultural properties and features of the products are managed. Culture world "givens" such as these complicate analysis of the global marketplace and at the same time open up new conceptual and theoretical possibilities for analysis of the structure of the global media industry.

We have also called for increased attention to empirical evidence that serves as appropriate measures for conceptual categories and goals. These logics, complicated by cultural considerations, moreover, redirect attention to the inherently complicating factors of the study of global media industry that are too often overlooked, that is, as a culture industry producing and distributing a *cultural* product. Conventional approaches in our discipline of sociology would focus on such important questions as the industrial organization of media companies, the composition of boards of directors and overlaps in membership across companies, or the cycles of diversity or homogeneity of cultural products. These are important research projects, but they are not sufficient.

We argue that it is important to bring evidence of organizational, institutional, or economic issues into *cultural* explanations, a reasoning that is consistent with the "cultural turn" in sociology, which has led to just such a revisioning within sociological approaches to organizations.

> Organizational theorists, long comfortable with conceptualizations of structure and strategy that depended on objective notions of resource relations, understood through competition and conflict, differentiation, and symbiosis, have increasingly recognized the conventional, and indeed fictional, quality of many organizational forms and strategies. Institutionalists who once looked to culture as a rushing in where rationality failed, where means-ends relations were uncertain or technologies untried, now increasingly recognize that culture plays a constitutive role in shaping organizational structure, strategies, and technologies. (Friedland & Mohr, 2004, pp. 1-2)

Organizational-behavior scholars have called for increased attention to "industry cultures," which are the "shared mindsets" that underpin perception, thought, feeling, and behavior of people who are members of groups in ways that are not directly obvious to either themselves or observers (Phillips, 1994). This turn to examining industry cultures is yet another call for attention to conceptual categories that elucidate institutional and production logics. The overtime accumulation of such middle-range, "culture-world"-pertinent evidence will contribute to moving beyond the theoretical impasse that has gripped the analysis of global media and its consequences for so long. Calls for such work have emerged from diverse disciplinary orientations, both in the social sciences and the humanities: "global transformations of television . . . cannot be analyzed in terms of an overarching theory that is applied from afar. Instead they can only be understood through close examination of the television industries, programs, technologies, and audiences in specific cultural contexts" (Parks & Kumar, 2003, p. 6).

A further challenge to the study of global media concentration is the necessity of moving beyond static descriptive models of the industry. This industry is especially dynamic (even "chaotic"), making the marshalling of evidence and the consistency of its interpretation a particular challenge. We note, for example, that even the evidence we attend to throughout our discussion is itself a "snapshot" of the worldwide syndication process, given the frequency and rapidity with which the global television industry changes. Furthermore, the unceasing transformation of this industry that makes it especially difficult to study may obscure relevant affirming or disconfirming evidence at a particular point in time. Media economists who have attempted to specify the structure of the media industry have called for analysis of the context of the media market, which entails attending to the market situation, external forces, strategic behavior of media companies, and public policy (Hendriks, 1995). Identifying key elements of its market structure is a necessary step, but extending a middle-range conceptual approach, as we have called for, directs attention to the institutional factors that shape its business models, which opens analysis of production cycles, styles, import and export trends, and monitoring the impact of changing regulatory environments.

We observe further that recognizing just how dynamic the global media industry is necessitates attention to the matter of constraints on this industry. Some years ago, sociologist Richard Peterson (1982) elaborated five such constraints on the production of culture more generally, in order to call attention to what is problematic in cultural production. These include law, technology, industry structure, organizational structure, occupational careers, and the market. Peterson's intent was to problematize the production-of-culture perspective, organizational analysis, and institutional or economic perspectives, which do not always conceptualize such limits or constraints as important to the analysis of production in culture industries. But by injecting production, organization, institutions, and economics into cultural explanations, constraints become apparent. Such thinking underscores the conceptual importance of Curtin's (2005) analysis of the consequences of "institutional logics" of politics, market, and production of television in China. Such findings highlight the relevance of formalizing analytical approaches to understanding constraints on the production of culture as part of global media concentration. A culture-world approach, which includes audiences and their input, would also permit the relevance of local perceptions of the quality of the product, either actual or perceived, as they affect the marketplace—hence the downturn of the popularity of U.S. products in the United Kingdom, or elsewhere.

In closing, studying trade flows is an important beginning in the project of grappling with pressing global media issues and concerns, but it is necessary to regard such information as only a beginning. We would hope that our cultural-sociological vantage point facilitates productive dialog in future scholarly analysis and debate about global media concentration.

ACKNOWLEDGEMENT

The research reported here was supported by the Academic Senate and the Institute for Social, Behavioral, and Economic Research, University of California, Santa Barbara.

REFERENCES

Antola, L., & Rogers, E. M. (1984). Television flows in Latin America. *Communication Research, 11*, 183-202.

Becker, H. S. (1982). *Art worlds*. Berkeley: University of California Press.

Bielby, D. D., & Harrington, C. L. (2002). Markets and meanings: The global syndication of television programming. In D. Crane, N. Kawashima, & K. I. Kawasaki (Eds.), *Global culture: Media, arts, policy, and globalization* (pp. 215-233). New York: Routledge.

Bielby, D. D., & Harrington, C. L. (2004). Managing culture matters: Genre, aesthetic elements, and the international market for exported television. *Poetics: Journal of Empirical Research on Literature, the Media and the Arts, 32*, 73-98.

Bielby, D. D., Moloney, M., & Ngo, B. Q. (2005). Aesthetics of television criticism: Mapping critics' reviews in an era of industry transformation. *Research in the Sociology of Organizations: Transformation in Cultural Industries, 23*.

Bielby, W. T., & Bielby, D. D. (1999). Organizational mediation of project-based labor markets: Talent agencies and the careers of screenwriters. *American Sociological Review, 64*, 64-85.

Bielby, W. T., & Bielby, D. D. (2003). Controlling primetime: Organizational concentration and network television programming strategies. *Journal of Broadcasting & Electronic Media, 47*(4), 573-596.

Biltereyst, D., & Meers, P. (2000). The international telenovela debate and the contra-flow argument: A reappraisal. *Media, Culture and Society, 22*, 393-413.

Brennan, S. (2003, March 18-24). All the TV in China. *The Hollywood Reporter*, pp. 18-19.

Brennan, S. (2005, May 26). They love L.A. screenings. *The Hollywood Reporter*, pp. 1, 31.

Cantor, M. G., & Cantor, J. M. (1986). Audience composition and television content: The mass audience revisited. In S. J. Ball-Rokeach & M. G. Cantor (Eds.), *Media, audience and social structure* (pp. 214-225). Newbury Park, CA: Sage.

Cavallin, J. (1998). European policies and regulations on media concentration. *International Journal of Communications Law and Policy, 3*(1), www.digital-law.net/IJCLP/index.html.

Chadha, K., & Kavoori, A. (2000). Media imperialism revisited: Some findings from the Asian case. *Media, Culture and Society, 2,* 415-432.

Chung, W-K., & Hamilton, G. (2001). Social logic as business logic: Guanxi, trustworthiness, and the embeddedness of Chinese business practices. In R. Appelbaum, W. Felstiner, & V. Gessner (Eds.), *Rules and networks: The legal culture of global business transactions* (pp. 325-346). Oxford: Hart Publishing.

Colitt, R., Fifield, A., Johnson, J., Marcelo, R., & Larsen, P. T. (2003, July 21). The United States of television. *Financial Times*, p. 8.

Collins, S. (2005, July 13). TV vets wind down a once powerhouse shop. *Los Angeles Times*, p. E-1.

Crane, D. (1992). *The production of culture: Media and the urban arts.* Newbury Park, CA: Sage.

Crofts, S. (1995). Global neighbours. In R. C. Allen (Ed.), *To be continued: Soap operas around the world* (pp. 98-121). London: Routledge.

Cunningham, S., & Jacka, E. (1994). Neighbourly relations? Cross-cultural reception analysis and Australian soaps in Britain. *Cultural Studies, 8,* 509-526.

Curtin, M. (2005). Murdoch's dilemma, or 'What's the price of TV in China?' *Media, Culture and Society, 27*(2), 155-175.

Douglas, M. (1986). *How institutions think.* Syracuse, NY: Syracuse University Press.

Dowd, T. J. (2004). Concentration and diversity revisited: Production logics and the U.S. mainstream recording market. *Social Forces, 82*(4), 1411-1455.

El-Magd, N. A. (2002, June 1). Cairo's "Millionaire" unabashedly takes Palestinian side in conflict. *Los Angeles Times,* p. F12.

Fahri, P., & Rosenfeld, M. (1998, October 25). American pop penetrates worldwide. *Washington Post,* p. A1.

Feigenbaum, H. (2004). Is technology the enemy of culture? *International Journal of Cultural Policy, 10*(3), 251-263.

Freedman, D. (2003). Who wants to be a millionaire? The politics of television exports. *Information, Communication & Society, 6*(1), 24-41.

Friedland, R., & Mohr, J. (2004). The cultural turn in sociology. In R. Friedland & J. Mohr (Eds.), *Culture matters* (pp. 1-68). Cambridge: Cambridge University Press.

Friedman, W. (2005, May 31). The upfront race nearing homestretch. *TV Watch,* http://publications.mediapost.com/index.cfm?fuseaction=Articles.showArticle&art_aid=30694.

Gomes, L. (2002, December 9). Globalization is now a two-way street—Good news for the U.S. *Los Angeles Times,* p. B-1.

Griswold, W. (2000). *Bearing witness: Readers, writers and the novel in Nigeria.* Princeton, NJ: Princeton University Press.

Griswold, W. (2004). *Cultures and societies in a changing world* (2nd ed.). Thousand Oaks, CA: Pine Forge Press.

Griswold, W., & Bastian, M. (1987). Continuities and reconstructions in cross-cultural literary transmission: The case of the Nigerian romance novel. *Poetics, 16,* 327-351.

Guider, E. (2005, May 20). Yanks bank on hot series: U.S. series find renewed favor abroad. *Variety,* p. 1.

Havens, T. J. (2003). Exhibiting global television: On the business and cultural functions of television fairs. *Journal of Broadcasting & Electronic Media, 47*(1), 18-35.

Hendriks, P. (1995). Communications policy and industrial dynamics in media markets: Toward a theoretical framework for analyzing media industry organization. *Journal of Media Economics, 8*(2), 61-76.

Herman, E. S., & McChesney, R. W. (1997). *The global media: The new missionaries of corporate capitalism.* London: Cassell.

Higham, G. (2003, March 14). U.S. programming losing its grip on Europe's prime-time audiences. *Broadcasting & Cable's TV International, 11,* 1.

Hirsch, P. M. (1972). Processing fads and fashions: An organization-set analysis of cultural industry systems. *American Journal of Sociology,* 77(4), 639-659.

Hirsch, P. M. (1980). An organizational perspective on television (Aided and abetted by models from economics, marketing, and the humanities). In S. Withey & R. Abeles (Eds.), *Television and social behavior: Beyond violence and children* (pp. 83-102). Hillsdale, NJ: Erlbaum.

Hofmeister, S. (2005, June 15). Viacom OKs plan to split, but 1 man will still run the show. *Los Angeles Times,* p. C-1.

Hoskins, C., & Mirus, R. (1988). Reasons for the US dominance of the international trade in television programmes. *Media, Culture and Society, 10,* 499-515.

Hubka, D. (2002). Globalization of cultural production: The transformation of children's animated television, 1980-1995. In D. Crane, N. Kawashima, & K. I. Kawasaki (Eds.), *Global culture: Media, arts, policy, and globalization* (pp. 233-255). New York: Routledge.

Iwabuchi, K. (2002). From western gaze to global gaze. In D. Crane, N. Kawashima, & K. Kawasaki (Eds.), *Global culture: Media, arts, policy, and globalization* (pp. 256-274). New York: Routledge.

Johnson, R. (2005, March 13). If the schtick fits. *Los Angeles Times,* p. E1.

Labaton, S. (2003, May 12). F.C.C. prepares to loosen rules on media ownership. *The New York Times.* http://www.nytimes.com/2003/05/13/business/media/13FCC.html?ex=1368158400&en=4f67a23f455753bc&ei=5007&partner=USERLAND

Litman, B. R., Shirkhande, S., & Ahn, H. (2000). A portfolio theory approach to network program selection. *Journal of Media Economics, 13*(2), 59-79.

Littleton, C., & Andreeva, N. (2005, May 18). A threepeat for Warners. *The Hollywood Reporter.* http://www.hollywoodreporter.com/hr/search/article_display.jsp? vnu_content_id=1000929140

Lloyd, R. (2003). Will sex sell this show? Based on a British series, "Coupling" hits a new level of network raunchiness. *Los Angeles Times,* p. E5.

McChesney, R. W. (2004). *The problem of the media: U.S. communication politics in the twenty-first century.* New York: Monthly Review Press.

Miller, T., Govil, N., McMurria, J., & Maxwell, R. (2001). *Global Hollywood.* London: British Film Institute.

NATPE Education Foundation. (2001). *Doing business with the U.S.: A format for success.* Videotaped Proceedings from NATPE Convention, New Orleans, LA.

Nordenstreng, K., & Varis, T. (1974). *Television traffic: A one-way street? A survey and analysis of the international flow of television programme material* (No. 70). Paris: UNESCO.

OECD. (2001). *Communication outlook 2001.* Paris: OECD.

Orwall, B., & Peers, M. (2002, May 10). The message of media mergers: So far, they haven't been hits. *Wall Street Journal,* pp. 1, 5.

Owen, B. M., & Wildman, S. S. (1992). *Video economics.* Cambridge, MA: Harvard University Press.

Parks, L., & Kumar, S. (Eds.). (2003). *Planet TV: A global television reader.* New York: New York University Press.

Paxman, A. (2003). Hybridized, glocalized and hecho en Mexico: Foreign influences on Mexican TV programming since the 1950s. *Global Media Journal,* 2(2).

Peterson, R. A. (1982). Five constraints on the production of culture: Law, technology, market, organizational structure and occupational careers. *Journal of Popular Culture, 16,* 143-153.

Peterson, R. A., & Berger, D. G. (1975). Cycles in symbol production: The case of popular music. *American Sociological Review, 40,* 158-173.

Phillips, M. E. (1994). Industry mindsets: Exploring the cultures of two macro-organizational settings. *Organizational Science, 5*(3), 384-402.

Rajagopal, A. (2002). Mediating modernity: Theorizing reception in a non-Western society. In T. Miller (Ed.), *Television: Critical concepts* (Vol. 1, pp. 308-336). New York: New York University Press.

Segrave, K. (1998). *American television abroad: Hollywood's attempt to dominate world television.* Jefferson, NC: McFarland.

Sinclair, J., Jacka, E., & Cunningham, S. (1996). Peripheral vision. In J. Sinclair, E. Jacka, & S. Cunningham (Eds.), *New patterns in global television* (pp. 1-32). Oxford: Oxford University Press.

Sorensen, A. B. (1998). Theoretical mechanisms and the empirical study of social processes. In P. Hedstrom & R. Swedberg (Eds.), *Social mechanisms* (pp. 238-266). New York: Cambridge University Press.

Spanish-lingo programmers eager to tap exports. (2003, February 28). *Broadcasting & Cable's TV International, 11,* 12.

Straubhaar, J. D. (1991). Beyond media imperialism: Asymmetrical interdependence and cultural proximity. *Critical Studies in Mass Communication, 8,* 39-59.

Thorton, P. H. (2004). *Markets from culture: Institutional logics and organizational decisions in higher education.* Stanford, C A: Stanford University Press.

Tracey, M., & Redal, W. W. (1995). The new parochialism: The triumph of the populist in the flow of international television. *Canadian Journal of Communications, 20*(3), 1-17.

TV International Sourcebook 1998. (1998). Torrance, CA: Baskerville Communications Corporation.

Untangling the Telecom Act. (2007). http://www.freepress.net/telecom/

U.S. Department of Commerce. (2002). *Survey of current business.* Washington, DC: Author.

Varis, T. (1985). *The international flow of television programmes: Reports and papers on mass communication* (No. 100). Paris: UNESCO.

Winslow, G. (2004). Power brokers. WorldScreen.com. http://www.worldscreen.com/featurescurrent.php?filename=0604agents

Whitley, R. (2001). The institutional structuring of business transactions. In R. Appelbaum, W. Felstiner, & V. Gessner (Eds.), *Rules and networks: The legal culture of global business transactions* (pp. 73-99). Oxford: Hart Publishing.

Wong, E. (2005, August 28). In war's chaos, Iraq finds inspiration for reality TV. *The New York Times.* http://www.nytimes.com/2005/08/28/international/middleeast/28television.html?ex=1282881600&en==b3af8927364796

IV

REGULATORY AND LEGAL ISSUES

14

MEDIA OWNERSHIP AND DIVERSITY ASSESSMENT

A Social Science Research Agenda

Philip M. Napoli

Nancy Gillis

Media policymaking appears to be undergoing an expansion in its analytical orientation. The questions being asked and the issues being raised by policymakers and the courts more frequently delve into a wider range of areas of social scientific inquiry than has characterized the past two decades of policymaking and policy analysis. Policymakers, policy analysts, and the courts now are raising issues and questions pertaining to how citizens use different media technologies to obtain information (Waldfogel, 2002); if/how media content varies in accordance with variations in market and ownership conditions (Federal Communications Commission [FCC], 2002, 2004; Spavins, Denison, Frenette, & Roberts, 2002), what factors contribute to biased or ideologically slanted news content (Balan, DeGraba, & Wickelgren, 2004); and what criteria should be employed in defining a media outlet and the magnitude of its impact (*Prometheus Radio Project v. Federal Communications Commission,* 2004).

These areas of inquiry delve into disciplinary areas (most notably communications) that traditionally have remained tangential at best to communications policymaking and policy analysis (Mueller, 1995; Noam, 1994). A number of explanations for this disconnect have been offered, including, on the supply side, intradisciplinary conflict, narrowness of research focus (Mueller, 1995), and resistance among scholars to engaging in "applied" research (Noam, 1994); and, on the demand side, a lack of disciplinary diversity within the policymaking sector that has prioritized legal and economic analysis over other potentially relevant research traditions (Napoli, 1999).

An examination of the communications policymaking environment in recent years, however, suggests that this state of affairs is beginning to change. The focal point for this change has been the issue of media ownership. Given this change of climate, and the broader range of questions now being asked by policymakers, this chapter attempts to identify existing areas of scholarly inquiry that may prove fruitful to decision makers; and to suggest extensions of these lines of inquiry that may more directly serve the needs of policy decision makers. This chapter presents a rather narrowly focused effort at such an assessment, via an examination of the lines of social scientific inquiry that can inform recent efforts to assess media ownership policies, particularly in terms of the FCC's recent effort to construct a rigorous and reliable Diversity Index to be used in the assessment of individual media markets and the effects of modifications of existing ownership regulations on levels of diversity within these markets.

The first section of this chapter provides a brief overview of the media ownership issue, in terms of the process undertaken by the FCC, the outcome of this process, and its ramifications. The second section examines the issues raised by the FCC, the courts, and other stakeholders in regards to the ownership proceeding for which areas of social scientific research not typically integrated into policymaking and policy analysis may be able to make a significant contribution. As this section illustrates, the ongoing discussion over how best to assess viewpoint diversity in local media markets touches on a host of research areas that have been explored in significant detail. This section offers suggestions as to how future research in these areas can more directly address the specific needs of policymakers. The final section summarizes the key points of this chapter and offers some concluding observations about the state of the relationship between social science research and communications policymaking.

MEDIA OWNERSHIP, DIVERSITY POLICY, AND THE DIVERSITY INDEX

Section 202(h) of the Telecommunications Act of 1996 required the FCC to biennially review its media ownership rules "to determine whether any of such rules are necessary in the public interest as the result of competition" (Section 202(h), 110 Stat. at 111-112). The Act also required the FCC to "repeal or modify any regulation it determines to be no longer in the public interest (Telecommunications Act of 1996, Section 202(h), 110, Stat. at 111-112). For the 2002 review, FCC Chairman Michael Powell initiated what he described as the "most comprehensive look at media ownership regulation ever undertaken by the FCC" (FCC, 2002, p. 1). Powell specifically emphasized the importance of developing the necessary factual record to guide the FCC's decision making in this area. As he stated in 2001, at the beginning of the media ownership inquiry, "Rebuilding the factual foundation of the Commission's media ownership regulations is one of my top priorities" (FCC, 2001, p. 1).

As part of this process, the FCC created a Media Ownership Working Group, which conducted internally—and commissioned from external sources—a series of research projects on a wide range of issues relevant to the media ownership proceeding (FCC, 2002). Included in the FCC's research record were studies that examined the relationship between ownership of media content/outlets and the diversity of content provided (Einstein, 2002; Pritchard, 2002), as well as studies examining how people use different media (Nielsen Media Research, 2002), and the extent to which different media technologies serve as substitutes for each other for consumers seeking news and information (Waldfogel, 2002).

The end result of this process was a June 2, 2003 Report and Order that substantially relaxed a wide range of media ownership regulations, including the local television ownership rule (limiting the number of television stations a single entity could own within a single market), the broadcast–newspaper cross-ownership rule (limiting broadcast–newspaper combinations within a single market), and the national television ownership rule (limiting the national audience reach of a television station owner; FCC, 2003).

Included in this decision was a newly developed Diversity Index, which the FCC created internally to serve as a guide in assessing the status of local media markets and the appropriateness of permitting further ownership consolidation within these markets. The FCC's Diversity Index had as its basis the well-known Herfindahl-Hirschman Index (HHI), which measures market concentration via summing the squared market shares of each firm in a market. The Diversity Index extended and modified the HHI in a number of ways. The FCC (2003) described the methodology as follows:

> In terms of calculating the Index, within each medium we combine commonly-owned outlets and calculate each owner's share of the total availability of that medium. We then multiply that share by the share of the medium in question in the total media universe (television plus newspaper plus radio plus Internet). Once these shares in the overall "diversity market" have been calculated, we add together the shares of properties that are commonly-owned (a newspaper and a television station), square the resultant shares, and sum them to get the base Diversity Index for the market in question. (p. 13790)

This description merits some clarification. First, as is indicated by the description already given, an owner's holdings in each medium (television, newspapers, radio, and the Internet)[1] are first computed separately, with an owner's "share" of a particular medium calculated in terms of the proportion of the available outlets that the owner controls (thus, e.g., a firm owning two of the six television stations in the market would have a 33.3% share). Then, the owner's share for each medium is weighted separately, with the weights being derived from a consumer survey (Nielsen Media Research, 2002) in which respondents were asked to identify their primary sources of local and national news, and to assess the importance of different media as news sources. Specifically, the weights were derived from a survey question in which respondents were asked which types of media they had utilized for obtaining news and current affairs within the past 7 days. Thus, for instance, 57.8% of respondents claimed to have utilized television to obtain news and current affairs within the past 7 days. Based on these responses, the weighting scheme employed for the media included in the Diversity Index was as follows: television—57.8% newspapers—25.8%, radio—10.3%, Internet—6.1% (FCC, 2003).

The weighting process would then be employed, with, for example, the 33.3% share of the broadcast television market already described multiplied by 57.8% (the weighting score for television)—and so on for each medium—to determine that firm's share of the "diversity market." Once this process is completed for all of the holdings of each firm in the market, each firm's total share is squared, then summed (following the HHI methodology) to produce the Diversity Index for that market.

The FCC's key objective in creating this index was to have a mechanism for empirically assessing viewpoint diversity within local media markets that could provide guidance as to when to allow mergers to take place within these markets. To the extent that data on media usage are incorporated into the calculus, the index represents a significant—and largely unprecedented—move forward in terms of factoring the dynamics of media consumption into media ownership policy analysis. Based in part on the conclusions derived from employing this index in the analysis of a sample of media markets, the FCC concluded that the bulk of its ownership regulations could be substantially relaxed (FCC, 2003).

The ramifications of the FCC's decision have been well documented (e.g., Scott, 2004). Amidst a substantial outcry from both the public and members of Congress (see Scott, 2004), the FCC's media ownership decision was challenged in court by public interest advocates arguing that the rule changes had gone too far in allowing increased consolidation, and by industry participants who argued that the relaxation of the ownership rules did not go far enough (see *Prometheus Radio Project v. Federal Communications Commission,* 2004, pp. 388-389). After a jurisdictional lottery, the case was tried in the U.S. Court of Appeals for the Third Circuit in Philadelphia. The court remanded the bulk of the FCC's ownership decision, due in large part to the perceived failure of the FCC to justify its particular chosen ownership limits (*Prometheus Radio Project v. Federal Communications Commission,* 2004, p. 382). The FCC is now, consequently, revisiting the ownership decision, and also working on strengthening the analytical work underlying its decision-making process.

MAKING AND REMAKING MEDIA OWNERSHIP POLICY: POINTS OF ENTRY FOR SOCIAL SCIENCE RESEARCH

On the basis the Court of Appeals for the Third Circuit's decision directing the FCC to reconsider and better support its decision making, and considering the fact that the FCC is required to reassess its ownership regulations every 4 years, the question of the most effective analytical approach to formulating and assessing media ownership policy is likely to remain a prominent one for years to come, even if a revised version of the Diversity Index is not employed by the FCC .Examining the FCC's analytical process pertaining to the media ownership rules in greater detail, and examining the specific concerns raised by the Court of Appeals for the Third Circuit, helps to illustrate the various areas of inquiry in which social science research may be able to make contributions to both ongoing and future efforts to assess diversity in media markets and craft effective media ownership regulations.

Many of these areas of inquiry stem from the central issue of how best to assess the "diversity importance" (FCC, 2003, p. 13777) of different information sources. This was the key objective of the FCCs Diversity Index—to develop a reasonably sensitive measure of the impact on viewpoint diversity of ownership changes within and across multiple media technologies in local media markets in a way that could account for variations in the importance, from a diversity standpoint, of different information sources (FCC, 2003). Moreover, it is around this issue of how to best assess the diversity importance of different information sources that the Court of Appeals for the Third Circuit seemed to have taken the greatest issue with the FCC's work (*Prometheus Radio Project v. Federal Communications Commission,* 2004; see below). Consequently, this issue area represents the line of inquiry around which the greatest amount of additional work needs to take place—and also the line of inquiry toward which existing lines of social science research may be able to make a significant contribution from a number of perspectives.

This section begins, then, with an examination of a number of different conceptual elements of potential relevance to assessing the importance of different information sources from a diversity assessment standpoint. In each case, the element under discussion is identified (although in some cases, somewhat indirectly) from the FCC's or the court's decision making. Figure 14.1 provides a visual map of the discussion that is to follow. In this figure, the central concept of the "diversity importance" of different information sources is surrounded by the various

FIGURE. 14.1. Potential elements of the "diversity importance" of different information sources and associated areas of research.

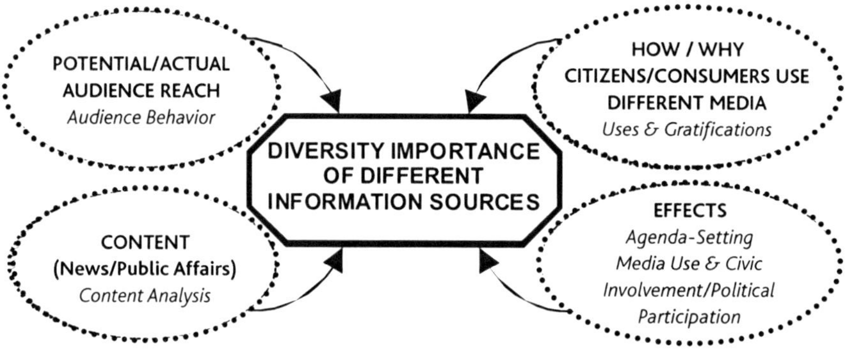

potentially relevant conceptual elements, which are, in turn, accompanied by listings of the areas of research that are of potential use in clarifying and applying these conceptual elements. The purpose of this overview is to identify relevant research issues that are raised and the possible areas of research that could inform policy decision making in relation to assessing the diversity importance of different information sources and formulating effective media ownership policies.

Content

The first element of potential relevance in assessing the importance of different media in relation to viewpoint diversity is content-related. Specifically, the amount of time/space devoted to local news or public affairs by a particular media technology or outlet could potentially function as a significant factor in determining the importance of different media outlets or technologies from a diversity standpoint. As the FCC stated in its June 2003 Report and Order, "we find that viewpoint diversity is most easily measured through news and public affairs programming. Not only is news programming more easily measured than other types of content containing viewpoints, but it relates most directly to the Commission's core policy objective of facilitating robust democratic discourse in the media" (p. 13631). Later in this decision, the FCC is even more direct in this regard, stating, "News and public affairs programming is the clearest example of programming that can provide viewpoint diversity" (FCC, 2003, p. 13776). The Commission related such content directly to its Diversity Index when it noted that, "Our Diversity Index focuses on availability of sources of local news and current affairs" (p. 13780).

Such statements suggest that content differences between individual media outlets or technologies—in terms of their provision of news and public affairs—should be factored into the calculus of assessing viewpoint diversity in media markets. The fact that the FCC produced such an assessment in conjunction with the work of its Media Ownership Working Group (Spavins et al., 2002) further suggests that attention to this issue may be appropriate. Nonetheless, the FCC (2003) elected not to integrate a content element into its Diversity Index, in part because of the following belief:

> this type of content analysis would present both legal/Constitutional and data collection problems. News and current affairs content is not necessarily limited to regularly-scheduled news programs. So we could be faced with deciding which other programs were news and current affairs, whether some portion of a program not primarily news should count as news, and, indeed, whether portions of a news report devoted, e.g., to movie reviews should count as news. (p. 13787)

The Court of Appeals for the Third Circuit, however, found this logic unpersuasive, describing the FCC's reference to data collection problems as "vague and unexplained" (*Prometheus Radio Project v. Federal Communications Commission,* 2004, p. 409). Moreover, the court expressed concern about any analytical approach that did not contain some content indicator when it stated that "assigning equal market shares to outlets that provide no local news almost certainly presents an understated view of concentration in several markets, thus contravening the Commission's goal of making 'the most conservative assumption possible' about viewpoint diversity" (*Prometheus Radio Project v. Federal Communications Commission,* 2004, p. 408, citation omitted).

If, then, news and public affairs content is a *potentially appropriate* indicator for determining the diversity importance of different information sources, then such differences in content provision across information sources should be assessed in a systematic and rigorous manner, something toward which the content analysis methodology employed by researchers across a wide range of social science disciplines (sociology, communication, political science) could make a valuable contribution.[2] We see, for instance, the important insights that can be derived from rigorously analyzing media outlets' local news and public affairs programming output in the work of Singleton and Rockwell (2003). Their study illustrates that the extent to which the FCC has traditionally relied on the provision of local news and public affairs in its definition of a "voice" in a local media market is, in fact, undermined by the fact that many media outlets provide no such content. Their analysis of television stations finds that, "In the nation's top fifty markets there were 162 television stations that apparently did not provide any local news or public affairs programming. Forty-nine of the top fifty television markets had at least one 'silent voice'" (Singleton & Rockwell, 2003, p. 398). Findings such as these[3] suggest that allocating equal weight to all outlets of a certain type in a market may severely misrepresent the extent to which these outlets are serving as comparable sources of local news and public affairs, and are thus of comparable diversity importance. Moreover, to the extent that the FCC has emphasized the importance of *locally oriented* news and information (FCC, 2003, 2004), research that has examined local news content to determine the true extent of its local orientation (E. Adams & Baldasty, 2001; W. Adams, 1978, 1980; Slattery, Hakanen, & Doremus, 1996)[4] may provide useful empirical findings or methodological suggestions for future efforts to assess the diversity importance of different information sources from a content standpoint.

Issues such as these were at the core of the Court of Appeals for the Third Circuit's rejection of the FCC's decision to include the Internet as a local information source in its viewpoint diversity calculus (*Prometheus Radio Project v. Federal Communications Commission,* 2003). In support of this decision, the court noted that "The Commission does not cite, nor does the record contain, persuasive evidence that there is a significant presence of independent local news sites on the Internet.... And the examples the Commission does cite—the Drudge Report and Salon.com—have a national, not local, news focus" (*Prometheus Radio Project v. Federal Communications Commission,* 2004, p. 406).

As this statement suggests, in terms of formulating and justifying its ownership policies to the satisfaction of the courts, the FCC may need to devote greater attention not just to the availability of different media outlets, but also to the nature of the content that they provide. Any effort to assess, from a content standpoint, the contribution of individual information sources to viewpoint diversity in the process of crafting a rigorous and reliable Diversity Index may need to be more intensively grounded in the extent to which different media technologies or outlets provide news and public affairs content—and the extent to which such content is indeed local in its orientation. Further exploration of the available content analysis tools and methods developed across the social sciences likely would prove useful in any such endeavor.

How and Why Citizens/Consumers Use Different Media

At the core of the FCC's proposed mechanism for weighting different information sources is the extent to which citizens/consumers report relying on different media for news and current affairs. In describing its methodology for computing the Diversity Index, the FCC (2003) stated that, "First, we premise our analysis on people's actual usage patterns across media today. Nonetheless, our method for measuring viewpoint diversity weights outlets based on the way people actually use them rather than what is actually available as a local news source" (p. 13787).[5]

Thus, how citizens/consumers use different media has played—and likely will continue to play—a significant part in assessing the importance of different media from a diversity policy standpoint. The logic here stems in part from the content-focused element discussed earlier, in which the FCC articulated its belief that the extent to which different media serve the news and public affairs needs of the citizenry should be at the core of determining their importance in assessing viewpoint diversity in local media markets (FCC, 2003). It is important to emphasize that this analytical orientation represents a virtually unprecedented step forward by policymakers in terms of attempting to integrate the dynamics of citizens' media usage behavior into formulating and analyzing media ownership policy.

What is vital, of course, in this situation, is that the FCC accurately reflect citizens/consumers' media usage behaviors in its calculus. Whether the FCC succeeded in this regard was a significant point of contention both during and after the June 2003 decision. Critics of the media usage survey that provided the foundation for the usage weighting noted that, even though the FCC was primarily concerned with media usage for local news and information, the survey question that formed the basis of the weighting measure was worded in such a way as to ask the respondents which sources of local or national news they had consulted within the past 7 days, thereby undermining the data as an effective means of assessing the diversity importance of different information sources at the local level—a weakness acknowledged by the FCC (2003, summarizing comments of the Consumer Federation of America; see also Cooper, 2004).

Moreover, these same critics argued that the underlying dynamics of media usage are more complex than the FCC's usage survey would suggest, with different media serving distinctly different functions for citizens/consumers—for example television news broadcasts serving as the primary source of "breaking news," newspapers providing a follow-up and analysis function, and talk radio programs serving primarily to disseminate viewpoints (FCC, 2003, summarizing comments of the Consumer Federation of America). From this perspective, the extent to which the FCC (2003) operated under the assumption that "consumers use multiple media as sources of news and current affairs, and hence that different media can be substitutes in providing viewpoint diversity" (Federal Communications Commission, 2003, p. 13779) might be inappropriate, or might not effectively capture the complexity inherent in the dynamics of media usage.

Critiques such as these point to the potentially greater level of sophistication that can (and perhaps should) be brought to bear on the process of assessing how and why citizens/consumers use different information sources in a way that most effectively captures variations in their diversity importance. The current state of affairs, then, would seem to be one in which it would be enormously valuable for policymakers to become more deeply versed in the literature on how and why citizens/consumers use different media, and thereby develop a more theoretically and empirically robust approach to integrating media usage data into media ownership policymaking.

The extensive uses and gratifications literature (e.g., Blumler & Katz, 1974; Rayburn, 1996; Ruggiero, 2000) would seem to provide a natural starting point. As

Mendelsohn (1974) noted in an early exploration of possible applications of uses and gratifications research to communications policy, "the integration of generalized social science intelligence regarding specific audiences, together with media-related information about them, allows for the development of policies for filling important gapes in the media enterprise. Thus, social science research can provide a basis for changing media policy in line with the media-related needs, expectations, uses, and gratifications of particular audiences" (p. 312).

Research in the uses and gratifications tradition can potentially inform the largely unresolved policy question regarding if—or to what extent—different information sources serve as substitutes or complements for each other, in ways that account for both how and why citizens/consumers use different information sources (e.g., Kaye & Johnson, 2002). Work on how citizens obtain political information from the media has mapped out the different ways that citizens/consumers use different information sources to obtain information, as well as the different types of news consumers served by these different media (e.g., Chaffee & Frank, 1997)[6]—work that would seem to bear directly on policymakers efforts to effectively calibrate the different levels of importance of different information sources. Additionally, some recent studies have addressed the issue of the substitutability of information sources, particularly in terms of the relationship between old and new media. For instance, recent research has failed to find particularly strong evidence that the Internet serves as a meaningful substitute for traditional news media (Althaus & Tewksbury, 2000; Dimmick, Chen, & Li, 2004), suggesting that a policy approach that considers the Internet as comparable to existing information sources from a diversity importance standpoint may not accurately reflect the dynamics of how citizens/consumers are integrating new media into their more established array of information sources.

In summary, as policymakers presumably move forward in developing more robust mechanisms for gauging the extent to which citizens/consumers rely upon different information sources for local news and information, the uses-and-gratifications perspective can serve as a potentially valuable source of theoretical, empirical, or methodological guidance.

Potential or Actual Audience Reach

Another concept of potential relevance in assessing the diversity importance of different information sources is the potential or actual audience reach of each source. As the FCC (2003) noted in expressing the nature of the concerns that were central to its ownership policy analysis, "we retain our emphasis on the citizen/viewer/listener and on ensuring that viewpoint proponents have opportunities to reach the citizen/viewer/listener" (p. 13775). As this statement suggests, assessing the potential of different viewpoints to reach audiences is a key concern from an ownership policy standpoint. Despite this concern regarding the potential reach of different viewpoints, the FCC did not go particularly far in its initial iteration of the Diversity Index in terms of integrating this concern into its analytical framework. As the FCC (2003) stated:

> Having decided on relative weights for the various media, we next confront whether and how to weight different media outlets within each category. The decision of whether to do weighting turns on whether our focus is on the availability of outlets as a measure of potential voices or whether it is on usage (i.e., which outlets are currently being used by consumers for news and information). We have chosen the availability measure, which is implemented by counting the number of independent outlets available for a particular medium and assuming that all outlets within a medium have equal shares.... The underlying assumption here is that all outlets have at least similar technical coverage characteristics. (p. 13786)

The FCC then noted that this assumption holds up relatively well within the context of television stations and newspapers, but not particularly well within the context of radio stations (where the different signal strengths of different stations lead to dramatically different levels of potential audience reach (FCC, 2003). Nonetheless, the FCC concluded that "we believe the assumption to be reasonable across all cases" (p. 13786).

Presumably, the FCC could have taken an approach that would have weighted media outlets by their true potential to reach viewers/listeners/readers within a particular media market. Thus, for instance, a full power VHF television station (presumably with cable carriage on all cable systems in a market) would have been weighted more heavily than a relatively low-power FM radio station with sufficient signal strength only to reach half of the listeners in a market. To the extent that the diversity importance of an information source is at least in part a function of its ability to reach citizens/consumers, then this more nuanced—although certainly much more complex and labor-intensive—approach would more accurately reflect the importance of different information sources.

Of course, one could also take this approach a step further, and go beyond the potential audience reach of an information source to its actual audience reach. Under this scenario, it is the size of the audience that the information source actually attracts that becomes the foundation for weighting its importance from a diversity assessment standpoint.

This approach was the path that the Court of Appeals for the Third Circuit seemed to view as most appropriate. The court took issue with the FCC's strategy of accounting for differentials in the reach/impact of different media at the level of technology but then ignoring such potential differentials at the level of individual outlets. According to the court: "The Commission's decision to assign equal market shares to outlets within a media type does not jibe with the Commission's decision to assign relative weights to the different media types themselves" (*Prometheus Radio Project v. Federal Communications Commission,* 2003, p. 408). The court pointed out the FCC's own statement that "'Not all voices, however, speak with the same volume,'" going on to argue that the FCC's approach "negates [its] proffered rationale for using the HHI formula in the first place—to allow it to measure the actual loss of diversity from consolidation by taking into account the actual 'diversity importance' of the merging parties" (*Prometheus Radio Project v. Federal Communications Commission,* 2003, p. 408, citations omitted). The court went on to say that "the Commission's attempt to justify its failure to consider actual market share of outlets within a media type is not persuasive. It suggests that actual-use data is not relevant because 'current behavior is not necessarily an accurate predictor of future behavior.' But this truism did not prevent the Commission from preferring actual-use data in assigning relative weight to the different media types" (*Prometheus Radio Project v. Federal Communications Commission,* 2004, pp. 408-409).

It would appear, then, that any revised efforts to assess the diversity importance of individual media outlets may need to address the actual audience reach of these outlets. Such an approach would seem to fit squarely within the realm of the field of audience research, as a number of audience research questions naturally arise once an audience reach element is incorporated into the calculus. One key question would involve the best approaches to aggregating and analyzing the relevant audience data in ways that would accurately differentiate the audience reach of different outlets. Answering this question may, on the surface, appear simple, but as audience behavior research tells us, audience exposure can be assessed at either the individual (micro) level or at the more aggregate (macro) level (Webster, Phalen, & Lichty, 2000), with the different approaches potentially providing significantly different portraits of the distribution of audience attention.

Another relevant question involves the extent to which any analysis of actual audience reach should be integrated with an assessment of potential audience reach, given the fairly extensive body of audience behavior research indicating that audience size is, to a large degree, a product of structural media factors such as signal strength, household penetration, channel capacity, and so on (Neuman, 1991; Webster & Phalen, 1997). Economist Bruce Owen (2004) noted the potential danger of confusing success with access—that is, of taking evidence of a heavy concentration of audience attention around a select few media sources (e.g., Hindman, 2007; Webster, 2007; Webster & Lin, 2002; Yim, 2003) as evidence that a media market is necessarily heavily concentrated, or possesses structural characteristics that are limiting access to a wide range of content options and viewpoints. To the extent that future efforts at assessing the diversity importance of different information sources may need to consider actual audience reach, audience behavior research could prove enormously useful in helping to disentangle the complex knot of factors that determine the audience size for a particular media outlet.

Effects

Finally, there is the question of whether any assessment of the diversity importance of different information sources should delve into the realm of media effects. To a certain degree, implicit in the often-used term *importance* by both the Court of Appeals for the Third Circuit and the FCC is some notion of the magnitude of the effect that a particular information source can have on an individual or a community, but this connotation is perhaps indirect at best. However, some other statements by the FCC in its June 2003 Report and Order suggest a more direct relationship between assessing viewpoint diversity and accounting for the magnitude of the effect of a particular information source. As the FCC noted in relation to its media usage survey (Nielsen Media Research, 2002), "The differences in usage across media documented in MOWG Study No. 8 are in part reflections of the differential impact on the user of television, radio, newspapers, etc. We believe that the overall impact of a medium is substantially determined by the physical attributes of its distribution technology, along with user preferences" (p. 13787).

Here, with the FCC's use of term *impact*, the relevance of considering the media effects issue becomes more pronounced. In gauging the diversity importance of different information sources, it would therefore seem of potential relevance to more directly examine the nature and magnitude of the effects that different information sources have on the citizenry. Given the enormity, and the theoretical and methodological diversity of the media effects literature, the question then becomes, within the context of media ownership policy, which realm(s) of media effects research are of the greatest relevance?

For guidance on this question, it is useful to consider the FCC's statement in its June 2003 Report and Order, in which it stated that "owners of media outlets clearly have the ability to affect public discourse, including political and governmental affairs, through their coverage of news and public affairs. . . . We believe sound public policy requires us to assume that power is being, or could be, exercised" (p. 13630). From this statement, it would seem that it would be best to focus on media effects issues in relation to political and governmental affairs. Two potentially useful bodies of research—and potentially fruitful lines of future inquiry—would be agenda-setting research (e.g., McCombs & Shaw, 1972; Wanta, 1997), as well as research on the relationship between media usage and civic involvement (e.g., Emig, 1995; Friedland, 2001) and political participation (Scheufele, 2002; Shah, McLeod, & Yoon, 2001).[7]

Considering first the potential relevance of agenda-setting research to media ownership policy, it is important to emphasize the central objective of the Diversity Index of providing a means of assessing the diversity importance of individual information sources. As was noted earlier, this importance may be appropriately conceptualized at least in part in terms of the magnitude of the effects of different information sources. Thus, as the FCC seeks to determine how best to weight the importance of radio versus television versus newspapers versus the Internet, and so on, it may be useful to have as the foundation of such an assessment a detailed understanding of the extent to which these different media technologies appear capable of influencing the citizenry's perceptions of the most important issues facing their communities. Such an approach would represent a more direct approach to the differential degrees of "impact" of different information sources that the FCC is attempting to tap via its current methodology of assessing reasons for usage and market shares.

Agenda-setting research typically has had a national orientation (due, in large part, to the greater availability of national media content for analysis). However, research examining the agenda-setting process at the local level—perhaps in conjunction with analysis of media usage patterns—could provide valuable baseline data in terms of which information sources truly exert the greatest impact on the citizenry in terms of their cognitions of issue salience. Previous research, for example suggests that newspapers exert the strongest agenda-setting effect, followed by broadcast television news (Wanta, 1997). Findings such as these, and/or the research designs from which they were derived, could potentially be integrated into future efforts by policymakers to differentiate information sources in terms of their diversity importance.

Moreover, the related agenda-setting work that has examined the intermedia agenda-setting process (wherein certain information sources appear to systematically impact the content output of other information sources; see Boyle, 2001; Roberts & McCombs, 1994; Trumbo, 1995) may contribute to an even more finely calibrated measure of the importance of different information sources. Thus, as the FCC decides the relative weights to assign broadcast television, versus radio, versus newspapers, the relative importance/impact of these different media may be better captured by not only assessing variations in how and to what extent they are used by the citizenry to perform different functions, but also how and to what extent some media exert significant influence over others. For instance, some scholars have found that daily newspaper coverage appears to be an important factor in determining which stories broadcast news programs devote attention to (Boyle, 2001; Ku, Kaid, & Pfau, 2003; Trumbo, 1995). To the extent that such phenomena may be prevalent at the local level, then a weighting system that assesses the relative importance of different media only via assessing the extent to which consumers report on relying on one versus the other may, in fact, underestimate the full importance of certain information sources.

Another possible avenue of inquiry from an effects standpoint would involve incorporating—and building on—the extensive literature that has examined the relationship between media usage and civic involvement or political participation. Such an approach would involve conceptualizing the diversity importance of different information sources at least in part in terms of their role in informing and mobilizing the citizenry, and facilitating citizens' involvement with their communities. To the extent that ownership policy is motivated in large part by the diversity and localism principles, and the long-standing relationships between these principles and the effective functioning of the democratic process (see FCC, 2003), then an analytical approach grounded at least in part in this body of literature would seem to be of significant potential value.

Research on the relationship between media usage and civic involvement and political participation shows, for example, stronger relationships between newspaper consumption and civic involvement/participation than between television news consumption and civic involvement/participation (Shah et al., 2001; see also Jeffres, Dobos, & Sweeney, 1987). Studies also have shown a significant relationship between local media usage and the strength of community ties (Emig, 1995), suggesting that the extent to which an information source is local in its orientation may bear directly on the extent to which it is able to have an impact on civic engagement. Much recent work has focused in the Internet, and the relationship between Internet usage and civic involvement (e.g., Bimber, 2003). Work in this vein has suggested that there is a positive relationship between Internet usage and community and civic engagement (see Katz & Rice, 2002).

Findings such as these could guide efforts to effectively calibrate the diversity importance of different information sources in a way that considers the extent to which these sources are capable of impacting political and civic involvement/participation at the local level. Moreover, contained within this body of literature may be methodological approaches that could be applied to systematic national or local level assessments of the relationship between the usage of different information sources and their effects on the political behavior variables that presumably underlie policymakers' concerns with preserving diversity and localism in local media markets.

CONCLUSION

This chapter attempted to explore areas of intersection between existing lines of social science research and the FCC's efforts to effectively assess viewpoint diversity in local media markets. As this chapter indicated, well-established areas of research, such as agenda-setting, uses and gratifications, audience behavior, and media usage and political participation—which traditionally have not had much bearing on communications policy—now appear to be much more directly relevant to issues and concerns being raised by policymakers in relation to their work on media ownership.

Ideally, this brief overview sheds light on some areas of research that could help guide policymakers and policy analysts in their decision making, in their identification of key research questions, and in their conceptualization and design of future research endeavors. Moreover, this overview hopefully will help scholars engaged in the lines of inquiry reviewed here to see more clearly the potentially significant policy ramifications for their work—something they more than likely have not been inclined to focus on in the past. More important, in now thinking this way, ideally these scholars can design studies that more directly address ongoing policy concerns than work in these areas generally has in the past. To the extent that past work in these areas has not been conducted with media policy applications in mind, there likely is substantial potential for researchers in these areas to conduct work that more directly addresses the FCC's media ownership policy concerns and informational needs than the sampling of research discussed in this chapter.

The likelihood of such work then having an impact on policy decision making seems much greater today than in years past, as policymakers are today not only more receptive to such work, but are actively seeking and engaging in such work. The era of the "toaster with pictures"[8] approach to communications policymaking appears to be passing. As policymakers devote greater attention to the political and cultural dimensions of media production and media usage, the opportunities for a broader range of social scientists to make a meaningful contribution to the policymaking process increase. And although economics is likely to remain the primary *lingua franca* of communications policy, to the extent that the relevant work from other disciplines at least gets digested into—and consequently informs—this more traditional analytical perspective (e.g., Brennan, 1992), then the likelihood of communications policymaking effectively accounting for the full range of policy priorities hopefully will increase.

NOTES

1. Magazines were discarded by the FCC as a meaningful source of local news and information on the basis of the data gathered in the Nielsen Media Research (2002) study, as well as on the basis of data from other sources. Cable television was excluded for similar reasons (see FCC, 2003).
2. In light of the heavy methodological criticisms leveled against the FCC's own effort at content analyzing broadcast stations' news and public affairs programming (Spavins et al., 2002; for an overview of the specific criticisms, see Napoli, 2004), it would also seem that the content analysis methodology is one in which the FCC could use substantial external guidance and assistance.
3. For analyses that reach similar conclusions within the narrower context of the provision of public affairs programming, see Napoli (2001) and Yan and Napoli (2004).
4. In these studies, methods are developed for determining the extent to which local news content is indeed "local" in its orientation, with the results often showing that a large percentage of "local" news does not address stories or issues related to the local community, and in fact has a more national or regional orientation.
5. This description of the Diversity Index would seem to contradict the FCC's (2003) earlier statement that, "Our Diversity Index focuses on availability of sources of local news and current affairs" (p. 13780). Usage and availability are very different concepts, and the extent to which the FCC seems to conflate the two suggests that perhaps both concepts should be incorporated into the next iteration of the Diversity Index—and, perhaps more important, that further FCC work in this area should be more thoroughly grounded in the existing research on the production and the consumption of news and public affairs content.

6. Chaffee and Frank (1997), for instance, illustrated how television and newspapers serve complementary functions in providing political information, and how newspapers primarily serve the more politically active component of the citizenry.
7. Certainly, there are other potentially relevant bodies of literature relevant to an effects-grounded approach to assessing the diversity importance of different information sources. Space limitations prevent further exploration of these possibilities.
8. This phrase refers to a statement by Reagan-era FCC Chairman Mark Fowler, who referred to television as a "toaster with pictures" (cited in Nossiter, 1985, p. 402) in order to illustrate his argument that the media industry is an industry like any other, requiring no special consideration of its political or cultural impact. Instead, the traditional, economically oriented marketplace analysis applied to other product markets would suffice.

REFERENCES

Adams, E.E., & Baldasty, G.J. (2001). Syndicated service dependence and a lack of commitment to localism: Scripps newspapers and market subordination. *Journalism and Mass Communication Quarterly, 78*(3), 519-532.

Adams, W.C. (1978). Local public affairs content of TV news. *Journalism Quarterly, 55*, 690-695.

Adams, W.C. (1980). Local television news coverage and the central city. *Journal of Broadcasting, 24*(2), 253-265.

Althaus, S., & Tewksbury, D. (2000). Patterns of Internet and traditional news media use in a networked community. *Political Communication, 17*(1), 21-45.

Balan, D.J., DeGraba, P., & Wickelgren, A.L. (2004). *Media mergers and the ideological content of programming.* Unpublished working paper, Bureau of Economics, Federal Trade Commission, Washington, DC.

Bimber, B. (2003). *Information and American democracy: Technology in the evolution of political power.* New York: Cambridge University Press.

Blumler, J.G., & Katz, E. (Eds.). (1974). *The uses of mass communication: Current perspectives on gratifications research.* London: Sage.

Boyle, T. (2001). Intermedia agenda setting in the 1996 presidential election. *Journalism and Mass Communication Quarterly, 78*(1), 26-44.

Brennan, T.J. (1992). Integrating communication theory into media policy: An economic perspective. *Telecommunications Policy, 16*(6), 460-474.

Chaffee, S., & Frank, S. (1997). Learning about politics from the mass media. *Political Communication, 14*, 421-430.

Consumer Federation of America. (2003). *Comments of Consumer Federation of America, Consumers Union, Center for Digital Democracy, Media Access Project, Federal Communications Commission Biennial Media Ownership Review.* Accessed January, 2, 2005, from http://gullfoss2.fcc.gov/prod/ecfs/retrieve.cgi?native_or_pdf=pdf&id_document=6513400922

Cooper, M. (2004, October). *When law and social science go hand in glove—Usage and importance of local and national news sources: Critical questions and answers from media market analysis.*

Paper presented at the Telecommunications Policy Research Conference, Arlington, VA.

Dimmick, J., Chen, Y., & Li, Z. (2004). Competition between the Internet and traditional news media: The gratification-opportunities niche dimension. *Journal of Media Economics, 17*(1), 19-33.

Emig, A. G. (1995). Community ties and dependence on media for public affairs. *Journalism & Mass Communication Quarterly, 72*(2), 402-411.

Einstein, M. (2002). *Program diversity and the program selection process on broadcast network television.* Federal Communications Commission Media Ownership Working Group Study #5. Accessed July 7, 2004, from http://hraunfoss.fcc.gov/edocs_public/attachmatch/DOC-226838A10.pdf

Federal Communications Commission. (2001). *FCC Chairman Powell announces creation of media ownership working group.* News release. Available: http://www.fcc.gov/Bureaus/Miscellaneous/News_Releases/2001/nrmc0124.html (Accessed December 27, 2004).

Federal Communications Commission. (2002). FCC *releases 12 studies on current media marketplace.* News Release. Accessed December 27, 2004, from http://hraunfoss.fcc.gov/edocs_public/attachmatch/DOC-226838A1.pdf

Federal Communications Commission. (2003). *2002 Biennial Regulatory Review, Report and Order and Notice of Proposed Rulemaking.* 18 FCC Rcd. 13620.

Federal Communications Commission. (2004). *Broadcast Localism.* Notice of Inquiry, 19 FCC Rcd. 12425.

Friedland, L. (2001). Communication, community, and democracy: Toward a theory of the communicatively integrated community. *Communication Research, 28*(4), 358-391.

Hindman, M. (2007). A mile wide and an inch deep: Measuring media diversity online and offline. In P. M. Napoli (Ed.), *Media diversity and localism: Meaning and metrics* (pp. 327-348). Mahwah, NJ: Erlbaum.

Jeffres, L. W., Dobos, J., & Sweeney, M. (1987). Communication and commitment to community. *Communication Research, 14*(6), 619-643.

Katz, J. E., & Rice, R. E. (2002). *Social consequences of Internet use: Access, involvement, and interaction.* Cambridge, MA: MIT Press.

Kaye, B., & Johnson, T. (2002). Online and in the know: Uses and gratifications of the Web for political information. *Journal of Broadcasting and Electronic Media, 46*(1), 54-71.

Ku, G., Kaid, L., & Pfau, M. (2003) The impact of web site campaigning on traditional news media and public information processing. *Journalism and Mass Communication Quarterly, 80*(3), 528-547.

McCombs, M. E., & Shaw, D. L. (1972). The agenda-setting function of mass media. *Public Opinion Quarterly, 36*, 176-187.

Mendelsohn, H. (1974). Some policy implications of the uses and gratifications paradigm. In J.G. Blumler & E. Katz (Eds.), *The uses of mass communication: Current perspectives on gratifications research* (pp. 303-318). Beverly Hills, CA: Sage.

Mueller, M. (1995). Why communications policy is passing "mass communication" by: Political economy as the missing link. *Critical Studies in Mass Communication, 12*(4), 457-472.

Napoli, P. M. (1999). The unique nature of communications regulation: Evidence and implications for communications policy analysis. *Journal of Broadcasting & Electronic Media, 43*(4), 565-581.

Napoli, P. M. (2001). Market conditions and public affairs programming: Implications for digital television policy. *Harvard International Journal of Press/Politics, 6*(2), 15-29.

Napoli, P. M. (2004). Television station ownership characteristics and news and public affairs programming: An expanded analysis of FCC data. *Info: The Journal of Policy, Regulation, and Strategy for Telecommunications, Information, and Media, 6*(2), 112-121.

Neuman, W. R. (1991). *The future of the mass audience.* New York: Cambridge University Press.

Nielsen Media Research (2002). *Consumer survey on media usage.* Media Ownership Working Group Paper #8. Available: http://hraunfoss.fcc.gov/edocs_public/attachmatch/DOC-226838A17.pdf (Accessed January 3, 2005).

Noam, E. (1994). Reconnecting communications studies with communications policy. *Journal of Communication, 43*(3/4), 207-214.

Nossiter, B.D. (1985, October 26). The FCC's big giveaway show. *The Nation*, pp. 402-404.

Owen, B. (2004). *Confusing success with access: "Correctly" measuring concentration of ownership and control in mass media and online services.* Stanford Law School, John Olin Program in Law & Economics Working Paper 284. Accessed January 1, 2005, from http://ssrn.com/abstract=545302

Pritchard, D. (2002). *Viewpoint diversity in cross-owned newspapers and television stations: A study of news coverage of the 2000 presidential campaign.* Media Ownership Working Group Paper # 2. Accessed January 1, 2005, from http://hraunfoss.fcc.gov/edocs_public/attachmatch/DOC-2268 38A7.pdf

Prometheus Radio Project. v. Federal Communications Commission. (2004). 373 F. 3d 372.

Rayburn, J. (1996). *Uses and gratifications: An integrated approach to communication theory and research.* Mahwah, NJ: Erlbaum.

Roberts, M., & McCombs M. (1994). Agenda setting and political advertising: Origins of the news agenda. *Political Communication, 11*, 249-262.

Ruggiero, T. (2000). Uses and gratifications theory in the 21st century. *Mass Communication & Society, 3(1)*, 3-37.

Scheufele, D. (2002). Examining differential gains from mass media and their implications for participatory behavior. *Communication Research, 29*(1), 46-65.

Scott, B. (2004). The politics and policy of media ownership. *American University Law Review, 53*, 645-677.

Shah, D., McLeod, J., & Yoon, S. (2001). Communication, context, and community: An exploration of print, broadcast, and Internet influences. *Communication Research, 28*(4), 464-506.

Singleton, L.A., & Rockwell, S.C. (2003). Silent voices: Analyzing the FCC "media voices" criteria limiting local radio-television cross-ownership. *Communication Law & Policy, 8*, 385-401.

Slattery, K.L., Hakanen, E.A., & Doremus, M.E. (1996). The expression of localism: Local TV news coverage and the new video marketplace. *Journal of Broadcasting & Electronic Media, 40*(3), 403-413.

Spavins, T.C., Denison, L., Frenette, J., & Roberts, S., (2002). *The measurement of local television news and public affairs programs.* Federal Communications Commission Media Ownership Working Group Study #7.Accessed December 27, 2004, from http://hraunfoss.fcc.gov/edocs_public/attachmatch/DOC-226838A12.pdf

Trumbo, C. (1995). Longitudinal modeling of a public issue: An application of the agenda-setting process to the issue of global warming. *Journalism & Mass Communication Monographs, 152.*

Waldfogel, J. (2002). *Consumer substitution among media.* Federal Communications Commission Media Ownership Working Group Study #3. Accessed December 27, 2004, from http://hraunfoss.fcc.gov/edocs_public/attachmatch/DOC-226838A8.pdf

Wanta, W. (1997). *The public and the national agenda: How people learn about important issues.* Mahwah, NJ: Erlbaum.

Webster, J.G. (2007). Diversity of exposure. In P.M. Napoli (Ed.), *Media diversity and localism: Meaning and metrics* (pp. 309-326). Mahwah, NJ: Erlbaum.

Webster, J.G., & Lin, S. (2002). The Internet audience: Web use as mass behavior. *Journal of Broadcasting & Electronic Media, 46*(1), 1-12.

Webster, J.G., & Phalen, P.F. (1997). *The mass audience: Rediscovering the dominant model.* Mahwah, NJ: Erlbaum.

Webster, J.G., Phalen, P.F., & Lichty, L.W. (2000). *Ratings analysis: The theory and practice of audience research.* Mahwah, NJ: Erlbaum.

Yan, M.Z., & Napoli, P.M. (2004, October). *Market structure, station ownership, and public affairs programming on local broadcast television.* Paper presented at the Telecommunications Policy Research Conference, Arlington, VA.

Yim, J. (2003). Audience concentration in the media: Cross-media comparisons and the introduction of the uncertainty measure. *Communication Monographs, 70*(2), 114-128.

15

FROM SCARCITY TO MARKET POWER QUID PRO QUO

Toward a Workable Right of Access in U.S. Media

Michael M. Epstein

ACCESS, DIVERSITY AND OWNERSHIP: AN EVOLVING DEBATE

U.S. broadcast regulation over the last eight decades was built on three interrelated issues: access, diversity, and ownership of scarce broadcast frequencies. Reflecting on this regulatory history, this chapter revisits the origins of the U.S. government's mandate to regulate access, diversity, and ownership, and engages in the modern debate over the sustainability of this mandate in light of technologies that have transformed the communications industry in recent years. Although many, including the Federal Communications Commission (FCC),[1] cite these transformative technologies as a justification for abandoning broadcast regulation altogether, this chapter looks beyond the scarcity rationale to an alternative view: one that assesses the scarcity issue not in terms of technology or channels, but rather in terms of market power conferred by government on private parties.

Indeed, if one accepts the premise that broadcasters, irrespective of numerical or allocational scarcity, are recipients of a scarce government benefit, it makes sense to argue that broadcasters should offer something of value in exchange for that benefit. Thus, this chapter proposes a *quid pro quo*[2] in which broadcast conglomerates can opt to increase their ownership of broadcast properties by granting local and diverse entities content-neutral access to airtime. With this proposal, all parties come out ahead—broadcasters can acquire more licenses, the public gets access to broadcast channels heretofore unavailable to them, and the government achieves the goals of diversity and localism through access regulations that have proven elusive through its regulation of ownership.

Foundations of Broadcast Regulation

The FCC was created pursuant to the Communications Act of 1934, foundational legislation that, although amended, still governs civilian electronic media to this day.[3] As a New Deal initiative, the Federal Communications Act (FCA) was not a revolutionary example of new federal regulation; it was, in fact, legislation that strengthened and expanded the power of the government to control electronic media, including use of public airwaves for broadcasting.[4] In large measure, the FCA and its less comprehensive predecessor, the Federal Radio Act of 1927, were passed to address a void in technology management in the 1920s that had led to a "wild west" atmosphere in broadcasting, as fledgling commercial radio stations overpopulated the airwaves, and radio receiver manufacturers developed incompatible transmission formats.[5] Under the FCA, the FCC was given a broad mandate to manage broadcast technology, a mandate including frequency allocations designed to clear signal interference, ensuring compatible formats and equipment, developing infrastructure such as transmission towers and signal relays, and administering a licensing regime whereby the FCC would choose who could become a broadcaster.[6] Radio broadcasters of the time, both independent and those affiliated with fledgling networks, largely viewed the FCC's role as akin to a police officer assuring the smooth operation of traffic on a public right of way. Under this traffic cop approach, the FCC could regulate the number of users of the airwaves, and even require those users to use the technology in a particular way, but the agency could not regulate the nontechnological aspects of a station's operations or regulate the broadcaster's substantive speech.[7]

However, the FCC, even in its earliest days, did not share the industry view that the FCA limited its function to policing technology. Although the FCC recognized that FCA Section 326 prohibited censorship of a broadcaster's speech content by government,[8] the agency used its licensing authority to deny renewals or revoke licenses to radio stations that they deemed to be acting contrary to the public interest.[9] To support this broader, more intrusive mandate, the FCC turned to a number of FCA sections that charged the agency with the authority to require licensees to act in the "public interest, convenience, and necessity."[10] This language, borrowed without debate from World War I-era railroad legislation, provided the FCC with the legal justification to consider programming content in the administration of its licensing power.[11]

In its first uses, against a snake-oil salesman involved in a kickback scheme and an early radio evangelist, the FCC's public interest mandate did not generate a lot of controversy among broadcasters.[12] But by the end of the 1930s, when the FCC used the FCA's public interest mandate to promulgate a series of agency initiatives known as the "Chain Broadcasting Regulations," the broadcast industry was ready to challenge the FCC in the courts. With the advent of the Chain Broadcasting Regulations, a series of limitations on the mounting power of networks to control the operation and programming of local stations, the agency was no longer advocating the public interest against licensees on the fringe.[13] As a regulatory matter, the FCC used its public interest mandate against national programming networks, the vanguard of the broadcast community, to advance a policy that promoted local and diverse licensees and programming.

In passing the FCA, Congress contemplated that licensees would operate in local communities, airing local programming; nowhere did the Act envision the primacy of national programming distribution in American broadcasting.[14] To prevent network owners such as NBC and CBS from using the leverage of their highly popular national radio programs—the "must-hear" programs of the golden age of radio—to effectively take over the programming of local stations around the country, the FCC greatly curtailed the ability of networks to enter into affiliation agreements with its licensees, requiring licensees, for example, to have the final say on the programs they aired, to limit network programming to specific times of the day, and to avoid long-term affiliations.[15]

By the spring of 1943, the networks' challenge to FCC authority was before the U.S. Supreme Court. The case, *National Broadcasting Co. v. U.S.*,[16] is a landmark in American broadcasting. In a 5 to 2 decision, the Court upheld the authority of the FCC to regulate broadcast licensees, and, by extension, networks, in the public interest.[17] Justice Felix Frankfurter, writing for the majority, sided with the FCC on both statutory and constitutional grounds. In assessing the FCC's authority under the Communications Act, Frankfurter expressly rejects the networks' traffic cop characterization. According to the Court, because the FCC is burdened to determine the traffic itself, the Act requires the agency to exercise subjective judgments.[18] The opinion also determines that the Chain Broadcasting Regulations are consistent with the public interest objectives of the FCA, providing the FCC with statutory cover to promote local and diverse broadcasting.[19]

Scarcity as the Central Regulatory Principle

At a Constitutional level, however, the Court's decision in *National Broadcasting* established precedent that reverberates to this day. Rejecting the contention that the FCA itself infringed on broadcasters' First Amendment rights, the Court held that the FCC's policy of denying licenses or renewals to broadcasters did not amount to a denial of free speech.[20] The First Amendment was not implicated, according to Frankfurter, because "[t]here is a fixed natural limitation on the number of stations that can operate without interference with one another."[21] The Supreme Court thus introduces the concept of *scarcity*. Because there are more applicants for licenses than available frequencies, Congress and the FCC can justifiably impose special obligations on those fortunate enough to receive a license.[22] This type of scarcity, based on the limited allocation of a renewable but finite natural resource—the frequencies contained in the electromagnetic spectrum—embodies what today would be called allocational scarcity.

In the aftermath of the NBC decision, Congress and the FCC used allocational scarcity to justify a series of structural and content regulations that applied only to broadcast media, including special rules for access by political candidates, indecency restrictions, community program responsibilities, and, perhaps most controversially, a general right of public access known as the "Fairness Doctrine."[23] Enforced by the FCC from 1949 to 1987, the Fairness Doctrine imposed two distinct access obligations on broadcasters. The first, an affirmative responsibility to air controversial issues of significant public importance,[24] was rarely invoked, the most notable case involving a group that petitioned a television station to engage a controversy over strip-mining.[25] The second part of the doctrine required broadcasters to give members of the public a right to respond to specific broadcasts, including political editorials and personal attacks.[26] Broadcasters, and especially television journalists, were strongly opposed to the Fairness Doctrine in any context, because it created a chilling effect on the airing of certain content and, when enforced, effectively allowed the government to make editorial judgments on behalf on broadcasters.[27] By the mid-1960s, the Radio and Television News Directors Association was ready to mount a legal challenge to the doctrine.[28]

In 1969, the U.S. Supreme Court delivered a decision in *Red Lion Broadcasting v. FCC*.[29] Although the specific controversy involved the validity of the Fairness Doctrine, the more general principle of scarcity as a justification for a variety of special broadcast content regulations was the key issue at stake. The Supreme Court again ruled for the FCC, despite the fact that in the quarter century since the NBC case, radio and television broadcasters had become more plentiful both nationally and locally.[30] Justice White, writing for the Court, made it clear that because there are not enough frequencies available to grant licenses to all applicants, broadcasters do not enjoy the same First Amendment rights of nonbroadcasters.[31] White also said that, because of this inherent scarcity, the First Amendment rights of the public take precedence over the broadcasters' speech rights.[32] In addition to reaffirming the scarcity concept, the Court posited a second justification for government regulation in the public interest: Because licensees operate over airwaves that belong to the people, broadcasters have a fiduciary duty to act on behalf of the public.[33]

Abundance of Media

Although little attention—perhaps too little—has been given to the public trustee theory for broadcast regulation in the years since *Red Lion*, challenges to the scarcity justification remained very much in the forefront. With the advent of proliferating electronic channels of discourse in the 1970s and 1980s, including cable, satellite, and microwave relays, broadcasters and some in government began to question whether broadcast frequencies could still be characterized as "scarce", while conveniently ignoring the public trustee component of the *Red Lion* argument.[34]

Cable and satellite reception in homes meant that, for the first time since the dawn of radio, an alternative to over-the-air broadcast programming was available to the American public. Television broadcasters, who had enjoyed dominant market power for decades because of their access to a government resource, saw their audience begin to erode, and their market power become less dominant, in a more competitive marketplace, a trend that continues to this day. By the early 1980s, Mark Fowler, the FCC chairman during the Reagan Administration, was calling for the deregulation of broadcasting, and, in particular, the repeal of the Fairness Doctrine. Fowler and other deregulatory advocates essentially argued that, since electronic media were no longer scarce, a free market would promote competition and foster diversity both in terms of license ownership and programming.[35] Although the Supreme Court has acknowledged the debate over the continued saliency of scarcity, it has not abandoned its holding in *Red Lion*.[36] The same cannot be said for the Fairness Doctrine, which the Court permitted the FCC to abandon in 1987 for reasons unrelated to scarcity.[37]

In recent years, Congress and the FCC have continued to deregulate electronic media, citing many of the same arguments from the 1980s. In the mid-1990s, for example, Congress lifted caps on radio station license ownerships, relaxed limits on broadcast ownership, and deregulated cable.[38] In 2003, the FCC proposed a blunderbuss review of all remaining ownership regulations for television.[39] Chairman Michael Powell, a long-time advocate of marketplace diversity, argued forcefully that new technologies like the Internet make the scarcity rationale obsolete. Breathing new life into the numerical scarcity approach, Powell believes that Americans live in a world where there is an "abundance of media," and, in a literal sense, he is right.[40] Compared with the 1970s, there are many more channels—broadcast, cable, satellite, Internet—competing for consumers of content (see chapter by Aufderheide, Compaine, Lievrouw, and Noam, this volume).

Free Markets, Diversity, and Localism

Media channels, in fact, may have become more numerically abundant, but are they more diverse? In trying to answer that question, advocates of greater government intervention have focused on the issue of media ownership from an antitrust perspective. In the context of broadcast ownership, a free market system does not bring about diversity, either in terms of the multiplicity of sources or programming content. Indeed, the lesson of broadcast deregulation is that market power left unchecked leads to three distinct concerns that curb both source and programming diversity: concentration, conglomeration, and the "corporatization" of media. For broadcasting, *concentration* of ownership has meant fewer companies now own more licenses.[41] The result, a reduction in the numbers of independently owned and community-managed stations, has increased profits for broadcasters even as it hurts diversity.[42] *Conglomeration* refers to the advent of large companies who, in the wake of market deregulation, have amassed sizable holdings in a variety of different media, such as broadcast, cable, satellite, content producers, and Internet. Again, the result, although profitable, harms diversity, as the conglomerate consolidates previously independent media outlets to reduce operational redundancies and cross-promote its businesses.[43] *Corporatization* of broadcasting is what happens when the free market drives up the costs of licenses to the point that only corporations can afford to be broadcasters. In the private marketplace, this type of price inflation is desirable, as sellers should be able to profit from the capital appreciation of their assets. But broadcasting is not a private marketplace, and diversity suffers qualitatively when only one segment of society—corporate America—controls what gets seen and heard on the public's airwaves.

In 2004, a group of re-regulation advocates, led by public interest law firm Media Access Project, successfully challenged the FCC's proposals to relax or eliminate ownership restrictions in the courts by emphasizing the impact of this deregulation on source diversity and program diversity, especially in news coverage.[44] Although this challenge, heard in the Third Circuit Court of Appeals, may have slowed the deregulation juggernaut at the FCC, it is unclear whether the more influential DC Circuit Court of Appeals, known for opinions advocating deregulation, will agree with the concerns espoused by its sister court. Whether it does or does not, the issue of numerical scarcity in light of an Internet-fueled "abundance of media" will likely wind its way into the Supreme Court by the end of the decade.

Throughout all this, Congress has remained largely acquiescent to the political power of media conglomerates, even though they have proven to be less willing to accommodate industry interests than the FCC on key issues such as broadcast television ownership caps,[45] children's television,[46] indecency,[47] and vertical and horizontal channel ownership limitations for cable systems.[48] In enacting cable caps, Congress recognized that unregulated market power would lead to continued consolidation of cable system ownership and control of channel programming in an industry that is already dominated by Time-Warner and Comcast.[49] At the same time, Congress has also been willing to enact structural regulation to protect broadcast licensees from the more dominant market power of cable, giving broadcasters the option of requiring local cable systems to retransmit station signals at no cost.[50] These "must-carry" rules, upheld as constitutional by the Supreme Court in 1997,[51] were designed not only to address market competition, but also to promote programming diversity in broadcasting, which Congress specifically found to be at risk by the early 1990s.[52] At least with respect to cable regulation, Congress has recognized a nexus between program diversity and consolidated ownership.[53]

Still, when it comes to broadcasting, Congress evidently agrees with the FCC that allowing concentrated ownership of broadcast licenses will keep over-the-air broadcasting viable and relevant in a market growing increasingly competitive with cable, satellite, and Internet.[54] The same can be said of Congress's failure to legislatively overrule the FCC's repeal of the Financial Interest and Syndication-Rules (fin-syn) in 1996, which, since 1970, had prevented broadcast networks from vertically consolidating with studios to produce and syndicate programming.[55] These deregulatory positions, consistent with the free market mandate of numerical scarcity advocates, may possibly help bolster broadcasting in the marketplace. Broadcast networks, after all, no longer attract the huge audience shares of the 1960s, and the synergy of owning both stations and studios allows broadcasters—or at least the conglomerates that own them—to cut programming costs and participate in the hugely profitable syndication environment.

What Congress and the FCC have failed to take into account is the extent of the negative impact on programming caused by the easing of structural broadcast regulation. Government policymakers need to recognize that numerical scarcity reduces not only diversity of ownership, but also the diversity of the programming aired by those owners, especially when the "abundance" of media outlets are controlled by a few powerful corporations. Local musicians already complain that they cannot get airtime on radio stations unless they have a business relationship with Clear Channel or Viacom, the two conglomerates that now dominate radio in the wake of its deregulation in 1996.[56] In the decade since the fin-syn rules were repealed, every major independent producer of television entertainment in Hollywood, save one (which has only two programs on air currently), has either gone out of business or been absorbed into a conglomerate.[57] Indeed, from a programming diversity standpoint, consolidation of radio and television ownership has meant consolidation of programming decisions, as corporate owners combine once-independent news operations, remove station management from the localities they purport to serve, and adopt more risk-averse rules to rein in controversial or edgy programming that might negatively affect shareholder return.

If one accepts the premise that scarcity cannot exist when media are numerically abundant, then it becomes difficult to justify regulating television and radio broadcasting any differently than other media. Except for antitrust issues relating to abuse of market power—laws that in recent years have been relaxed across industries—there is little that diversity advocates can do to stop the current trend toward consolidation and merger among broadcasters. But this premise need not be accepted. In fact, the better argument is that the numerical abundance of electronic media channels opens the door to justifying greater regulation of broadcast ownership and programming diversity. Because broadcasting is now a significantly smaller piece of the media pie, the use of public airwaves to distribute program content should be less important to these conglomerates. Yet, the large media companies continue to lobby for the opportunity to increase their broadcast holdings, and pay millions to acquire new licenses despite the fact that broadcasting may be subject to content regulations á la *Red Lion.*

Why do these companies refuse to rid themselves of the specter of intrusive regulation by simply migrating their distribution operations to cable or the Internet? The answer lies in a reality that numerical scarcity advocates seem unwilling to recognize: Numerical abundance fails to account for qualitative disparities among media outlets. Put simply, not all media channels are the same. Conglomerates are eager for more broadcast stations precisely because broadcasting gets special treatment from the government; regulations have essentially privileged broadcasting in the marketplace for decades, and continue to do so. In addition to protective structural regulations such as the "must-carry" rules, broadcasting is the only medium with universal access—all televisions, for example, are capable of receiving broadcast signals. The conglomerates who own incumbent broadcasters are keenly aware of the competitive advantages of the broadcast medium. As the market has become more competitive, companies with diverse media holdings have become increasingly willing to leverage their incumbency as broadcasters to extend their audience reach through cross-promotion on nonbroadcast outlets and by requiring affiliates to carry strategically significant programming.[58]

In an abundant media marketplace, the owners of a transmission conduit should be completely free to impose whatever conditions they choose on those seeking to make use of that conduit's market power. Since the U.S. government "owns" the airwaves in public trust for the people, Congress and the FCC thus have the right to impose conditions that may be quite onerous on a broadcaster's speech rights without trampling on the First Amendment.[59] Congress affirmed this right on behalf of government expressly in the FCA itself, providing "for the use of such channels, but not the ownership thereof, by persons for limited periods of time, under licenses granted by Federal authority, and no such license shall be construed to create any right, beyond the terms, *conditions*, and period of the license" (italics added).[60] If a conglomerate does not like the condition, it can take its programming elsewhere.[61]

THE CURRENT STATE OF ACCESS RIGHTS

Although the Rehnquist Court voiced support for the allocational scarcity arguments of *Red Lion* in three of its decisions,[62] if the FCC and many broadcasters have their way, it may only be a matter of time before the "abundance of media" argument gains ground in the reconstituted Roberts Court.[63] And if, as a result, content regulations are deemed unconstitutional, then the implications on rights of access to broadcasting may be significant.

In discussing the current state of these rights, it is important to distinguish between a right of *reply* and one of *access*. A right of reply becomes operative as a result of a speaker's content. This would encompass the right of someone to respond to a personal attack, as in the *Red Lion* case, or the right to air an opposing view in a public controversy. A right of access, on the other hand, may become operative for reasons wholly unrelated to speech content, including the nature of a media technology or the status of the initial speaker. The difference is important under U.S. Constitutional law, as rights of access may be devised as content neutral structural regulations, while rights of reply are necessarily content regulations. In the aftermath of the FCC's repeal of the Fairness Doctrine, there is currently no right of reply or access available to the American public. A limited right of access exists for qualified candidates for federal office, and a content-triggered general right of reply is available to all legally qualified candidates, but these statutory provisions do not apply to the general public, or indeed to anyone but opposing candidates.[64]

For print media, the U.S. Supreme Court ruled, in *Miami Herald v. Tornillo* (1974), that the First Amendment forbids the government from legislating rights of reply to newspapers.[65] Justices were specifically concerned with the chilling effect a right of reply statute would have on robust debate because editors would have an incentive to avoid news coverage of controversies.[66] *Tornillo* means that if the courts were to treat no-longer-scarce broadcasting as indistinguishable from print media, then a right of reply such as the Fairness Doctrine would no longer be constitutional for the airwaves, even if Congress were to re-enact it.

However, whether it effectively bans all rights of access to media is a matter open to debate. With the exception of advertisements that discriminate on the basis of gender,[67] the Supreme Court has not validated any rights of access for print media. Thus, in articulating a workable right of access to government-owned media technology, it is important to address the First Amendment concerns articulated in *Tornillo*. *Tornillo* must not be read broadly as a bar to all rights of access to media. Justice Brennan, in his concurring opinion, suggested as much when he questioned the scope of the principal opinion's broad ruling.[68] Still, even if the Supreme Court were to hold that the First Amendment strictly prohibits government-mandated access to newspapers, broadcasting is a different medium, with different characteristics, even in the absence of scarcity.[69] To some extent, courts have already recognized that medium-specific differences can act to limit *Tornillo*'s application to non-scarce media access. Instructive on this point is the willingness of courts to uphold the constitutionality of public access rights to cable television systems, even in the absence of allocational or numerical scarcity.[70]

PUBLIC ACCESS IN EXCHANGE FOR MARKET POWER: TWO APPROACHES BASED ON QUID PRO QUO

As increased bandwidth, digital compression, and alternate content sources make it possible for more companies to own more media outlets, policymakers need to look beyond scarcity if they want to foster a true marketplace of ideas. Courts and policymakers must look at market power and public access in order to find a viable solution to the lack of diverse programming content. The debate is not about whether a regional corporation should be permitted to compete against a conglomerate in the broadcast marketplace, but whether broadcast media with market power, in return for the access granted to them by the government,[71] have an obligation to air both local and nationally diverse views of the public to whom the airwaves belong. This chapter proposes that Congress revisit the concept of a right of access, expanding it through a system whereby the greater a conglomerate's audience reach and market power, the more time that conglomerate must provide to citizens with diverse and/or local content.

The access regimes proposed in this chapter are each based on a theory of *quid pro quo*:[72] a bargained-for-exchange in which broadcasters would trade media access for market power. Under this *quid pro quo*, media owners would be required to provide the general public with a right of access to the media in exchange for the government permitting the broadcasters to increase their financial interest in media properties. One possible application of the *quid pro quo* theory would be to look to a time-based right of access that is accepted voluntarily by media companies in exchange for the government granting increased electronic media ownership caps.

Although this chapter focuses on a specific proposal for "zone-based" public access, which is defined later, there are two distinctively different approaches that would legally permit the government to trade market power for access even if one were to abandon the scarcity justifications presented in *Red Lion*. The first is an expansive approach to *quid pro quo*, which engages existing jurisprudence relating to contractual waiver of constitutional rights and implements the licensee as public trustee justification that is also proffered in *Red Lion*, albeit with less notoriety. Building on existing Supreme Court contracts cases and applied in a media context in *Erie Telecommunications v. City of Erie*,[73] this contractual *quid pro quo* approach allows the government to enact content regulations not subject to strict scrutiny.

The second approach is a more narrowly construed form of *quid pro quo* that would subject "zone-based" public access to the same type of intermediate scrutiny used by the Supreme Court to validate structural cable regulation in *Turner Broadcasting Co. v. FCC*.[74] In this second approach, the *quid pro quo* is not contractual, but rather is present in the manner in which a court would evaluate the speech burden incidental to the structural regulation.

APPLYING THE QUID PRO QUO ARGUMENT TO ELECTRONIC MEDIA

Quid Pro Quo in Lieu of Scrutiny

Erie's Expansive Quid Pro Quo Approach. Some courts have employed a more expansive approach to *quid pro quo*, arguing that as a matter of contract, a company can waive constitutional rights,[75] including fundamental First Amendment rights that would otherwise be subject to a level of constitutional scrutiny. With respect to First Amendment rights, the expansive *quid pro quo* argument finds its first articulation in Justice Rehnquist's dissent in *First National Bank of Boston v. Bellotti.*[76] Although the Court held that Massachusetts could not proscribe the political speech rights of a corporation created under state charter, Justice Rehnquist disagreed on the principle that corporations, as persons created pursuant to state law, can have additional regulations placed on them, even at the cost of rights that would otherwise be available to natural persons.[77] Because a state "confers special privileges and immunities different from those of natural persons" when it creates a commercial entity such as a corporation, the entity "would be subject to like regulation."[78]

In *Erie Telecommunications*, the Third Circuit used an expansive *quid pro quo* approach to uphold an agreement between the City of Erie and its cable franchisee that imposed fees and public access requirements that would otherwise have been subject to constitutional scrutiny as a content regulation.[79] Although the court made no reference to Justice Rehnquist's implied social contract reasoning, the rationale for waiver, as in the *Bellotti* dissent, was rooted in contract principles.[80] After reviewing a long line of Supreme Court precedents that permit waiver of constitutional rights in general, the Third Circuit articulated a broad rule that First Amendment rights may be bargained away when "the party foregoing its rights has done so of its own volition, with full understanding of the consequences of its waiver."[81]

In arriving at this holding, the Circuit panel reviewed two prior Supreme Court cases in which a contracting party alleged that the other party waived fundamental constitutional rights. In the first Supreme Court case, *D.H. Overmyer v. Frick Co.*,[82] the Court held that a warehousing company waived procedural due process rights of notice and counsel when it entered into, and subsequently breached, an installation contract with a refrigerator manufacturer.[83] In the second case, *Fuentes v. Shevin*,[84] the justices ruled that a consumer who purchased household goods did not waive due process rights in a dispute with a retailer, despite the fact that an adhesion contract[85] stipulated such a waiver.[86]

The distinction between the two cases is set forth in a passage from *Fuentes* that is cited in the *Erie* opinion, which explains that the two cases are distinguishable because of the parties' relative bargaining ability and actual knowledge of waiver.[87] Although the Supreme Court in *Fuentes* made reference to unequal bargaining power, the Court focuses on the limited context of adhesion contracts.[88] The more fundamental issue, as it turns out, is whether the waiving party was "aware of the significance" of the waiver provision.[89] Indeed, under the specific facts of *Fuentes*, the court determined that the appellant homemaker did not waive her constitutional rights because the consumer product manufacturer "made no showing whatever that the appellants were actually aware or made aware of the significance of the fine print now relied on as a waiver of constitutional rights."[90]

For the *Erie* court, this reliance on actual awareness as a condition to waiver offers specific guidance into what the Supreme Court would require for a valid contractual waiver in the context of constitutional free speech rights.[91] The principal focus for the Third Circuit was on whether the waiving party did so with "volition and understanding," which the court explained is deemed to be present "where the parties to the contract have bargaining equality and have negotiated the terms of the contract, and where the waiving party is advised by competent counsel and has engaged in other contract negotiations."[92]

Although the court did not say it in so many words, ultimately it comes down to a question of whether the party is sophisticated and well-advised. If so, the Third Circuit will deem that the party entered into the contract with the volition and understanding necessary to make a waiver of constitutional rights valid.[93] *Fuentes* was a homemaker who entered into a contract with fine print in a consumer context.[94] *Overmyer*, on the other hand, concerned a corporation both represented by counsel and who had negotiated a deal in a business context.[95] Although the focus on negotiation and bargaining equality might have proven to be a problem in a contract in which a government dictates terms to a private business entity in exchange for media rights or licenses, the *Erie* court, by making the real focus of its analysis sophistication and legal counsel, was able to apply the Supreme Court's prior jurisprudence on contractual waivers of rights to the case of a cable company suing over content regulations imposed by a city government.[96]

It may be interesting to note that although the circuit court affirmed the lower court's summary judgment decision, its decision to apply a *quid pro quo* analysis as an alternative to constitutional scrutiny represented a conscientious departure from the trial judge's approach.[97] Although the district court found for the City of Erie, it was not on the basis of contractual waiver.[98] In concluding that the city's access requirements were constitutional, the district court rejected the applicability of a waiver rationale because the cable company "lacked knowledge of its rights."[99] The court explained that "in light of the ambiguity and uncertainty" of the law with respect to cable companies' First Amendment rights at the time of the contract, Erie Telecommunications Inc. could not have waived a right of which it was not actually aware.[100] The fact that the appellate court dispensed with a requirement of actual knowledge in favor of deemed "volition and understanding" is significant, because under the Third Circuit's reasoning, a sophisticated, well-advised party may be deemed to have waived rights even if the right had not fully emerged as a matter of law, or if the party was unaware that the right, in fact, existed.[101]

Expansive Approach to Quid Pro Quo After Erie. *Erie* is still good law in the Third Circuit, but it remains the only case to expressly apply a contractual waiver of First Amendment rights in a dispute involving media access.[102] Although the Seventh Circuit makes reference to *Erie* in *Chicago Cable Communications v. Chicago Cable Commission*,[103] a case that also engaged the validity of local cable ordinances, the *Chicago Cable* court's reference to contractual waiver is limited to an introductory footnote in which the court determined that it will not decide the issue "since neither party advanced the argument here or in the district court."[104] Even so, the footnote contains enough analysis to offer readers some sense of how the Seventh Circuit panel would have considered the issue had they ruled it justiciable.[105] In the first part of the footnote, the panel raises the waiver issue as "conceivabl(e)," and focuses on the fact that Erie's contract was a consent decree to prior litigation, as opposed to an initial franchising agreement between the government and a cable company, which was at issue in *Chicago Cable*.[106] The differences in the type of contract turns out to be a "significant distinction" for the court, because the footnote states that "serious reservations arise when a local government with a virtual monopoly on cable access conditions a franchise contract with a cable programmer on the stipulation that the programmer waive certain constitutional rights."[107]

Although one must be careful not to draw too many conclusions from a single footnote, the Seventh Circuit's brief treatment of the waiver issue suggests that its view would be in conflict with the Third Circuit's holding in *Erie*. In emphasizing "virtual monopoly power"[108] on the part of the city, the *Chicago Cable* court seems to be suggesting that the two parties to a cable franchising agreement have unequal bargaining power.[109] Although the court did not say so expressly, the connection to the Supreme Court's contractual waiver precedent is evident.[110] The difference between the outcomes of *Overmyer* and *Fuentes*, as the Third Circuit pointed out, was that Mrs. Fuentes signed an adhesion contract as a condition to a consumer purchase.[111] Whereas, Overmyer's contract was a negotiated business transaction between parties with equal bargaining power.[112] Although it is true that some contracts involving monopolistic power may involve unequal bargaining power, the Seventh Circuit panel would be at odds with the Third Circuit if it suggested that such unequal bargaining power would be present in franchise agreements merely because the government controls access.[113]

Expansive Quid Pro Quo and the Nature of Cable Franchise Agreements. First, the reality of a cable franchise agreement is that it is a contract that effectively grants monopoly power *to* a cable company.[114] Because local rights of way and other access benefits belong to the government, the government is not using unequal bargaining power to condition access; it is merely retaining some of the power it would otherwise have the right to reserve for its own use or could bargain away freely to another party. In granting benefit that is not available to everyone, the government should be permitted to condition its grant on a waiver of rights that the cable company would otherwise *not* have. Additionally, prohibiting the government from placing conditions on its grant of a monopoly would *significantly reduce* a municipality's power to negotiate on behalf of its citizens. Cities would effectively be limited to an all-or-nothing approach to bargaining.[115] Government would either have to give up all of its access rights without limitation, or refrain from entering into these franchise agreements altogether, because cities are not required to contract with cable companies. The result would be less bargaining power for both sides, as there would be virtually no room for negotiation on the issue.

Moreover, a franchise agreement between a city and a cable provider is not an adhesion contract. Unlike a household consumer, a cable company is expected to be aware of the rights it is giving up in exchange for the benefits it will receive under a franchise agreement. This is not a question of an unsophisticated party not reading the fine print of a form contract. In every instance, cable companies seeking to enter into a franchise agreement are familiar with the issues, have the capability to conduct negotiations, and are advised by competent counsel.[116] Indeed, this is precisely why the Third Circuit deemed "volition and understanding" to encompass a totality of circumstances that included bargaining equality, the presence of actual negotiation, and the availability of advice of counsel, and why the *Erie* court held that Erie Telecommunications Inc., had acted with volition and understanding when it entered into a contract that required it to waive certain First Amendment rights.[117]

Although the decision in *Erie,* that volition and understanding includes consideration of the totality of the circumstances, is based on a fair reading of the Supreme Court's decisions in *Fuentes* and *Overmyer,*[118] other circuits, including the Seventh, remain free to develop their own approach to these precedents. Unfortunately, the *Chicago Cable* court did not engage either of these cases, choosing instead to list, without elaboration, older Supreme Court cases with less relevance to contractual waiver jurisprudence.[119] However, even if another circuit were to reject *Erie's* concept of volition and understanding, the court would have to undertake the same type of thorough analysis of *Fuentes* and *Overmyer* that the Third Circuit undertook.[120] In analyzing these two cases, a court would have to discuss why the Supreme Court decided for waiver in *Overmyer* but against waiver in *Fuentes.* Because *Fuentes* expressly embraces lack of "awareness of the significance" of the waiver as its standard for invalidating contractual waiver, merely demonstrating that the parties had unequal bargaining power would therefore not be consistent with the Supreme Court's treatment of the issue.[121]

Some might complain that a content-based right of access places too great a burden on broadcasters' speech rights; however, the reality is that there is no burden under the First Amendment, only a conditional government benefit at stake. If a particular speaker does not want to accept the speech limitations required by government as a *quid pro quo* for receiving this benefit, that speaker is just as free as anyone else in this country to exercise its free speech rights. The speaker has merely chosen not to avail itself of a government benefit on which the government has a right to place restrictions. The plausibility of this argument is suggested in *Red Lion* itself, not in the scarcity rationale, but in the broadcaster as public fiduciary analysis, a justification for content regulation that receives far less attention by the courts and policymakers than scarcity has in the last three decades.[122] Broadcasters are like private companies hired to manage a public housing project, or concessionaires in a public park—the government does not deny rights by imposing conditions to those seeking to profit from government-owned resources.[123]

The Scrutiny Alternative: Quid Pro Quo Under the O'Brien Test

In cases involving structural cable regulation, including cable access cases, courts have addressed *quid pro quo* arguments through an application of the *O'Brien* test of intermediate scrutiny. Under the *O'Brien* test, a "content-neutral regulation will be sustained under the First Amendment if it advances important governmental interests unrelated to the suppression of free speech and does not burden substantially more speech than necessary to further those interests."[124] In those cases, the *quid pro quo* argument is used more narrowly in order to address the issue of burden under the *O'Brien* test's fourth prong.[125]

Although *quid pro quo* is no longer applied expressly in the context of contractual bargaining, courts employ a *quid pro quo* analysis within the framework of the *O'Brien* test's scrutiny of substantial governmental interest and more importantly, narrow tailoring.[126] The end result is that subsequent courts have justified government regulations because, first, the government has a substantial interest in regulating rights of a private party that would not have existed but for the government's decision to grant a special benefit not available to everyone, and second, the burden on cable's speech rights is low because it benefits from a government monopoly.[127]

In *Chicago Cable*, the Seventh Circuit adopted the intermediate scrutiny approach to *quid pro quo* by applying the *O'Brien* test to the city's cable access regulations.[128] The court characterized its decision to use *O'Brien* as a "novel question" because the Supreme Court, at least as of 1989, had not yet determined "the applicable standard of review" for a First Amendment challenge to municipal cable regulations.[129] The Court's choice in applying the *O'Brien* test is significant because it presages the approach to constitutional scrutiny that the Supreme Court later applied to structural cable regulation, which impacted cable speech rights in *Turner II*.[130]

Chicago Cable applied the *O'Brien* test to a city regulation requiring locally originated programming and focuses on the economic scarcity of cable as a medium under the fourth prong of the *O'Brien* test, which, as interpreted in *Fox* and

Clark,[131] requires that the "means must be narrowly tailored but not the least restricted alternative."[132] Although the case cited *Red Lion* as an example of justified government regulation of television and distinguished *Red Lion* from *Tornillo*, the opinion makes an *economic scarcity* argument quite different from the *allocational scarcity* justification offered for access under the Fairness Doctrine.[133] It is an argument about market power that applies to all electronic media in which access is controlled: "Cable programming, like other forms of the electronic media, is an economically scarce medium. Unlike the traditional forms of print media, a cable programmer enjoys a virtual monopoly over its area, without the threat of an alternative provider."[134]

What is most significant about *Chicago Cable* is its willingness expressly to engage a *quid pro quo* argument to justify content neutral regulation as narrowly tailored under the *O'Brien* test.[135] As the court explained: The government "is duty-bound to recognize the effects of 'medium scarcity' by ensuring that the few programmers who are granted a franchise make optimum use of it."[136] Therefore, because the government has granted the cable system economic rights that are not available to all, the government can require the cable entity to give up some of its speech rights on behalf of cable customers.[137]

Although the *Chicago* opinion focuses on a single franchise with monopoly power over access, *Turner II* expands the economic scarcity argument to encompass restricted competition.[138] In *Turner II*, the Supreme Court held, 5 to 4, that federal legislation requiring cable systems to retransmit local broadcast signals did not violate the First Amendment rights of cable interests.[139] In validating these so-called "must-carry" rules, five justices determined that regulations designed to foster diversity, localism, and certain types of programming were content-neutral regulations subject to a highly deferential application of intermediate scrutiny under the *O'Brien* test.[140]

Justice Breyer determined that the issue of speech burden can be mitigated under the fourth prong of the *O'Brien* test of intermediate scrutiny when media entities have benefited from a grant of monopoly power from government.[141] In his pivotal concurrence, Justice Breyer recognized that governmental regulations suppressing cable speech rights are justified because cable systems "face little competition" and, therefore, act as a "bottleneck that controls the range of viewer choice."[142] The same logic would apply to broadcast licensees because they also face little economic competition and control what viewers can see in a medium with great market power. If a local cable system can be required to give the public access to its conduit, so too should broadcasters be required to provide the general public with a right of access to the conglomerate's media holdings in exchange for the government permitting the conglomerate to increase its financial interest in media properties that receive a government benefit, such as a license or infrastructure rights of way.

Turner II is also instructive because the Supreme Court determined that the *1992 Cable Act*'s "must-carry" rules were narrowly tailored because they scaled the burden in accordance with an Multi System Operator's (MSO) channel capacity.[143] In sections 614 and 615 of the Communications Act of 1934, Congress very specifically set forth an escalating must-carry obligation: Systems with fewer than 12 channels had a low burden, systems with 12 to 36 channels had an intermediate burden, and systems with more than 36 channels had the greatest burden.[144] The Act also exempted the smallest systems—those with fewer than 300 subscribers—from the rules entirely.[145] This scaling scheme, absent in previous codifications of the must-carry rules, was key to the *Turner II* Court's determination that the 1992 rules were narrowly tailored.[146] Indeed, in the 1985 *Quincy* case, the DC Circuit struck down must-carry rules precisely because they applied to all MSOs equally, and thus were too burdensome on the smallest cable systems.[147, 148] Moreover, any burden is further reduced by current technological innovations, such as digital compression and signal multiplexing, because public access could be limited to one programming stream, permitting broadcasters to control content on the remaining streams of their signal bandwidth.[149]

In a *quid pro quo* access system, scaling a court-determined burden is relatively easy to accomplish. Because the amount of time or space devoted to public access is scaled to the amount of additional market power from which the conglomerate is seeking to benefit, any burden would be mitigated for media companies with smaller market reach. In that respect, public access would operate in a manner similar to the *1992 Cable Act's* narrowly tailored must-carry rules. In any event, this narrow tailoring analysis would only be necessary if the courts determine that a *quid pro quo* access system is a content-neutral structural regulation that burdens the conglomerate's speech rights, a proposition that would not be easy to make because the *quid pro quo* is voluntarily entered into. Thus, even if courts were to determine that a market power *quid pro quo* would impose a real speech burden on media conglomerates that choose to avail themselves of the bargain, the government could make a strong argument that any burden would be narrowly tailored under the *O'Brien* test.

Structural regulation also has the additional benefit of being content-neutral. In its analysis of burden, *Chicago Cable* focused on the fact that the regulation is minimally burdensome because it does not require the cable system to air any specific program or editorial viewpoint.[150] *Turner II*, as previously explained, similarly extended the content-neutrality analysis to types of programming, including local and diverse.[151] The idea here is that although the government cannot require specific content to be aired, it can require that cable companies air types of programs that further governmental interests without sacrificing content-neutrality.[152] This would surely encompass regulations requiring electronic media—including broadcasting—to give access to local or diverse content.

Finally, with respect to both *quid pro quo* approaches, the diversity justifications for regulation of the cable medium would apply even more directly to broadcasting, because the reasons for the legislation relate to the protection of government-owned airwaves. Putting aside the now controversial scarcity justifications of the *Red Lion* case,[153] although still good law, the diversity and localism justifications found in *Turner II*[154] make even more sense in the context of broadcasting. Broadcast frequencies, unlike cable transmissions, are a medium that by their very nature belong to the public. Broadcasters, of course, operate on these frequencies only to the extent that Congress permits them, through licensing and regulatory oversight.[155] As such, requiring broadcasters to give up rights of access—even content-based ones—as a *quid pro quo* for receipt or continuance of an operating license is quite reasonable. The *Chicago Cable* and the *Turner II* decisions thus effectively open the door to a regulatory approach that, with government action, would overcome constitutional concerns that heretofore prevented or hindered Congress from enacting, and the FCC from enforcing, a general right of access to media.

IMPACTS

Recognizing the Grant of a Government Benefit

The salience of the theory of broadcaster as public trustee is such that it should stand alone as a justification for broadcast access regulation, divorced from the imperiled scarcity justification of *Red Lion*.[156] Even considering the scarcity argument, in an expansive media market, recall that speakers who do not wish to accept the conditions that attach to the government-granted benefit of a broadcast license now have viable alternatives such as cable, satellite, and the Internet.

To a great extent, the major media conglomerates have already become dominant owners in these media alternatives, partly in an effort to provide programming not subject to the lowest common denominator of the highly regulated broadcast market, but mainly as part of a strategy to promote increased viewership of the conglomerate's flagship broadcast outlet.[157] For example, Fox will use its Fox Sports Net and FX cable stations to air Major League baseball games with limited appeal, or even minor playoff games, when a game with greater audience appeal is scheduled to air on the Fox Broadcast network, but the World Series games always air on the Fox Broadcast network, just as the Super Bowl always airs on a broadcast station. The fact that major television events air on licensed broadcast properties establishes that the government benefit conferred on broadcasters—market power—remains just as strong, if not stronger, in an era of rapidly expanding multichannel video programming distribution (MPVD) alternatives.

Media conglomerates recognize that they are receiving this government benefit when they choose to continue to operate broadcast licenses, when they produce original news and entertainment programs for broadcast, and when they buy or sell broadcast licenses for millions of dollars. Conglomerates may not thank the government for the benefits bestowed on them by the grant of a broadcast license, but the marketplace speaks for itself. After all, if broadcasting were a bad bargain, licenses would presumably be plentiful and inexpensive.

Scaled Zone-Based Access Formula

One formula, noted in Figure 15.1, would be to scale the amount of access, in this case, time as a function of the percentage of national audience a broadcaster's licenses reach, as measured by the FCC. Thus, a conglomerate with 50% of national reach could be responsible for 50 minutes of access time each week (or other such period). A greater or lesser amount of time could also be achieved by adjusting the conversion factor up or down. Setting a value for the conversion factor would be a matter for Congress or the FCC to determine. Congress could presumably develop more complicated formulae that could account for characteristics of the speakers granted access. For example, Congress could adjust the conversion factor based on the degree of diversity (see Napoli, this volume) or localism among the users. Access could be based on the status of a petitioner as "local," and possibly even the local nature of the speech itself, because *Turner II* held that laws that promote locally based programming and quality programming were not impermissible content regulations.[158]

In the context of the digital broadcast television, the access unit of time could be replaced by bandwidth, because digital signals can be "multiplexed." With multiplexing, a digital broadcast signal can be divided into a number of programming streams that on digital television tuners would appear as subchannels.[159] Because multiplexing would essentially permit licensees to transmit programming to as many as 12 subchannels simultaneously, the zoned-access formula could be modified to account for proportionate access by requiring a broadcaster to dedicate all or part of a subchannel, or perhaps even more than one subchannel, to public access, depending on the amount of additional market power granted to the broadcaster.[160] From a policy standpoint, requiring licensees to "give back" some of the benefit of increased programming capacity now possible with digital transmissions seems to make sense. In this respect, setting aside one or two programming streams out of a multiplexed broadcast signal for public access is not that much different than the public access set asides already in place for cable[161] and direct broadcast satellite[162] systems under the 1992 Cable Act. Those

FIGURE. 15.1. **Formula for scaled zone-based access.**

$$z = p \times \frac{u}{f}$$

where:

z = zone of access based on time (or bandwidth)
p = market reach, as measured by FCC
u = unit of time or space measurement (e.g., minutes or sq. inches)
f = factor of conversion to be determined by Congress

set-asides, which scale the number of public access channels proportionally to a system's total channel capacity, were upheld as constitutional structural regulation by the DC Circuit in *Time-Warner Entertainment Co., L.P. v. FCC.*[163] Broadcast set-asides would also be consistent with the position of the Gore Commission, which in the late 1990s recommended that digital broadcasters be required to dedicate some of their more efficient bandwidth to public interest programming.[164]

Moreover, the notion that digital broadcasters can earmark a multiplexed channel for transmissions not related to their over-the-air broadcasting service has already been recognized by the FCC in its regulations governing digital television. Section 73.624(g) of the Code of Federal Regulations allows licensees effectively to sublease a multiplexed subchannel to themselves or third parties for private use, provided that the licensee tenders an annual fee representing 5% of the revenue received from these ancillary services.[165] Permitting digital broadcasters to privatize public airwaves in return for a fee seems very different from requiring broadcasters to dedicate a programming stream to public access in return for market power, but the logic is the same. In both cases, the licensee would essentially be "spinning off" a portion of its bandwidth to operations not related to its use of the airwaves for broadcast programming. The only difference is that in a zoned-access regime, the beneficiary would be the public, and the content transmitted on any dedicated subchannel would not be subject to the licensee's control or gatekeeping.

Additionally, such a system would be relatively easy to measure and monitor because it essentially creates a relationship between the market power given and the time or bandwidth yielded. Since the government gives broadcasters measurable market power in granting a broadcast license, such a scaled right of access would be constitutional as a structural regulation, so long as access were determined through content-neutral means such as a lottery, first-come-first-serve basis, or random selection rather than through a content-driven selection process.[166]

EXTENDING THE QUID PRO QUO TO ALL CONGLOMERATE-OWNED MEDIA

Given that a broadcast license grants a great government benefit of market power to a broadcaster, the questions remains, what should the government require in exchange for this benefit of market power? The answer is simple—diversity and localism.[167] The government should require a broadcaster to provide access to the general public in order to ensure local and diverse content. Based on the expansive contractual *quid pro quo* theory outlined here, a content-based approach such as the Fairness Doctrine—independent of scarcity justifications—would be constitutionally permissible. If the government were to bargain with broadcasters over market power, it might be possible to extend broadcast access regulations to nonbroadcast media owned by broadcasters.

Waiving First Amendment Rights in Return for Increased Market Power

The fundamental question here is not whether the First Amendment rights of non-scarce print publishers would be infringed on, but rather whether the *quid pro quo* can be sufficiently broad to include a waiver of the publisher's free speech rights for nonbroadcast media. Again, under the Supreme Court's reasoning in the contractual *quid pro quo* cases, a party to contract can waive constitutional rights as a condition for receiving the benefit offered in the bargain.[168] Thus, just as the government can impose conditions on the grant of a broadcast license, it follows that the government can broaden the condition to address market power even to print. *Tornillo* would not be implicated by such a broadly constructed condition agreed to by a conglomerate as part of a voluntary contract.[169] As was the case with cable owners, the owners of print publications are not required by the government to give up a portion of their publication, here measured as space and location, as opposed to time; the conglomerates remain free to exercise their full First Amendment rights by choosing not to accept the additional broadcast market power being offered to them.

Although the contractual waiver approach may open the door for a host of content regulations in which a conglomerate's ownership of broadcast properties could be leveraged against the acceptance of content regulations affecting their nonbroadcast properties, such a revolutionary approach would be unprecedented and, in light of current law, unnecessary. Under the more narrow *O'Brien quid pro quo* approach, the market power for timed access *quid pro quo* also could effectively provide the public with structural access to newspapers and other publications owned by broadcast conglomerates.[170] In addition to being more politically palatable, such a regulation, in which the government would require a content-neutral unit-based approach to media access, makes more sense for four reasons.

First, requiring media conglomerates to accept a structural regulation in return for market power avoids the enforcement problems that plagued the Fairness Doctrine, including concerns over the definition of controversial issues and the specter of a chilling effect on broadcaster speech.[171] Second, a structural regulation that grants local and diverse access without regard to content helps foster diversity without making content judgments that might overlook similar, but not identical views, or views clearly out of the mainstream. Third, it also takes the broadcaster out of the gatekeeping process, removing any question of editorial favoritism or bias based on political or social agenda. Fourth, as important as the aforementioned benefits are, what makes structural regulation as a *quid pro quo* for media market power especially attractive (compared with a scarcity argument) is its potential application to nonbroadcast media such as cable—where *Turner II* and the *O'Brien* test already apply—and in some cases, even to print media owned by media conglomerates, *Tornillo* notwithstanding.

This broader scope of regulation would be constitutional under *O'Brien* as applied in *Turner II* because the broadcasters effectively can be required to give up their First Amendment rights in return for a government benefit of being a broadcaster.[172] As was the case with the contractual approach, the only question would be whether the government could extend the burden it is already permitted to exact against a broadcaster to First Amendment rights that would otherwise be beyond the government's ability to regulate, absent a compelling interest. In that sense, it is not much different from content-neutral, market-based logic that led the Supreme Court to validate broadcast cross-ownership regulations in 1978.[173] Cross-ownership rules, although not currently in favor at the FCC, require companies who own different media properties to divest nonbroadcast properties in local markets with excessive ownership concentration.[174]

This proposed *quid pro quo* approach, although consistent with Justice Marshall's thinking in *National Citizens Committee for Broadcasting v. FCC*, would be far less burdensome (if at all) on broadcasters than cross ownership regulations.[175] Here, if broadcasters want access to licenses with greater market reach, they must give back some of that power by providing time-based diverse and local access, not only on their broadcast stations, but also on their cable channels. As is often the case in contract law, the broadcaster does not have to accept the government's offer.[176] These broadcasters may be permitted to broadcast, but they would not be permitted to amass the type of media ownership that many media conglomerates seek today. Moreover, they may not be permitted to own other nonbroadcast properties in markets where cross ownership rules are needed to guard against local concentration.

Content-Neutral Right of Access and Editorial Discretion

Constructing the public's right not as a right of *reply* but rather as a content-neutral right of *access*, makes it possible to overcome *Tornillo*'s concern that the government must not intrude into the discretion of the editorial boardroom. The corporate ownership, choosing to avail itself of a government incentive, would be the one to make the business decision to provide an access corner or access page in the publication. A governmental system can simply be designed to permit structural access based on the type of content—local and diverse—just as a majority of the Supreme Court did in *Turner II* when it determined that the *1992 Cable Act's* must-carry rules were not a content regulation, even though they promoted diverse and local programming over other types of programming.[177]

By making the access decisions nondiscretionary for editors, the *quid pro quo* system would be the type of noneditorial structural access that the Supreme Court upheld as constitutional in *Pittsburgh Press v. Pittsburgh Commission on Human Relations*.[178] In *Pittsburgh Press*, the Court held that a government regulation requiring gender-equal access to classified job listings did not infringe on the First

Amendment rights of newspapers.[179] In that case, the court reasoned that the compelled access did not infringe on speech rights because it lay "outside the editorial sphere."[180] Critical to the court's finding was that advertising staff, not editors, were charged with making the access decisions for classified listings. Decisions on how to apportion access to public speech zones would be even further outside the editorial sphere, as they would be business decisions made in the board room of the publication's corporate parent, and not by the staff of the newspaper. All the staff of the newspaper needs to do would be to implement a corporate policy that would provide access to local and diverse sources without regard to specific content.[181]

Making access a business decision implemented by noneditorial staff also addresses one of the *Tornillo* court's concern that permitting government to mandate access would have a chilling effect on journalists, because they would avoid controversial topics or criticism.[182] This fear of a chilling effect is also a principal objection lodged against the Fairness Doctrine, and, to some extent, was borne out during the 25 years that the doctrine was enforced by the FCC.[183] The beauty of a structural access zone approach is that it has virtually no editorial impact on journalists. Unlike the Florida reply statute found unconstitutional in *Tornillo*, and unlike the no-longer enforced Fairness Doctrine, access is not triggered by specific content or lack of it.[184] Access is merely granted in proportion to the market power granted by government.

Finally, it is important to emphasize that companies owning broadcast licensees with little market power would not be covered by the proposed zone-access system. Independent, "mom-and-pop" operations would have no reason to enter into the *quid pro quo* bargained-for exchange, unless, of course, they were interested in strengthening their market power through horizontal acquisitions in broadcast television and radio. The same is true for independent newspapers and magazines. Without concentrated broadcast ownership, the government benefit essential to a *quid pro quo* analysis is absent. In such a case, print publications, even ones owned by powerful corporations, would not be subject to any type of government-sponsored public access system, in accordance with the holding in *Tornillo.*

CONCLUSION

If Congress, the FCC, and the courts accept the idea that broadcasters can be required to give up some of their speech rights in return for market power based on a government grant of access, the contentious discussions about the nexus of ownership and programming content, market-based content preferences, and multiplicity of sources become moot.[185] Debates over compulsory speech and chilling effect would no longer exist, as timed access would not be tied to any particular speech content, and would not require broadcasters to air particular viewpoints. In the event that a content regulation under *Red Lion* and ownership caps no longer proves viable, this proposal allows Congress and the courts to engage diversity and localism as structural broadcast regulation, with obligations predicated on the public interest mandate of the Communications Act and consistent with the reasoning of the U.S. Supreme Court in *Turner II.*

Moreover, even if such an access system were ultimately determined to be a content regulation, Congress and the FCC may nonetheless be able to avoid strict scrutiny by extending to broadcasting the same *quid pro quo* theory used by courts in cable access cases. It may also be worth considering whether the government has the right to require broadcasters to waive certain rights contractually—including speech rights—as a *quid pro quo* for the receiving a government benefit, such as a broadcast license and relaxed ownership caps—to which they otherwise would have no right.[186] Such an approach would not only provide constitutional justification for intrusive content regulations like the repealed Fairness Doctrine, but may also provide a constitutional means for extending access rights to print properties owned by media conglomerates, notwithstanding *Tornillo.*

In the end, media ownership is not a right; it is a government (e.g., public) benefit. Thus media companies can be required to give up rights in order to access this benefit. And although attention must be given to the burden on speech, the burden would be insignificantly low. Such a plan meets the FCC's public interest mandate by providing a benefit both to society and to the broadcasters,[187] while fulfilling a common goal, increasing localism and diversity. Increasing access to the public is a small price to pay for media conglomerates further increasing their power in the market.

NOTES

1. *See* In Re Complaint of Syracuse Peace Council, 2 FCC Rcd. 5043 (1987). The FCC determined that the "comprehensive study of the telecommunications market in the 1985 Fairness Report has convinced us [the FCC] that [the scarcity] rationale that supported the doctrine in years past is no longer sustainable in the vastly transformed, diverse market that exists today." *Id.* at para. 64. *See also* FCC Commissioner Michael Powell Advises Investment Analysts to Look For Evidence Of Regulators Promoting Innovation and Competition, 1998 WL 110174 (FCC 1998). Quoting Albert Einstein, Powell said, "The unleashed power of the atom has changed everything save our modes of thinking and we thus drift toward unparalleled catastrophe." Reflecting on where the country is in the communications revolution, Powell remarked "I often think of Einstein's warning," and questioned whether the industry is "truly changing enough, and quickly enough to accommodate" the transformation to a pro-competitive, deregulatory communications regime. Powell set forth several tenets of "new regulatory thinking" that should guide policy leaders in light of the pro-competitive,

de-regulatory environment and the immense power and potential of communications technology unleashed:

"1) Faith in competition, and courage to cede control to the marketplace;
2) Focus on innovation when dealing with network industries that are driven by technology;
3) Prepare for the regulatory implosion of traditionally distinct technologies and services;
4) Strive for regulatory efficiency. In this regard, regulatory agencies must make timely decisions, be sensitive to business realities and capital markets, and shift their efforts to enforcement."

2. "An action or thing that is exchanged for another action or thing of more or less equal value; a substitute." *Quid pro quo*, BLACK'S LAW DICTIONARY (Bryan A. Garner et al. eds., 8th ed. 2004). "Lat. 'what for what' or 'something for something.' The concept of getting something of value in return for giving something of value. For a contract to be binding, it usually must involve the exchange of something of value." *Quid pro quo*, THE 'LECTRIC LAW LIBRARY'S LEXICON, *available at* http://www.lectlaw.com/def2/q003.htm. (last visited on March 13, 2007).
3. *See* Purposes of Chapter; Federal Communications Commission Created, 47 U.S.C. § 151 (2005) (entitled "Purposes of chapter; Federal Communications Commission created") ("For the purpose of regulating interstate and foreign commerce in communication by wire and radio so as to make available, so far as possible, to all the people of the United States, without discrimination on the basis of race, color, religion, national origin, or sex, a rapid, efficient, Nation-wide, and world-wide wire and radio communication service with adequate facilities at reasonable charges, for the purpose of the national defense, for the purpose of promoting safety of life and property through the use of wire and radio communications, and for the purpose of securing a more effective execution of this policy by centralizing authority heretofore granted by law to several agencies and by granting additional authority with respect to interstate and foreign commerce in wire and radio communication, there is created a commission to be known as the "Federal Communications Commission", which shall be constituted as hereinafter provided, and which shall execute and enforce the provisions of this chapter.")
4. *See supra* n. 3
5. *Id.; see also generally* Federal Radio Act of 1927, Pub. L. 632, 47 U.S.C. § 81-83 (1927) (repealed 1934).
6. *See generally*, Nat. Broad. Co. v. United States, 319 U.S. 190 (1943) (*National Broadcasting*) (noting that the FCC's purpose under the 1934 Telecommunications Act was to "make available, so far as possible, to all the people of the United States a rapid, efficient, Nationwide, and world-wide wire and radio communication service with adequate facilities at reasonable charges."); United States v. Sw. Cable Co., 392 U.S. 157 (1968) (noting the Commission's "broad responsibilities for the orderly development of an appropriate system of local television broadcasting"); James S. Rivers, Inc. v. FCC, 351 F.2d 194 (D.C. Cir. 1965) (denial of request to operate a radio station at night as well as daytime is not abuse of discretion).
7. *See National Broadcasting*, 319 U.S. at 215-16. ("The Act itself establishes that the Commission's powers are not limited to the engineering and technical aspects of regulation of radio communication. Yet we are asked to regard the Commission as a kind of traffic officer, policing the wave lengths to prevent stations from interfering with

each other. But the Act does not restrict the Commission merely to supervision of the traffic. It puts on the Commission the burden of determining the composition of that traffic. The facilities of radio are not large enough to accommodate all who wish to use them. Methods must be devised for choosing from among the many who apply. And since Congress itself could not do this, it committed the task to the Commission.")

8. Censorship, 47 U.S.C. § 326 (2005) ("Nothing in this chapter shall be understood or construed to give the Commission the power of censorship over the radio communications or signals transmitted by any radio station. . . .")
9. KFKB Broad. Ass'n v. Fed. Radio Comm'n. (*KFKB*), 47 F.2d 670, 672 (D.C. Cir. 1931) ("In considering the question whether the public interest, convenience, or necessity will be served by a renewal of appellant's license, the commission has merely exercised its undoubted right to take note of appellant's past conduct, which is not censorship.")
10. *See Id.*; Licenses: Grant, 47 U.S.C. § 307(a) (2005) ("The Commission, if public convenience, interest, or necessity will be served thereby, subject to the limitations of this chapter, shall grant to any applicant therefore a station license provided for by this chapter."); Application for License, 47 U.S.C § 309(a) (2005) ("[t]he Commission shall determine, in the case of each application filed . . . , whether the public interest, convenience, and necessity will be served by the granting of such application, and, if the Commission, on examination of such application and on consideration of such other matters as the Commission may officially notice, shall find that public interest, convenience, and necessity would be served by the granting thereof, it shall grant such application.")
11. U.S. v. American Bond & Mortgage Co., 31 F.2d 448, 453 (N.D. Ill. 1929).
12. *See KFKB*, 47 F.2d at 672 (Court affirmed the Commission's denial of a license renewal; finding that the Commission's consideration of the appellant's past conduct while determining "whether the public interest, convenience, or necessity will be served by renewal of appellant's license" was not censorship.); *see also* Trinity Methodist Church, South v. Fed. Radio Comm'n., 62 F.2d 850, 852 (D.C.Cir 1932) (The Court affirmed the Commission's denial of a license renewal because broadcasting information "without facts to sustain or to justify them" was not in the public interest. The Court also noted that the Commission was free to "take notice of appellant's conduct in his previous use of the permit" under the public interest standard.)
13. *National Broadcasting*, 319 U.S. at 218 (In an effort to maximize the public interest within the limited number of radio channels, the FCC is given the power to refuse licenses to persons who continue to enter into contracts with network organizations. This practice "limits his ability to make the best use of the radio facility assigned to him.")
14. *See* Purposes, 47 U.S.C. § 521 (2005) ("The purposes of this subchapter are to . . . (2) establish franchise procedures and standards which encourage the growth and development of cable systems and which assure that cable systems are responsive to the needs and interests of the local community; . . . (4) assure that cable communications provide and are encouraged to provide the widest possible diversity of information sources and services to the public; . . . (6) promote competition in cable communications and minimize unnecessary regulation that would impose an undue economic burden on cable systems.") *See also* Congressional Findings and Policy: Cable Television Consumer Protection and Competition Act of 1992, Pub. L. 102-385 § 2(a) (1992) ("The Congress finds and declares the following . . . (16) As a result of the eco-

nomic incentive that cable systems have to delete, reposition, or not carry local broadcast signals, coupled with the absence of a requirement that such systems carry local broadcast signals, the economic viability of free local broadcast television and its ability to originate quality local programming will be seriously jeopardized . . . (21) Cable systems should be encouraged to carry low-power television stations licensed to the communities served by those systems where the low-power station creates and broadcasts, as a substantial part of its programming day, local programming.") *See also National Broadcasting*, 319 U.S. at 217 ("The avowed aim of the Communications Act of 1934 was to secure the maximum benefits of radio to all the people of the United States.")

15. Requirements as to Certain Applications in Broadcasting Service, 47 U.S.C. § 311 (2005); Definitions: Chain Broadcasting, 47 U.S.C. § 153(9) (2005) (defines chain broadcasting as the "simultaneous broadcasting of an identical program by two or more connected stations."); *see also National Broadcasting*, 319 U.S. at 218 ("In essence, the Chain Broadcasting Regulations represent a particularization of the Commission's conception of the 'public interest' sought to be safeguarded by Congress in enacting the Communications Act of 1934. The basic consideration of policy underlying the Regulations is succinctly stated in its Report: 'With the number of radio channels limited by natural factors, the public interest demands that those who are entrusted with the available channels shall make the fullest and most effective use of them. If a licensee enters into a contract with a network organization which limits his ability to make the best use of the radio facility assigned him, he is not serving the public interest. The net effect (of the practices disclosed by the investigation) has been that broadcasting service has been maintained at a level below that possible under a system of free competition. Having so found, we would be remiss in our statutory duty of encouraging 'the larger and more effective use of radio in the public interest' if we were to grant licenses to persons who persist in these practices.'")
16. *National Broadcasting*, 319 U.S. 190.
17. *Id.* at 190, 227 (Justices Black and Rutledge took no part in the consideration or decision of these cases).
18. *Id.* at 224-27.
19. *Id.* These rules upheld by the Supreme Court in 1943 still appear in the Code of Federal Regulations verbatim, despite uneven enforcement by the FCC. *See* Affiliation Agreements and Network Program Practices; Territorial Exclusivity in Non-Network Program Arrangements, 47 CFR 73.658 (2001).
20. *National Broadcasting*, 319 U.S. at 226-27.
21. *Id.* at 213.
22. A license for an available frequency is essentially a license to use a publicly owned resource; therefore, it is not just scarcity *per se*, but scarcity of something that belongs to the public. As a result of public ownership, there are social implications to allocation of available frequencies.
23. *See* Report of the Commission in the Matter of Editorialization by Broadcast Licensees, 13 FCC 1246 (1949) (*Report on Editorialization*) (The Fairness Doctrine is a former FCC rule that required the broadcast media owner to furnish a reasonable opportunity for discussion of conflicting views on issues of public importance. The FCC abandoned the fairness doctrine in 1987); *see also,* Candidates for Public Office, 47 U.S.C. § 315 (2005) (Fairness doctrine as embodied in this statute is described as "Equal opportunities requirement; censorship prohibition; allowance of station use;

news appearances exception; public interest; public issues discussion opportunities"); Administrative Sanctions: Revocation of Station License or Construction Permit, 47 U.S.C. § 312(a)(7) (2005) ("The Commission may revoke any station license or construction permit . . . for willful or repeated failure to allow reasonable access to or to permit purchase of reasonable amounts of time for the use of a broadcasting station, other than a non-commercial educational broadcast station, by a legally qualified candidate for Federal elective office on behalf of his candidacy").

24. Red Lion Broad. Co. v. FCC, 395 U.S. 367, 371 (1969)(*Red Lion*). *See also* Times-Mirror Broadcasting Co., 24 P & F Radio Reg. 404 (1962).
25. In Re Complaint of Representative Patsy Mink, Memorandum Opinion and Order, 59 FCC 2d 987, para. 29 (1976) ("It is our belief, as stated in the Fairness Report, *supra*, that the licensee could not reasonably fail to cover an issue which has tremendous impact within the local service area—that such failure would violate the fairness doctrine. We now reaffirm that principle. Where, as in the present case, an issue has significant and possibly unique impact on the licensee's service area, it will not be sufficient for the licensee as an indication of compliance with the fairness doctrine to show that it may have broadcast an unknown amount of news touching on a general topic related to the issue cited in a complaint. Rather it must be shown that there has been some attempt to inform the public of the nature of the controversy, not only that such a controversy exists."); *see also Report on Editorialization*, 13 FCC at 1249 ("It is axiomatic that one of the most vital questions of mass communication in a democracy is the development of an informed public opinion through the public dissemination of news and ideas concerning the vital public issues of the day. . . . The Commission has consequently recognized the necessity for licensees to devote a reasonable percentage of their broadcast time to the presentation of news and programs devoted to the consideration and discussion of public issues of interest in the community served by the particular station.")
26. *See* 47 CFR 73.123 (repealed 1987); *see also Report on Editorialization*, 13 FCC at 1249.
27. *See* Nat'l Broad. Co. v. FCC, 516 F.2d 1101 (D.C. Cir. 1975); Columbia Broad. Sys., Inc., v. Democratic Nat'l Comm., 412 U.S. 94 (1973).
28. Radio Television News Dirs. Ass'n. v. United States, 400 F.2d 1002 (7th Cir. 1968); Red Lion Broad. Co. v. FCC, 381 F.2d 908 (D.C. Cir. 1967).
29. *Red Lion*, 395 U.S. 367.
30. *Id.*
31. *Id.* at 386-389.
32. *Id.* at 386-387.
33. *Id.* at 389.
34. *See* FCC v. League of Women Voters, 468 U.S. 364, 376 n.11 (1984); *see also* Radio-Television News Directors Assoc. v. FCC, 229 F.3d 269 (D.C. Cir. 2000).
35. *See, e.g.*, Mark S. Fowler and Daniel L. Brenner, *A Marketplace Approach to Broadcast Regulation*, TEXAS LAW REVIEW, 60 Tex. L. Rev. 207, 221-26 (1982).
36. Turner Broadcasting System, Inc. v. FCC, 512 U.S. 622, 638-639 (1994) (*Turner I*) ("The broadcast cases are inapposite in the present context because cable television does not suffer from the inherent limitations that characterize the broadcast medium. Indeed, given the rapid advances in fiber optics and digital compression technology, soon there may be no practical limitation on the number of speakers who may use the cable medium. Nor is there any danger of physical interference between two cable speakers attempting to share the same channel. In light of these fundamental techno-

logical differences between broadcast and cable transmission, application of the more relaxed standard of scrutiny adopted in Red Lion and the other broadcast cases is inapt when determining the First Amendment validity of cable regulation.") *See also*, Action for Children's Television v. F.C.C., 58 F.3d 654, 674 n.8 (D.C. Cir. 1995) Interestingly, in responding to Government's argument that cable and broadcast are alike in that they both are beset by "market dysfunction," the *TBS* Court stated that "the special physical characteristics of broadcast transmission, not the economic characteristics of the broadcast market, are what underlies our broadcast jurisprudence" (citations omitted). Apparently, the Court is now prepared to abandon the *economic* scarcity theory.")

37. *See* Telecomms. Research and Action Ctr. v. FCC, 801 F.2d 501 (D.C. Cir. 1986), *cert. denied*, 482 U.S. 919 (1987); Meredith Corp. v. FCC, 809 F.2d 863 (D.C. Cir. 1987).
38. Pub. L. 104-104, Title II, § 202 (1996), *as amended* Pub. L. 108-199, Div. B, Title VI, §629 (2004). *See also* § 629(a) ("The Commission shall modify section 73.3555 of its regulations by eliminating any provisions limiting the number of AM or FM broadcast stations which may be owned or controlled by one entity nationally."); § 629 (c) (instructing the Commission to eliminate "the restrictions on the number of television stations that a person or entity may directly or indirectly own, operate, or control, or have a cognizable interest in, nationwide" and to increase the "national audience reach limitation for television stations to 39 percent."); § 629 (f) ("The Commission shall revise section 76.501 of its regulations to permit a person or entity to own or control a network of broadcast stations and a cable system.")
39. In the Matter of Review of the Commission's Regulations Governing Attribution of Broadcast and Cable/MDS Interests, Further Notice of Proposed Rulemaking, MM Docket No. 94-150, 11 FCC Rcd. 19,895 (1996).
40. *See* Larry Shaughnessy, FCC: Media Rules Will Change, CABLE NEWS NETWORK, June 2, 2003, *available at* http://www.cnn.com/2003/BUSINESS/06/02/fcc.media/index.html (last visited on Sept. 8, 2005) (this article discusses the merits proposed rules, which would allow "a single media company [to] own enough television stations to reach as much as 45 percent of the U.S. television market, up from the current ceiling of 35 percent. Companies would be allowed to own both television stations and newspapers in all but the smallest markets, and in large markets, individual companies could own several radio and television stations as well as the newspaper and cable outlet.")
41. *See* Who Owns the Media, FREE PRESS, *available at* http://www.freepress.net/content/ownership (last visited March 13, 2007).
42. In broadcast radio, for example, station ownership declined by 25%, from 5,100 owners prior to the 1996 Telecommunications Act to a mere 3,800 owners by the end of 2001. Anastasia Bednarski, From Diversity to Duplication: Mega-Mergers and the Failure of the Marketplace Model Under the Telecommunications Act of 1996, 55 Fed.Comm.L.J. 273, 287 (2003). As a result of consolidation in the radio market, there has been a rapid increase in syndicated programming.
43. *See* Charles Layton, News Blackout, AMERICAN JOURNALISM REVIEW, Jan. 2004, *available at* http://www.ajr.org/article.asp?id=3500 (last visited March 13, 2007); *see also* Robert Philpot, Evil Empires, FORT WORTH STAR TELEGRAM, April 2004, *available at* 2004 WLNR 1855709. *See also infra* n. 58.
44. Prometheus Radio Project v. FCC, 373 F.3d 372 (3d Cir. 2004) (*Prometheus*). Large companies that implement consolidation and synergy techniques are dropping original independent newscasts because they do not maximize corporate profits, opting

instead for national syndicated newscasts. Michael S. Schneider, Local Newscasts Fall Victim To Cost Cuts, Variety 21, (Jan. 28, 2002–Feb. 3, 2002); Dan Trigoboff, CBS Drops News In Detroit, Broadcasting & Cable (Nov. 25, 2002). As a result of increased acquisition of owned and operated stations by networks and fewer affiliate stations, affiliates are unable to exercise their right to preempt under network/affiliate rules. Preemption is a valuable right, since it allows a local affiliate to reject network programming. *Ex Parte* Presentation of NASA/NAB, MB Docket 02-277 at 3-5 (May 9, 2003) [JA5383-85]. Due to their increased power, networks are able to decline special news presentations, such as presidential debates. Paul B. Matay, Abundant Media, Viewer Scarcity: A Marketplace Alternative To First Amendment Broadcast Rights and the Regulation of Televised Presidential Debates, 36 Ind. L.Rev. 101, 127 (2003). Networks have also taken advantage of their consolidated market position to cuts costs of coverage by pooling their resources, resulting in erroneous results and one-sided newscasts. Robert H. Lande, Consumer Choice As The Ultimate Goal Of Antitrust, 62 U.Pitt. L. Rev. 503, 519 (2001).

45. Congress has repeatedly thwarted FCC initiatives to relax television ownership restrictions: in 1984, when, Congress enacted a 25% ownership limitation, effectively overruling then-FCC chairman Mark Fowler's attempt to sunset station ownership caps altogether; in 1996, when Congress maintained an increased ownership cap of 35% for television after deregulating radio ownership; and in 2003, by enacting legislation to raise the cap to 39%, despite an FCC effort to institute a 45% cap. *See* In the Matter of Amendment of Section 73.3444 [formerly Sections 73.35, 73.240, and 73.636] of the Commission's Rules Relating to Multiple Ownership of AM, FM, and Television Broadcast Stations, 100 F.C.C.2d 17 (July 1984) (Congress imposes 25% cap in 1984); Telecommunications Act of 1996, 47 U.S.C.A. § 202 (1996) (Congress maintains 35% cap for TV); In The Matter of 2002 Biennial Regulatory Review—Review of the Commission's Broadcast Ownership Rules and Other Rules Adopted Pursuant to Section 202 of the Telecommunications Act of 1996, 18 FCC Rcd. 13,620 (June 2003) (FCC sets cap at 45%); Prometheus Radio Project v. FCC, 373 F.3d 372, (3rd Cir. 2004) (response to 45% cap); Consolidated Appropriations Act of 2004, Pub. L. No. 108-199, sect. 629, § 202(c)(1)(B), 118 Stat. 3, 99 (2004) (Congress sets cap at 39% while Prometheus is still pending).
46. Children's Television Act of 1990, Pub. L. No. 101-437, 104 Stat 996 (1990).
47. *See generally* Broadcasting Obscene Language, 18 U.S.C. 1464 (1994); FCC v. Pacifica Foundation, 438 US 727 (1978); *see also* Enforcement of 18 U.S.C. 1464, 47 CFR 73.3999 (1995).
48. *See generally* Second Further Notice of Proposed Rulemaking, In the Matter of the Commission's Cable Horizontal and Vertical Ownership Limits, 20 FCC Rcd. 9374 (May 2005); Cable Television Protection and Competition Act of 1992, Pub. L. No. 102-385 (Oct. 1992).
49. *See* Time Warner Entertainment Co. v. FCC, 56 F.3d 151, 179 (D.C. Cir. 1995) ("Experience under the 1984 Cable Act's deregulatory regime led Congress to enact the 1992 Cable Act. . . . In 1992, Congress found that the deregulated cable industry had become the 'dominant nationwide video medium,' serving 'over 60 percent of the households with televisions.' The legislative record also showed that the industry was "highly concentrated." In addition, Congress found that 'most cable television subscribers have no opportunity to select between competing cable systems,' and that '[t]he result is undue market power for the cable operator as compared to that of con-

sumers.' . . . Congress also found that the average monthly cable rate had increased 'almost 3 times as much as the Consumer Price Index since rate deregulation.' Congress concluded that rate reregulation was necessary to ensure that cable operators would not exercise 'undue market power vis-à-vis video programmers and consumers.'") (quoting *1992 Cable Act* §2) (internal citations omitted). *See also* Turner Broadcasting System Inc. v. FCC, 520 U.S. 180, 197 (1997) (*Turner II*) ("Evidence indicated the structure of the cable industry would give cable operators increasing ability and incentive to drop local broadcast stations from their systems, or reposition them to a less-viewed channel.")

50. Carriage of Local Commercial Television Signals, 47 U.S.C. § 534 (2003); Carriage of Non-Commercial Educational Television, 47 U.S.C. § 535 (2003) (original version at Pub. L. 102-385; § § 4, 5; 106 Stat. 1471, 1477) (Congress enacted the Cable Television Consumer Protection Act of 1992 on October 5, 1992. This Act contained two provisions that require cable operators to carry local broadcast station signals without charge at the broadcaster's election. These provisions, sections 614 and 615 of the *1992 Cable Act*, apply to the signals of commercial and noncommercial educational stations respectively. Together, these sections are known as the "must-carry" rules.)
51. *Turner II*, 520 U.S. at 224-225 (The Supreme Court held that must-carry regulations were content-neutral and narrowly tailored to advance Congress' interests in preserving the benefits of free, over-the-air local broadcast television, promoting the widespread dissemination of information from a multiplicity of sources, and promoting fair competition in market for television programming.)
52. *See generally* Pub. L. 102-385 §2 (1992) (*1992 Cable Act*); *1992 Cable Act* § 2(a)(4) ("The cable industry has become highly concentrated. The potential effects of such concentration are barriers to entry for new programmers and a reduction in the number of media voices available to consumers."); *Id.* § 2(a)(6) ("There is a substantial governmental and First Amendment interest in promoting a diversity of views provided through multiple technology media.")
53. *See generally 1992 Cable Act* § 2 (1992).
54. *See generally Id.*
55. Prime Time Rules, 47 C.F.R. § 73.658 (j)(ii) (1994) (*Fin-Syn Rules*).
56. *See, e.g.* Ben Bagdikian, Grand Theft: The Conglomeratization of Media and the Degradation of Culture, MULTINATIONAL MONITOR, Volume 26, Issue ?, 2005 WLNR 4970719 (2005) ("For 25 years, a handful of large corporations that specialize in every mass medium of any consequence has dominated what the majority of people in the United States see about the world beyond their personal experience. These giant media firms, unlike any in the past, thanks to the hands-off attitude of the Federal Communications Commission (FCC) majority, are unhampered by laws and regulation. In the process, they have been major agents of change in the social values and politics of the United States.")
57. Hey Bud, Cut Us a Slice, FT REPORTS, February 5, 2002, 2002 WLNR 10635428; Independent Production Companies, THE MUSEUM OF BROADCAST COMMUNICATIONS, *available at* http://www.museum.tv/archives/etv/I/htmlI/independentp/independentp.htm (last visited on March 13, 2007).
58. The synergy that results from internal production, initial distribution, syndication, and repurposing allows these consolidated media giants to gain carriage on cable systems, enabling the parent corporations of broadcasters to capture a large share of the non-broadcast video market. Hearing on Concentration of Media Ownership Before

the S. Comm. On 43 Commerce, Sci. and Transp., 108th Cong., Oct. 2, 2003 (Testimony of Dr. Mark Cooper) *available at* http://commerce.senate.gov/hearings/testimony.cfm?id=950&wit_id=2679 (last visited March 13, 2007). For example, the General Electric merger permitted profitable "cross-promotion across the new company's extensive cable and broadcast networks," allowing the company to use its broadcast and cable properties to promote company business interests, rather than the goals of localism and diversity. General Electric Co., *available at* http://www.ge.com/pr/display.php?highlight=true&id=1443&keyword=vivendi (last visited March 13, 2007).

The vast consolidation of these media outlets results in efforts of cross-promotion, synergy and cost-saving techniques, which fail to promote the Commission's goals of localism and diversity. For example, independent local news stations are being acquired by conglomerates and managed by their local affiliates, effectively removing any independent local news coverage. Bill Moyers, On Big Media, Truthout, Oct. 10, 2003, *available at* http://www.globalpolicy.org/socecon/tncs/mergers/2003/1010bigmedia.htm (last visited March 13, 2007. Original programming has given way to syndicated programming, which is also increasingly owned by the conglomerate that controls the licensee. Edward D. Cavanagh, Symposium: Are New Media Really Replacing Old Media? Broadcast Media Deregulation and the Internet: De-Regulation of the Air Waves: Is Antitrust Enough?, 17 St. John's J.L. Comm 67, 72 n.42 (2003)(citing Polly Higgins, Tucson Radio Making Waves: Corporate Radio Moves In, Tucson Citizen May 3, 2002, at A1). Additionally, vertical integration allows conglomerates to own much of the content they air, making it more likely for them to exclude local content, which they do not have a financial stake in. FCC Broadcast Localism Hearing, Charlotte, North Carolina, Oct. 2003, *available at* http://hraunfoss.fcc.gov/edocs_public/attachmatch/DOC-242307A1.pdf (last visited March 13, 2007).

59. To some extent, Congress has exercised that right, using the scarcity argument, but not the public fiduciary argument, to justify specific burdens on broadcasters in the Children's Television Act, indecency legislation, and the repeated imposition of ownership caps. *See supra* n. 45-48.
60. License for Radio Communication on Transmission of Energy, 47 U.S.C. § 301 (2005).
61. Some deregulation advocates contend that onerous government regulation of broadcasters amounts to an unconstitutional taking of private property. *See, e.g.*, Satellite Broadcasting and Communications Ass'n v. F.C.C. 275 F. 3d 337 (4th Cir. 2001). Such a position is not supportable under the FCA, which requires licensees to waive any claim to a property interest in the broadcast spectrum. Waiver by License of Claims to Particular Frequency or of Electromagnetic Spectrum, 47 U.S.C. § 304 (2005) ("No station license shall be granted by the Commission until the applicant therefor shall have waived any claim to the use of any particular frequency or of the electromagnetic spectrum as against the regulatory power of the United States because of the previous use of the same, whether by license or otherwise.") *See also* F.C.C. v. Sanders Bros. Radio Station, 309 U.S. 470, 475 (1940). "The policy of the Act is clear that no person is to have anything in the nature of a property right as a result of the granting of a license."
62. *See generally Turner II*, 520 U.S. 180; Reno v. American Civil Liberties Union, 521 US 844 (1997) (*Reno*); McConnell v. FEC, 540 US 93 (2003).
63. *See e.g.*, FOX TV Stations, Inc. v. FCC, 280 F.3d 1027, 1046 (D.C. Cir. 2002) ("contrary to the implication of the networks' argument, this court is not in a position to reject the scarcity rationale even if we agree that it no longer makes sense. The Supreme

Court has already heard the empirical case against that rationale and still "declined to question its continuing validity."" (citing *Turner I*); Communications Corp. v. FCC, 835 F.2d 292, 294-295 (D.C. Cir. 1987) ("Wire-carried media like cable, of course, have no such limitations, and thus we found the "scarcity rationale" that the Supreme Court has used to justify broadcast television regulations to offer no succor to those seeking to establish the constitutional validity of cable television regulations.")

64. Candidate for Public Office, 47 U.S.C. § 315 (2002); Administrative Sanctions, 47 U.S.C. § 312 (2004).
65. Miami Herald v. Tornillo, 418 U.S. 241 (1974) (*Tornillo*). The Supreme Court, in the 1969 *Red Lion* decision, validated the constitutionality of a limited right of access for broadcast media, the FCC, and some courts question the continued validity of that decision because the Court based its right of access largely on the issue of scarcity. Four years later, the Supreme Court struck down a state's right of reply statute, *Tornillo*, again choosing to focus on scarcity—or the lack of it—as the pivotal issue in its review of the newspaper industry. *Id.* at 256-257.
66. *Id.* at 257.
67. *See* Pittsburgh Press v. Pittsburgh Commission on Human Relations, 413 U.S. 376, 387-388 (1973) (*Pittsburgh Press*).
68. *Tornillo*, 418 U.S. 241, 258-259 (1974) (Brennan, J., Concurring).
69. In Mt. Mansfield Television, Inc. v. FCC, for example, the Second Circuit alludes to incumbent broadcasters' market dominance as "the fruit of a preferred position conferred by the Government," stating that "(l)ong experience in broadcasting, confirmed habits of listeners and viewers, network affiliation, and other advantages in program procurement give existing broadcasters a substantial advantage . . ." Mt. Mansfield Television, Inc. v. F. C. C. 442 F.2d 470, 477 (1971) (citing *Red Lion*).
70. *See. e.g.* Time Warner Entertainment Co., L.P., v. F.C.C., 93 F.3d 957 (D.C. Cir.1996), *see infra* n.163.
71. The government grants access in the form of spectrum, bandwidth, and other valuable media.
72. *See supra* n. 2.
73. *See* Erie Telecommunications v. City of Erie, 853 F.2d 1084, 1094-1095 (3d Cir. 1988) (*Erie*).
74. *Turner II*, 520 U.S. 180.
75. *See Erie*, 853 F.2d 1084, 1096 (holding that as a matter of contract, a company can waive constitutional rights, including fundamental First Amendment rights that would otherwise be subject to a level of constitutional scrutiny).
76. First Nat. Bank of Boston v. Bellotti, 435 U.S. 765, 822 (1978) (Rehnquist, J., dissenting).
77. *Id.* at 827 (Rehnquist, J., dissenting).
78. *Id.*
79. *See generally Erie*, 853 F.2d 1084 at 1084.
80. *Id.* at 1090 (recognizing a party's freedom to contract).
81. *Id.* at 1096.
82. D.H. Overmyer Co. v. Frick Co., 405 U.S. 174 (1972) (*Overmyer*).
83. *Id.* at 187-188.
84. Fuentes v. Shevin, 407 U.S. 67 (1972) (*Fuentes*).
85. An adhesion contract is "[a] standard-form contract prepared by one party, to be signed by the party in a weaker position, usu. a consumer, who adheres to the contract

with little choice about the terms." *Contract*, n.8 *Adhesion Contract*, BLACK'S LAW DICTIONARY (Bryan A. Garner et al. eds., 8th ed. 2004).

86. *Id.* at 95-96.
87. *Erie*, 853 F.2d at 1096 (citing *Fuentes*, 407 U.S. at 95).
88. *Fuentes*, 407 U.S. at 95.
89. *See Erie*, 853 F.2d at 1096 (citing *Fuentes*, 407 U.S. at 95).
90. *Fuentes*, 407 U.S. at 95.
91. *See Erie*, 853 F.2d at 1096.
92. *Id.*
93. *Id.* at 1097
94. *Fuentes*, 407 U.S. at 70, 95.
95. *Overmyer*, 405 U.S. at 186.
96. *See Erie*, 853 F.2d at 1096.
97. *Id.* at 1096-97 (noting that although contractual waiver issues were not raised in the lower court, the failure to raise those issues should not have any bearing on determining whether or not a party waived its constitutional rights).
98. *See* Erie Telecomms., Inc. v. Erie, 659 F. Supp. 580, 605 (D. Pa. 1987).
99. *Id.* at 586.
100. *Id.* at 585. Because he views lack of actual knowledge as a bar to waiver, Judge Mencer relies instead on the *O'Brien* test of intermediate scrutiny to determine whether the access requirements were constitutional noncontent regulation. In turning to *O'Brien*, the district court engages the First Amendment implications of government access requirements in a manner that has, in the years since the *Erie* case, become the more typical approach of courts. *See also*, United States v. O'Brien, 391 U.S. 367 (1968) (*O'Brien*).
101. *Erie*, 853 F.2d at 1095-1096 (citing *Fuentes* and *Overmyer*).
102. *See Id.* at 1084.
103. Chicago Cable Commc'ns v. Chicago Cable Comm'n, 879 F.2d 1540 (7th Cir. 1989) (*Chicago Cable*).
104. *Id.* at 1548 n.6.
105. *See Id.*
106. *Id.*
107. *Id.*
108. *Id.*
109. *See Id.* at 1550.
110. *See Id.* at 1550-1551.
111. *Fuentes*, 407 U.S. 67.
112. *Overmyer*, 405 U.S. 364
113. *Compare Chicago Cable*, 879 F.2d at 1548 n.6 *with Erie*, 853 F.2d at 1101.
114. Although the *1996 Telecommunications Act* prohibits exclusive franchising by municipalities, permitting alternative cable providers to "overbuild" rights of way, such overbuilding is rare on account of cost and market saturation. Efforts to overbuild even in large markets, such as New York, Boston, Washington, and other cities, have failed to make an alternative cable provider competitive in those markets. *See* K.C. Neel, Fresh Start for RCN, MULTICHANNEL NEWS, Volume 26, Issue 16, 2005 WLNR 6152620 (April 18, 2005).
115. As this chapter goes to press, the U.S. Senate is considering the *Ensign Bill*, which would preempt municipalities from entering into these negotiations by making cable

franchising a matter of federal law. In the event that the *Ensign Bill* becomes law, the federal government would be responsible for negotiating quid pro quos with cable providers. By placing cable franchising under federal control, the relationship between cable companies and the government would be more like the relationship between broadcasters and the government, which has always been governed exclusively by federal law. Broadband Investment and Consumer Choice Act, S.1504, 109 th Cong. (2005).

116. *See Erie*, 853 F.2d at 1101.
117. *Id.* at 1096.
118. *See supra* n. 96.
119. *Chicago Cable*, 879 F.2d at 1548.
120. *Erie*, 853 F.2d at 1095-97.
121. *Fuentes*, 407 U.S. at 95.
122. *Red Lion*, 395 U.S. at 389.
123. On this point, see Rust v. Sullivan, which held that the government could impose conditions on recipients of a government benefit without violating the recipients' first amendment rights. Rust v. Sullivan, 500 U.S. 173, 193-198 (1991).
124. *See Turner II*, 520 U.S. at 189; *see also O'Brien*, 391 U.S. at 376-377 ("[A] government regulation is sufficiently justified if it is within the constitutional power of the Government; if it furthers an important or substantial governmental interest; if the governmental interest is unrelated to the suppression of free expression; and if the incidental restriction on alleged First Amendment freedoms is no greater than is essential to the furtherance of that interest.")
125. The fourth prong of the *O'Brien* test requires that the "incidental restriction on alleged First Amendment freedoms is no greater than is essential to the furtherance of that interest." *O'Brien*, 391 U.S. at 376. The *Turner* court found that the must-carry provisions met the fourth prong of the *O'Brien* test "[b]ecause the burden imposed by must-carry is congruent to the benefits it affords . . . [and] . . . must-carry is narrowly tailored to preserve a multiplicity of broadcast stations for the 40 percent of American households without cable." *Turner II*, 520 U.S. at 215-16. In applying the fourth prong, the *Turner II* court considered earlier precedent. *Id.* at 216. ("[T]he essence of narrow tailoring" is "focusing on the evils the [Government] seeks to eliminate . . . [without] significantly restricting a substantial quantity of speech that does not create the same evils") (citing Ward v. Rock Against Racism, 491 U.S. 781, 799, n.7 (1989).
126. *See Turner II*, 520 U.S. at 189, 215-216 ("Because the burden imposed by must-carry is congruent to the benefits it affords, we conclude must-carry is narrowly tailored to preserve a multiplicity of broadcast stations for the 40 percent of American households without cable.") *See also Chicago Cable*, 879 F.2d at 1549.
127. *See generally Turner II*, 520 U.S. 180 (upholding a city's "must-carry" provisions explaining that the ordinances did not violate the First Amendment).
128. *Chicago Cable*, 879 F.2d at 1549.
129. *Id.* at 1548.
130. *Compare Id.* at 1548-49 *with Turner II*, 520 U.S. at 189.
131. Board of Trustees of State Univ. of N.Y. v. Fox, 492 U.S. 469 (1989); Clark v. Community for Creative Nonviolence, 468 U.S. 288 (1984).
132. *Chicago Cable*, 879 F.2d at 1550.
133. *Id.* at 1548-1549; *compare Chicago Cable*, 879 F.2d at 1550 *with Red Lion*, 395 U.S. at 400-401.

134. *Chicago Cable*, 879 F.2d at 1550.
135. *See Id.*
136. *Id.*; *see also* Community Commc'ns Corp. v. City of Boulder 660 F.2d 1370, 1379 (10th Cir. 1981); Berkshire Cablevision of Rhode Island v. Burke 571 F. Supp 967, 986-87 (1st Cir. 1985).
137. *See Chicago Cable*, 879 F. 2d at 1550 (holding that the city's ordinance requiring time for local programming was valid because of the "virtual monopoly" over television media); *Turner II*, 570 U.S. at 215-16 (holding that the "must-carry" provisions were valid because the burden imposed was proportionate to the benefit afforded).
138. *Compare Chicago Cable* 879 F.2d at 1550 *with Turner II*, 520 U.S. at 185.
139. *Turner II*, 520 U.S. 180 (1997).
140. *Id.*; *see also O'Brien*, 391 U.S. 367 (1968).
141. *Turner II*, 520 U.S. 180 at 227-228 (1997) (Breyer, J., concurring).
142. *Id.*
143. *Turner II*, 520 U.S. 180 (1997).
144. *See Supra* n.50.
145. *Id.*
146. *Turner II*, 520 U.S. 180, 215-216 (1997) ("Because the burden imposed by must-carry is congruent to the benefits it affords, we conclude must-carry is narrowly tailored to preserve a multiplicity of broadcast stations for the 40 percent of American households without cable . . . Congress took steps to confine the breadth and burden of the regulatory scheme. For example . . . Congress exempted systems of 12 or fewer channels, and limited the must-carry obligation of larger systems to one-third of capacity . . . allowed cable operators discretion in choosing which competing and qualified signals would be carried . . . and permitted operators to carry public stations on unused public, educational, and governmental channels in some circumstances . . .") (citations omitted).
147. Quincy Cable TV, Inc. v. FCC, 768 F.2d 1434 (1985).
148. 47 U.S.C. § 534 (2005); *see Turner II*, 520 U.S. at 215-216 (concluding that the must-carry provisions were narrowly tailored). *See also* the cable public access set aside provisions *infra* n.161.
149. *See Infra* n. 159.
150. *Chicago Cable*, 879 F.2d at 1551.
151. See *Turner II*, 520 U.S. at 186.
152. *Id.* at 213.
153. *Red Lion*, 395 U.S. at 400-401.
154. *Turner II*, 520 U.S. at 226-227 ("[i]n justifying the First Amendment "price" suffered by cable operators and some viewers in implementing the must-carry rules, the court refers to the regulatory scheme's benefits, such as "prevent[ing] too precipitous a decline in the quality and quantity of programming choice for an ever-shrinking non-cable-subscribing segment of the public. . . . This purpose reflects what "has long been a basic tenet of national communications policy," namely, that "the widest possible dissemination of information from diverse and antagonistic sources is essential to the welfare of the public.") (citations omitted).
155. Dale Bickel, How to Apply for a Broadcast Station, FEDERAL COMMUNICATIONS COMMISSION, Mar. 27, 2002, *available at* http://www.fcc.gov/mb/audio/howtoapply.html (last visited March 13, 2007).
156. *Red Lion*, 395 U.S. at 400-401.

157. *See* Shaughnessy, *supra* n. 40.
158. *Turner II*, 520 U.S. 180 (1997).
159. *See* Modulation: Multiplexing, SEARCHNETWORKING.COM, *available at* http://searchnetworking.techtarget.com/sDefinition/0,,sid7_gci212586,00.html (last visited on March 13, 2007). *See also* Anthony E. Varona, Changing Channels and Bridging Divides, MINNESOTA JOURNAL OF LAW, SCIENCE & TECHNOLOGY, 6 Minn. J. L. Sci. & Tech. 1 (2004); 47 U.S.C. § 534 (b)(3)(A) (2000); TV Stereophonic Aural and Multiplex Subcarrier Operation, 47 C.F.R. § 73.669(b) (2005); In re Carriage of Transmission of Digital Broad. Stations, 13 F.C.C.R. 15092, 15125 (1998).
160. *Id.*
161. §532(b)(1) of the *1992 Cable Act* scales channel set asides into four tiers, as follows: "(A) An operator of any cable system with 36 or more (but not more than 54) activated channels shall designate 10 percent of such channels . . . not otherwise required for use . . . .
(B) An operator of any cable system with 55 or more (but not more that 100) activated channels shall designate 15 percent of such channels . . . not otherwise required for use. . . .
(C) An operator of any cable system with more than 100 activated channels shall designate 15 percent of all such channels.
(D) An operator of any cable system with fewer than 36 activated channels shall not be required to designate channel capacity for commercial use by persons unaffiliated with the operator, unless the cable system is required to provide such channel capacity under the terms of a franchise in effect on the date of the enactment of this title [Oct. 30, 1984]." Cable Channels for Commercial Use: Designation of Channel Capacity for Commercial Use, 47 U.S.C. §532(b)(1).
162. Direct Broadcast Satellite (DBS) Service Obligations, 47 U.S.C. § 335(b)(1)(2005). Although the statute does not expressly scale the DBS set aside obligation, Congress intended that the percentage of capacity to be allocated to noncommercial educational or informational programming should be considered in light of the total channel capacity of the DBS system in question, with larger capacity systems required to reserve a higher percentage of capacity than smaller DBS systems. *See* S. Rep. No. 92, 102d Cong., 2d Ses. 4 at 92 (1991).
163. Time-Warner Entertainment Co., L.P., v. F.C.C., 93 F.3d 957 (D.C. Cir.1996)
164. *See* Final Report of Advisory Committee on Public Interest Obligations of Digital Television Broadcasters (1998), *available at* http://www.ntia.doc.gov/pubintadvcom/piacreport.pdf (last visited March 13, 2007); *see also* The Gore Commission, THE MEDIA INSTITUTE (2002), *available at* http://www.mediainstitute.org/gore/gc1.html (last visited March 13, 2007).
165. Digital Television Broadcast Stations, 47 C.F.R. § 73.624 (g) (2005) ("Commercial and noncommercial DTV licensees must annually remit a fee of five percent of the gross revenues derived from all ancillary or supplementary services, as defined by paragraph (b) of this section.")
166. "First-Come First-Served" is the process of filling applications on a first come, first served basis. "Random Selection" is the process of placing applications meeting specifications in a random order, usually by a computer program, and then filling the applications in that random order. Judge Spatt, Court Declines to Enjoin Cable TV Operator's Changes to Public Access Channel Time Slots, N.Y.L.J. 24 (2005). "Lottery" is the process of distributing tickets to all applicants and then granting

applications to those who hold the winning number/s, which are drawn at random. *Lottery*, BLACK'S LAW DICTIONARY (Bryan A. Garner et al. eds., 8th ed. 2004).

167. *See e.g.*, FCC v. National Citizens Committee for Broadcasting, 436 U.S. 775, 777 (*National Citizens*) ("the FCC made a rational judgment in concluding that the need for diversification was especially great in cases of local monopoly . . . the Supreme Court held that the challenged regulations, which were designed to promote diversification of the mass media as a whole, were properly based on the "public interest" standard . . ." *See also*, *Id.* at 780 ("in setting its licensing policies, the Commission has long acted on the theory that diversification of mass media ownership serves the public interest by promoting diversity of program and service viewpoints, as well as by preventing undue concentration of economic power.")
168. *See generally Erie*, 853 F.2d 1084.
169. *Tornillo* 418 U.S. 241 (In 1974, the court held that a Florida statute requiring newspapers to provide right of reply for political candidates unconstitutional because newspapers are not a scarce resource. However due to consolidation of media outlets in recent years, and the scarcity of major newspapers even in big cities, it is conceivable that the court *would* include a conglomerate's newspaper entities, in addition to its broadcast and cable outlets, as part of an entire corporate media package from which the conglomerate is reaping a benefit.)
170. *O'Brien*, 391 U.S. 367 (holding that where a regulation furthers a substantial government interest, provided that it is narrowly tailored, the limitation on speech is constitutional).
171. *Red Lion*, 395 U.S. at 392-393 (In upholding the Fairness Doctrine, the *Red Lion* court stated, "station owners and a few networks would have unfettered power to make time available only to the highest bidders, to communicate only their own views on public issues, people and candidates, and to permit on the air only those with whom they agreed." Additionally the court agreed with station owners' arguments that "if political editorials or personal attacks will trigger an obligation in broadcasters to afford the opportunity for expression to speakers who need not pay for time and whose views are unpalatable to the licensees, then broadcasters will be irresistibly forced to self-censorship and their coverage of controversial public issues will be eliminated or at least rendered wholly ineffective.")
172. *Id.* at 400 (The *Red Lion* court recognized the advantages of viewer loyalty and network affiliation as the "fruit of a preferred position conferred [to the broadcasters] by the government.")
173. *National Broadcasting*, 436 U.S. 775 at 801-802.
174. *Id.* at 779, 783-784
175. Nat. Citizens Comm. for Broad. v. FCC, 567 F.2d 1095 (1977).
176. *See generally*, Final Written Expression: Parol or Extrinsic Evidence, UCC § 2-206 (2003).
177. *Turner II*, 520 U.S. 180 (1997).
178. *Pittsburgh Press*, 413 U.S. 376.
179. *Id.*
180. *Id.* at 386 ("Under some circumstances, at least, a newspaper's editorial judgments in connection with an advertisement take on the character of the advertisement and, in those cases, the scope of the newspaper's First Amendment protection may be affected by the content of the advertisement.") *See also Id.* at 391 ("we reaffirm unequivocally the protection afforded to editorial judgment and to the free expression of views on

these and other issues, however controversial. We hold only that the Commission's modified order, narrowly drawn to prohibit placement in sex-designated columns of advertisements for nonexempt job opportunities, does not infringe the First Amendment rights of Pittsburgh Press.")

181. *Id.* 386-387.

182. *Tornillo*, 418 U.S. at 257 ("Faced with the penalties that would accrue to any newspaper that published news or commentary arguably within the reach of the right-of-access statute, editors might well conclude that the safe course is to avoid controversy.")

183. In the Matter of Inquiry into Section 73.1910 of the Commission's Rules and Regulations Concerning the General Fairness Doctrine Obligations of Broadcast Licensees, 102 FCC.2d 142 (1985) ("On basis of a comprehensive record, Commission determined the fairness doctrine in operation thwarts laudatory purpose it is designed to promote. Instead of furthering discussion of public issues, Commission found the doctrine inhibits broadcasters from airing controversial issue programming. In addition, Commission expressed concern that administration of fairness doctrine has unintentionally resulted in stifling viewpoints which may be unorthodox, unpopular or unestablished.")

184. *Id.* ("The fairness doctrine, as developed by the Commission, imposes on broadcasters a two-pronged obligation. Broadcast licensees are required to provide coverage of vitally important controversial issues of interest in the community served by the licensees and to provide a reasonable opportunity for the presentation of contrasting viewpoints on such issues."); *Tornillo*, 418 U.S. at 258 ("The choice of material to go into a newspaper, and the decisions made as to limitations on the size and content of the paper, and treatment of public issues and public officials—whether fair or unfair—constitute the exercise of editorial control and judgment. It has yet to be demonstrated how governmental regulation of this crucial process can be exercised consistent with First Amendment guarantees of a free press as they have evolved to this time. Accordingly, the judgment of the Supreme Court of Florida is reversed.")

185. *See Chicago Cable*, 879 F.2d at 1550 (holding that the city's ordinance requiring time for local programming was valid because of the "virtual monopoly" over television media); *Turner II*, 520 U.S. at 215-16 (holding that the "must-carry" provisions were valid because the burden imposed was proportionate to the benefit afforded).

186. *See Erie*, 853 F.2d at 1097 (holding the cable company's waiver of its free speech rights was valid because the party was advised by counsel and the parent cable company had engaged in numerous other transactions).

187. *See* FCC Localism Task Force Mission Statement, available at http://www.fcc.gov/localism/ (last visited on March 13, 2007).

V

ETHICAL AND ACCESS ISSUES

16

MODEL MEDIA FIRMS

Kenneth Harwood

We ask media firms to behave ethically as they go about their work. What we ask might lead to some theory of ethics, theory of the firm, and theory of the media firm. We could build one or more model media firms from theory, compare and contrast what we see in some operating media firm with a model, and to the extent that we see differences between an operating firm and a model firm, perhaps find ways to bring the operating firm closer to the model. To this end, some specifications are sketched, and an operational transformation of them is indicated. A few implications are suggested for deregulation, consolidation, and globalization of model media firms.

MODEL MEDIA FIRMS

Concerns such as deregulation, consolidation, and globalization mark the work of many firms. Potential for decline in competition of ideas seems to be a concern marking the work of media firms particularly.

Preliminaries

1. *Aims.* To meet these concerns, one might look to some behavioral obligations of firms. An aim is to offer ways of perceiving some obligations of licensed media and unlicensed media firms, as well as media firms of public investors, private ones, and jointly public and private investors.
2. *Limitations.* The main attention is directed to recent years in the United States.
3. *Assumptions.* The view here is Western, and largely that in the United States, while honoring and respecting other views.
4. *Individual.* A natural individual is distinguished from a fictive one such as a firm.
5. *Firm.* A firm is an assuror of an economic good or service.

Theories

Philosophy

A central moral view is that developing the individual is the heart of what we should do in general, and especially through the media. Kant's (1956) practical imperative is this: "Act in such a way that you always treat humanity, whether in your own person or in the person of any other, never simply as a means, but always at the same time as an end." Rawls (Rawls & Kelly, 2001) offered a view of justice as fairness. Kant's practical imperative and Rawls' justice as fairness suggested that a media firm might benefit individuals and their societies through just behavior.

Firm

Behavioral theories of the firm ordinarily are of two broad kinds. One seeks to describe and predict behavior, often with the help of quantification. The other seeks to specify ideal behavior of a firm or preferred behavior. Penrose (1959) offered a theory of the firm Koenig (1993) reviewed several theories of the firm, and Jensen (2000) extended a neoclassical view of the firm.

The smallest firms put ownership and management in one person or a few persons. Larger firms often put ownership separate from management. Owners often seek maximal returns to the risks of their financial investments in a firm, whereas managers and nonsupervisory employees often seek maximum compensation for the risks of their work.

Owners or employees or both often adapt to wants and needs of customers, suppliers, creditors, neighboring publics, distant publics, allied firms, states, and others. Adapting usually is in hope of continued returns or larger returns to ownership. Real firms and ideal firms differ widely in the priority each kind of firm places on internal and external constituents.

Apart from the constituents of a firm are its parties, or its parties of interest. A party represents an oppositional group such as a collective bargaining unit or a competitor. Firms interest themselves in the kinds of parties and others such as possible candidates for purchase by the firm or merger with the firm. A party might also be a constituent such as employees who engage in collective bargaining.

A business alliance embraces firms seeking common benefits through business cooperation. A business alliance of large airline firms, for example, is to share passengers and passenger facilities to the mutual benefit of the firms. Each firm is an ally of the others. Cooperatives, nonprofit organizations, nongovernmental organizations, and state-owned or operated enterprises may be seen as firms.

An aside here is that philanthropy by firms is permitted and encouraged in the United States. Following a judicial decision in 1969, firms-for-profit are able to give reasonable amounts to charitable, religious, or civic organizations, considering net assets of the firm, and up to a limit to take tax deductions for some of the gifts. The gifts often are viewed as expenditures to further the purposes of business (Hamilton, 1991).

Media Firm

A media firm is the same as any other firm in most ways. A media firm differs from other firms politically in giving special attention and other resources to broadening the competition of ideas. Usual economic specifications of media firms are described in textbooks such as those by Picard (1989), Albarran (1996), and Doyle (2002). The United States protects media firms, among others, through the Constitution, which provides in its First Amendment that “Congress shall make no law . . . abridging the freedom of speech, or of the press. . . .” (Hickey et al., 1878). Perhaps an obligation of media firms, if not others, is to exercise and defend that special protection through offering a diversity of ideas.

Models

A Just Media Firm

Fair often is taken in ordinary language to mean impartial and honest. A just media firm is fair to its internal constituents, including its owners, managers, and nonsupervisory employees. A just media firm is fair to its external constituents such as collective bargaining units, the state, competitors, and candidates for purchase or merger.

Optimization of return to the firm often is an aim of a just media firm. An ideally just firm optimizes both return to the firm and the satisfaction of its constituents. In practice each constituent might have less than a complete array of satisfactions; yet the array as a whole is satisfying, if not entirely so. An ideal array of satisfactions is full and complete. A just media firm seeks to provide a fair share of satisfactions to all constituents. Comparing actual media firms with ideally just firms might suggest some ways in which to create more nearly just firms.

An Optimally Regulated Firm

An optimally regulated media firm makes just allotment of resources with minimal regulation by states, while optimizing return to the firm. A just media firm is regulated optimally if its inputs and outputs are regulated minimally. Minimal regulation of a just media firm is marked by special attention to minimizing regulation of the offering of diverse ideas by the firm.

Self-Regulation. Self-regulation means regulation of itself by a firm. Firms in general and media firms alike seem to prefer self-regulation to other regulation. Usually the administrative costs of self-regulation are low. Penalties often seem to be light or absent, whereas the power of a state to compel is not apparent.

Social Regulation. Social regulation, too, is less expensive and less compelling than regulation by a state. Social regulation ordinarily rests on a basis of general approval or disapproval mixed with greater or lesser demand for the output of a firm. Firms, including media firms, generally seem to prefer self-regulation, and then social regulation, to regulation by a state. Administrative costs and power to compel appear to put regulation by the state as the lowest preference of firms in general and media firms in particular.

Global Scale and Scope

Global scale of a firm often tends to lower costs and raise returns to a firm, beyond those of national or subnational scale. Media firms seem to share in this phenomenon. The cost of reporting an event, for example, may be divided among more recipients as more receive the same report. A similar phenomenon seems likely when a report is recycled or repurposed or revised to appear in two or more media, thus adding economy of scope to that of scale. In media firms as well as in others, some economies of scope are obtained in delivering diverse products or services from the same firm.

Neither vertical nor horizontal integration substitutes for economies of scale or economies of scope (Chandler & Hikino, 1994). Vertical integration usually refers to combining in one firm the work of different steps in production, instead of assigning different steps to different firms. For example, a newspaper might acquire another firm for its printing plant instead of relying on another firm to print what the acquiring firm readies for printing. Horizontal integration ordinarily means that two firms offering the same stage of production combine as one firm. Two weekly newspapers without a printing plant might merge, for example.

Local Scale and Scope

Local scale and scope bring to the fore some essentials of all firms, and of media firms especially. Local scale of firms tends to rise and fall with local population and depopulation of places. An example might be that a small place has perhaps one or two small retail establishments, whereas a large place has many shops, small and large. A very small place has no shop. Local scope refers to several different products being produced by the same firm. A long-time example in print media is the production of local billheads, handbills, and the like by a newspaper print shop. A local radio station might produce and market compact discs of local musicians fairly easily and cheaply. Firms in general and media firms in particular, tend to continue longer with economies of scale and scope than without such economies.

Local media firms, too, follow a pattern of rising and falling with population and depopulation. Some places are too small to sustain one local medium, although such places are served by media from larger places. Media from a large place tend to offer little or no content about the very small place.

Few places are for long stable in population. Most rise and fall with local opportunities for employment. Loss of local media to media in larger places tends to occur when local employment falls and depopulation occurs. Profit-seeking, unlicensed media such as local newspapers tend to disappear earlier than local media such as radio broadcasting stations, which are licensed. A local cable system goes dark when population density in an area falls below an economic minimum, typically taking with it one or more local community channels. Satellite delivery of local commercial radio and television seems to be in its infancy, as does delivery of local noncommercial media. A local cinema often goes dark before a sole radio station departs. Lest depopulation be taken as unworthy of attention, a recent study noted one in four of 3,141 rural counties to be depopulating, and 232 counties to be depopulating rapidly (Walser & Anderlik, 2004).

More generally, the demise of locally owned and operated media, because of overinvestment, sometimes may be stemmed by consolidating instead of closing. For example, consolidation brings economies of scale and scope to continue some local media where there would be silence. A just media firm seeks to disseminate diverse content where none would be otherwise. A locally owned combination of media in a community might be welcomed by most residents instead of having fewer media or none. Economies of scale and scope marked a widely accepted co-ownership in Lawrence, Kansas, during recent years, when the Simon family emphasized localism in its newspaper, cable system, and broadband service by accepting a smaller rate of return on investment than was typical of some other media owners (Folkenflik, 2005).

What remain, if local media remain, are noncommercial, often part-time, volunteer-staffed media. These include low-power television, low-power radio, community newsletters, volunteer film and video making, and local Internet and World Wide Web services in text, audio, still photography, and video.

Little noticed by many is the media organization of first and last resort in many communities. Often a library is in place before and after a local newspaper or a local radio station. A public library usually offers newspapers, magazines, audio- and videotapes and discs, and connection to the Internet and the Web. It may be seen as a public firm. When depopulation requires a public library to close, the remaining audiences might be served by a traveling public library, or they might go to another place for library services. Probably neighbors and friends continue to exchange with one another some books, videos, and other recorded media, as often they do when population is large enough to support public libraries or privately funded local media.

A MODEL MEDIA FIRM

To build a model media firm seems somewhat akin to building a model house. One is aware that other models could be on offer, and that structures stemming from a model might differ from each other in their details. The proposed model media firm is offered on the principle of taking individuals to be worthy ends, and not merely means. A moral condition of this model media firm is to be just, when that means to be fair, and to be fair means to be even handed and more than self-serving. The model media firm has a special obligation to offer diverse ideas when it competes with no other firm, a few others, or many.

Any media firm, whether in the private or public sector of the economy, may choose whether to compare itself with or draw close to the model. Some social regulators of media, or some official regulators, might choose to forward the ethical principle, moral condition, and special obligation of this model. To operationalize ethics, morals, and special obligations such as these, one might turn to owners and leaders of media firms as the individuals responsible for transforming theory to practice by developing the cultures of their organizations through encouraging just actions.

A just firm abides by its contracts inside and outside the firm. Examples of contracts within a firm are those with employees. Outside contracts are with customers, suppliers, and neighbors. A just media firm adds written or unwritten contracts with readers, viewers, and listeners as particular to media, as well as contracts with media regulators. A just media firm in the United States acts to preserve and enhance freedom of expression, and thereby fosters a diversity of ideas.

To compare a particular media firm with a model media firm requires analysis of the content and the context of its contracts, and analysis of adherence to internal and external contracts. Beyond contractual considerations of law and equity are those of ethics and morals. A particular media firm might measure itself by its own actions to develop just individuals who foster a just society.

CONCLUSION

It remains to say a few words on deregulation, consolidation, globalization of media, and examples of future studies.

Implications

What might be some aims of a just media firm in an environment of deregulation, consolidation and globalization?

Some deregulations could be accepted as changing the end point of fewest feasible interventions of the state in the just functioning of a firm. A just media firm offers special attention to deregulation, which comes from enhanced competition of ideas. It creates competition of ideas where none exists, if need be. Some consolidation could be accepted as a way in which a just media firm stimulates more local production in place of less. For example, a joint operating agreement could be accepted to permit two newspapers to share a printing plant.

A just media firm offers enhanced local content to places in which local content otherwise would grow smaller or less diverse. Distinctly local news is preserved when a larger media firm replaces a smaller one. Some globalization could be accepted if a just media firm enhances abilities of others to compete globally. A just media firm offers to people in more diverse places the fuller competition of ideas than would be available otherwise. Diverse local and nonlocal ideas are sought and transmitted.

Future Studies

Here are a few kinds or levels of possible future studies. To what extent a just media firm makes a good media firm should be worth some study. This chapter contains statements on what should be, whereas another might offer correlative statements of end philosophical goods to be sought. Specific media firms have highly variable life spans. Small firms tend to disappear in a few years. A very large firm might last a century or more. It might be worth studying life spans of just media firms to detect any change in life span with degree or duration of justness. Perhaps it will be undertaken to rank or rate media firms as being just. Regular publication of the data might forward the rise of just media firms.

REFERENCES

Albarran, A. B. (1996). *Media economics: Understanding markets, industries, and concepts* (1st ed.). Ames: Iowa State University Press.

Chandler, A. D., & Hikino, T. (1994). *Scale and scope: The dynamics of industrial capitalism.* Cambridge, MA: Belknap Press.

Doyle, G. (2002). *Understanding media economics.* London and Thousand Oaks, CA: Sage.

Folkenflik, D. (2005). Watchful eyes on Kansas media innovations. Retrieved August 16, 2005, from http://www.npr.org/templates/story/story.php?storyId=4599579

Hamilton, R. W. (1991). *The law of corporations* (3rd ed.). St. Paul, MN: West.

Hickey, W., United States, United States, & Cummings, A. (1878). *Hickey's constitution of the United States of America, with an alphabetical analysis: Proceedings of the Continental Congress* (New and enl. ed.). Baltimore, MD: J. Murphy.

Jensen, M. C. (2000). *A theory of the firm: Governance, residual claims, and organizational forms.* Cambridge, MA: Harvard University Press.

Kant, I. (1956). *Groundwork of the metaphysics of morals* (H.J. Paton, Trans.). New York: Harper & Row.

Koenig, G. (1993). *Les theories de la firme.* Paris: Economica.

Penrose, E. T. (1959). *The theory of the growth of the firm.* New York: Wiley.

Picard, R. G. (1989). *Media economics: Concepts and issues.* Newbury Park, CA: Sage.

Rawls, J., & Kelly, E. (2001). *Justice as fairness: A restatement.* Cambridge, MA: Harvard University Press.

Walser, J., & Anderlik, J. (2004, May 18, 2004). *Rural depopulation: What does it mean for the future economic health of rural areas and the community banks that support them?* http://www.fdic.gov/bank/analytical/banking/2005jan/article2.html

17

THE CONVERGING TRENDS OF MEDIA OWNERSHIP AND COPYRIGHT EXTENSION

The Impact on Access to Scholarly Information

Sarah M. Pritchard

American libraries, both public and academic, have a long tradition of offering free and equal access to information of all sorts, for all people. Although historically there have been libraries that were mechanisms of control (e.g., those of various rulers), or libraries only open to those who could pay (such as 19th-century subscription libraries), we have come to value the open and neutral character of most libraries and to assume that they will be able, through purchasing or borrowing, to acquire almost any information that a researcher may want. That breadth of access is now under threat from the convergence of price hikes and copyright controls that become ever more restrictive as the ownership of world publishing media becomes increasingly consolidated.

This chapter reviews the recent trends of consolidation and globalization in the publishing and media sectors that are most dominant in the production of scholarly information. The impact of this is shown in the actual content produced and, most alarmingly, in the sharp increase in prices, which in turn further reduces the amount that most libraries can purchase. The concurrent shift that is traced is the series of expansions in copyright protection that have been legislated in the United States (see chapters by Rose and Warner, this volume), and the accompanying increase in vigilance by publishers over perceived threats to their intellectual property. These directions are exacerbated by the monopolistic character of the industry, and their combined effect is shown on the collections and services of libraries. Finally, a brief description is made of new models of publishing that propose more open approaches both to pricing and to copyright, and some evidence will be given of the wide scholarly support of these emerging genres.

Definitions of the media as outlined by Aufderheide (Chapter 3, this volume) play out visibly in the specific case of access to scholarly information. Access, academic credibility, and pricing are different in "big media" and "little media"; the increased quantity and openness of the latter is welcome, but does not substitute for the high-quality but expensive information controlled by the former. Both copyright control and pricing barriers are heightened by the tendency to vertical integration in media production, which is especially visible in the electronic information world where content, aggregation, file storage, delivery, and user interfaces were originally separate businesses but now may all be provided by a single source. The control of these channels and filters, as Aufderheide explains, has great impact on the nature of the content, its penetration of the markets, who has access to it and what they are allowed to do with it.

CONSOLIDATION OF OWNERSHIP IN PUBLISHING AND MEDIA INDUSTRIES

The collapsing of many formerly separate publishing companies into a few large and diversified entities has a significant impact on diversity and creativity in output. Despite the proliferation of grassroots outlets via the Internet, these small niche products will take a long time, if ever, to achieve the scholarly credibility and production reliability that is necessary for the collecting of scholarly information. Libraries rely heavily on the reputation and past output of a publisher to determine whether the materials are of high academic quality and relevant to a university curriculum. Moreover, now that so much information is produced electronically and only "rented" by the library (e.g., through annual subscription access to a vendor's server), it is critical to know whether a publisher's electronic files will be persistently available and what might be the archivability of the digital content. Academic institutions continue to be reliant on the large well-established publishers for the core of their materials. This concern for long-term digital sustainability may be unique to research-level institutions and is unimportant to many Internet sites targeting fast-paced popular media topics.

Very small numbers of companies now dominate entire sectors of the market. Over the last 10 or 15 years, the pace of mergers has accelerated; for example, there were more than 60 publishing mergers in 1998 and 1999 alone (Monroe,

2000) and things have not slowed much in recent years. Although there are still several thousand publishers from whom academic libraries buy, the field is dominated by six to eight major players, especially for scientific and technical information (Monroe, 2005; for a similar discussion, see Holt, Chapter 6, this volume). One of these, Bertelsmann, a very large German company with $16.5 billion of business, now owns Springer, Random House, Knopf, Bantam, Crown, Dell, Ballantine, Barnes & Noble, and Doubleday publishers. This is not including many subimprints like Fodors, Pantheon, and Vintage Harmony, or its holdings in magazines, music, TV and radio, and book industry services like printing. We can see that it is not just mergers within media sectors like books, but across types of media so that software, music, entertainment, and scholarly or informational publishing may all be within the control of a single company like Sony or Time-Warner.

As publishers merge into large global corporations, we see less breadth of scope and content, and fewer "fringe" or controversial subjects, especially from the more highly regarded scholarly journals and publishers. There is a distancing between publishers and their markets as contrasted with what used to be a pattern of fairly close interaction that created a responsiveness to local topical needs. Anything thought less marketable to large numbers of customers is left to the "little media" that usually do not have the infrastructure and market penetration to have as much impact, reputation or durability.

Small presses and alternative Web sites thrive but do not get packaged in ways that make them easy to acquire and retain in libraries. Students, faculty, and the community who rely on libraries may not see some kinds of information at all, or may have to work much harder to find it if they even look in the first place. Libraries have expressly articulated these concerns in statements about media concentration and diversity, with professional associations encouraging a series of strategies to ensure better coverage; and media scholars have documented the lack of coverage of the mainstream media in such Web sites as that of Project Censored (American Library Association, 2005; Campbell, 2002).

The business models of very large publishers are such that they undertake many fewer reprints and keep books backlisted for a much shorter time, thus making it more difficult to purchase except immediately upon publication. Books are on the market more briefly and in smaller quantities, and there is more emphasis on topics that will sell heavily in their first issuance. Libraries and scholars who wait a year (e.g., until the more scholarly reviews have appeared or budgets improve) are often out of luck.

PRICE MANIPULATION AND CONTROL

The sharp rise in the prices of scholarly and technical periodicals over the last 15 years has been widely documented in the library press.[1] Prices have risen at a rate even exceeding that of health care costs. Although this rise is attributed to many factors (currency fluctuation, production costs), it has been noticeably accelerated as a result of media consolidations. Consolidation has brought the vertical integration of channels of production—a single corporation may often own the journal itself, the printing and electronic production operations, the linked assortment of hundreds of journals that together form a reference database, the search software that mediates the user's interaction with the database, and both wholesaling and retailing services to libraries and bookstores. This gives libraries fewer procurement options and locks them into escalating costs; corporations admit that integration improves profits (Milliot, Mutter, & Holt, 2004).

Monographs, serials, and databases are three areas of scholarly publishing that are each affected by publishing shifts in varying ways; they are different in pricing, use patterns, demand inelasticity, "half life," and reproduction convenience. In the *monograph* area, publisher consolidation has had a gradual if not well-documented impact on the numbers of titles published annually. This is difficult to chart because monographic publishing overall is on the decline as a result of external market factors. However, observers have noted drops in the total list of current production following many mergers, as the larger organization seeks to rationalize its many subsidiary specialties. Conversely, the effect of "capital deepening," where large conglomerates are able to expand market share at the direct expense of university presses and scholarly societies, results in fewer monographs from those other sources (Greco, 2001).

In the case of *periodicals*, publisher mergers create effective monopoly control because the submarkets are so specialized. For any given academic field, there are a very small number of leading journals to which the top researchers in that discipline will submit their articles and research reports. Over time, a given journal will become almost irreplaceable and irreproducible, in the sense that a different journal in exactly the same field may not have the same caliber of editors and authors. Scholars—and libraries—are hooked, with inelastic demand for the premier journals. In the past, these were typically produced by scholarly societies, but there has been rapid acquisition of such titles by commercial publishers, as well as launching of new titles as new academic fields have emerged. Libraries also must now negotiate for entire "bundles" of journals when seeking to gain access in electronic form. In this model, publishers market a large fixed set of titles with the option to gain access to more at a seemingly attractive initial price, but with ultimately less flexibility to cancel when budgets get tight, and the potential for steep hikes when contracts come up for renewal (Frazier, 2001). This model enables publishers to cross-subsidize less popular and lower quality journals by packaging them with the top sellers and making the entire bundle a take-it-or-leave-it proposition. In fact, the antitrust implications of these bundles are starting to be scrutinized (Edlin, 2004).

With these packages and with *reference databases*, purchasing now is basically renting. The library may not permanently own the back files, only the right to connect via the publisher's server. If the library stops paying or the publisher stops supporting a certain section of the information, the content is often lost with little advance notice. Contracts may require that the publisher be willing to provide the already paid back files if the library cancels, but this has rarely been tested. In the opposite scenario from bundling, disaggregation of the content may also increase the price. Under this model, access to the full text and the images and ancillary datafiles can be handled and charged separately. As all of these options are negotiated through contracts, there is little standardization across libraries as to the scope of the content, the restrictions embedded in the search interface or the retention and lending rights.

The upshot of consolidation in publishing and the resultant control over distribution and access models has resulted in price increases. Several studies have analyzed this in detail (see McCabe, 1999; Susman, Carter, Ropes, & Gray, 2003). Some examples are that the Elsevier merger with Pergamon resulted in an average price increase of 22% for former Pergamon titles and even an 8% increase for unchanged Elsevier titles; and the Lippincott-Kluwer merger resulted in a 35% price increase. Studies document that the reasons for the increases cannot be shown to be related to increased costs of editing or production. Some mergers have been challenged through the Federal Trade Commission or the Department of Justice but it is very difficult to claim monopoly, or barriers to entry, under current definitions because the market is not defined journal by journal.

The constant price increases have created unsustainable subscription models, and in academic libraries, the cost of journals is eating up the entire budget. Libraries are forced to cancel (when they can, depending on how journals are bundled), to eliminate paper-duplicate titles, and most worrisome, to cancel the small independent serials because those are the only ones outside the large publisher contracts. The number of actual subscriptions in libraries is fewer, but the unit costs are higher, which is counter to the fundamental concept of the information economy, in which we supposedly should see high first-copy costs, then very low marginal additional costs when production has become largely electronic. Overall, this trend is most striking in graphic form, as Figure 17.1 shows.

FIGURE. 17.1. Monograph and serial costs in ARL librarieS, 1986-2004.*

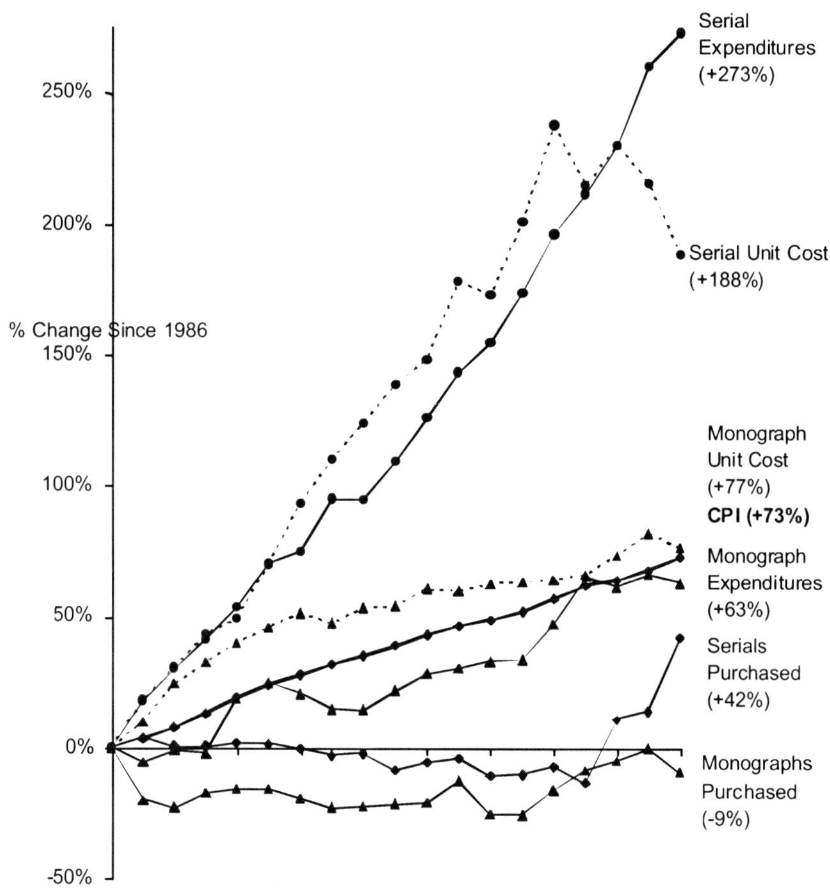

Source: *ARL Statistics 2002-03*, Association of Research Libraries, Washington, D.C.
*Includes electronic resources from 1999-2000 onward.

IMPACT ON USERS

The impact of these price increases on the users of libraries, on students and faculty and researchers and the community, is marked (see Fig. 17.2). Over the last 20 years, while publishing itself has continued to grow especially in countries other than the United States, the actual numbers of items acquired annually by research libraries has remained level or even slightly decreased.

The reduction in monographic purchasing is estimated to be more than 90,000 books foregone by each research library (of the approximately 110 reporting data) over the years 1986-2000. The choices are made differently at each library, but usually the result is that the books most needed for the immediate curricular needs are the highest priority. This may mean that library collections are becoming increasingly homogeneous and ahistorical, as libraries forego the specialty items in narrow subjects that may be related to their longer term research collecting policies but that have lower demand in any given year.

FIGURE. 17.2. **Resources per student in ARL university libraries, 1986-2004.**

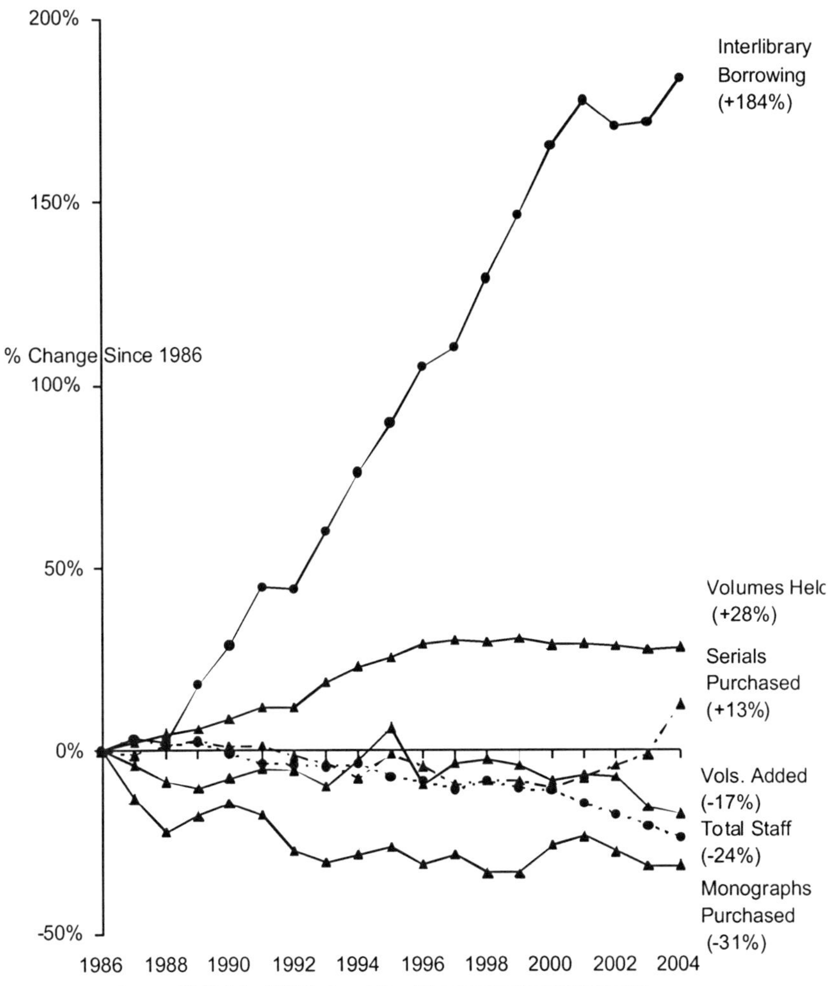

Source: *ARL Statistics 2003-04,* Association of Research Libraries, Washington D.C.

There are fewer volumes, smaller staffs, and interlibrary lending has skyrocketed. Although this is not necessarily bad and reflects a positive sharing of common resources, it begs the question: Whom do you borrow it from if fewer and fewer libraries can afford it in the first place? Where will we be able to rely on a long-term archival set? The Library of Congress is not the obvious answer, as is often assumed by the public.

Universities are not in a very influential economic position vis-à-vis the most expensive periodicals—those in the science, technology and medical fields (STM). Companies do not release their competitive customer data, but estimates are that in the STM sector, the market share of academic purchases is perhaps as low as 25%, not large enough to wield true consumer power with the largest multinational publishing conglomerates. Even for university press monographs, libraries and institutions are only about 30% of the market, and it is flat or gradually decreasing. Through banding together as consortia, universities have been able to negotiate favorably for some of the large database and journal packages, but the overall prices are still extremely high.

There is an indirect societal impact of this reduction in information. People who cannot buy their own subscriptions, are not affiliated with major academic institutions, or rely on public libraries may become information "have-nots." If libraries in schools and communities are central to literacy and an informed citizenry, the gap between sectors of society is being widened, paradoxically, by the enlarging media industries that purport to extend to all those sectors.

GLOBALIZATION

As media and publishing consolidation has progressed, one of the most salient characteristics of the resulting conglomerates is their international makeup. Print and electronic outlets that operate across many countries are under a single corporate umbrella, often outside the United States. Even if there are sales and distribution offices in many countries, policies and strategic decisions tend to be made at the primary location. This has noticeable impact on both pricing and copyright, and on responsiveness (if any) to customer pressures. A multinational company will generally set its prices in the currency that is demonstrating the most favorable exchange rate; in recent years, that has often been Dutch or Swiss currencies, with resulting disadvantages for U.S. and other customers. Traditional relationships with and understanding of customer expectations are eroded; the shared cultural context for business between, for example, American libraries and academic publishers, can no longer be assumed. Distribution, discounts, and bundling may be shaped based on channels and timing in other countries.

International copyright conventions to the contrary, there is little consistency in how concepts of intellectual property are interpreted and applied in a multinational context. Libraries may interpret fair use or academic freedom according to U.S. laws and then find that a transnational publisher disagrees and chooses to litigate using a German or French law; although ultimately the library may be cleared, these conflicts become protracted and have a chilling effect on other institutions.[2] Authors' rights, lending, copying, subsidiary and reprint rights, all vary from country to country, which in turn affects the availability of publications. Even when dealing with an international publisher, a library may have difficulty purchasing a particular version of a given title that has not yet reached U.S. distribution channels, unless it can do business directly in the other country. Globalization may help open up some niche markets but in combination with media consolidation it may reduce options related to pricing and usage rights.

COPYRIGHT CONTROL AS A DERIVATIVE OF ECONOMIC CONTROL

Media consolidation enables large publishing conglomerates to exert another form of economic control in addition to pure pricing: an indirect form of pricing through the control of intellectual property and thus control of the use and re-use of information. Copyright, originally a mechanism to encourage greater creativity and production (see Rose, Chapter 4 and Warner, Chapter 5, this volume) may be having the opposite effect as the balance has shifted from promoting access to maximizing economic return by controlling supply. Successive forms of legislation and business contracts have lengthened the term of copyright and increased limitations on copying, re-use, lending, and archiving particularly in digital media. Publishers have also become far more powerful agents of control than the original authors. Journal publishers, who offer a service highly desired by scholars, have often been able to require that authors sign over their copyrights to individual articles including spin-off rights such as inclusion in databases. Media and entertainment companies control coveted production and distribution channels and exert similar requirements over the creative community. As a result, the sector of copyright owners has grown, separate from either the creators or the end users who were the two parties originally targeted by, and relevant to, most laws.

Reflecting ongoing attempts to extend control by copyright owners, there has been a group of major new acts in the United States since 1998, including copyright term extension, the Digital Millennium Copyright Act (DMCA), the Uniform Computer Information Transaction Act (UCITA) and the Technology, Education and Copyright Harmonization Act (TEACH). Together, these have created enormous complexity especially when trying to determine the status of a given work such as an unpublished manuscript, or of a given use such as linking from a web page.[3] The DMCA covers web sites and peer-to-peer file sharing and has put universities in the position of *de facto* Internet service providers (ISPs) responsible for all the individual sites mounted by students and faculty on institutional servers. UCITA was a state-by-state initiative rooted in contract and commercial law; successful in some states and not others, it promoted "shrink wrap" and click-through licensing without the opportunity for user negotiation (Association of Research Libraries, 2000). TEACH has regulated much of what can be done with Internet and video uses in the classroom and via distance-education technologies. These and other initiatives that resurface with regularity have been lobbied for heavily by the media and information industries; library, education, and civil rights groups have battled back with mixed success. Generally, provisions allowed under fair use have shrunk, so that universities and libraries have fewer guaranteed rights. Meanwhile, fast-moving changes in technological approaches and media formats create a lot of gray areas and require constant effort to monitor permissions and oversee compliance, punctuated by occasional risk-taking and legal battles.

Attempts to seek balance do not necessarily have better results for opening up access to information. In the 2001 Tasini case, a free-lance writer took on The *New York Times,* which had incorporated his stories (as published in the paper) into their commercially distributed database without further permissions or fees beyond those for the original story. When the court found for the plaintiff, it stated that aggregators of information could not overlook the need for individual permissions for these re-uses. The result was a victory for author rights and was supported by libraries, but paradoxically has also resulted in a reduction of content in databases because several large publishers have chosen to delete content rather than deal with tracking and paying for many small individual items that may come from third-party providers and subaggregators (Quint, 2002).

Publishers have become increasingly more vigilant and litigious in monitoring how their products are being used. There is a distinct "intimidation factor" in selective threats and lawsuits that are well publicized. Libraries have been sued over providing access to supposedly unauthorized users, over mounting articles in electronic reserves systems, over streaming audio music files, infringing web links, and document delivery (copying of articles for a fee for users). Most of these uses are, ironically, responses of libraries and users to the high costs and limited access created by the publishers. Although there are legitimate concerns by publishers about "commercial" use within libraries, there are often disagreements over the interpretation of the coverage of fair use in these contexts, and over what behaviors constitute commercial (e.g., research conducted for profit-making businesses) as opposed to educational uses.

Licensing, and thus contract law, is taking precedence over statutory fair use rights in copyright law. If it is in your contract, you can do it, but otherwise not. This trend may narrow the legal definitions of acceptable use rather than open up new models, and brings increasing complexity in the layers of state and federal regulations and compliance.

COMPLIANCE BURDENS

All information collected and accessed in the research library has to be viewed through the lens of copyright, whether it is handwritten literary manuscripts or 1950s aerial photography or contemporary electronic journals. A complete chart of what the rules are stretches to almost 25 categories, including, for example, 5 date ranges that apply to unpublished works, 9 for works published in the United States, and 8 sets of dates and compliance statutes for works published outside the United States (Hirtle, 2005).

The need to observe a dizzying array of copyright coverages and licensing provisions has a daily impact on internal procedures, policies, and costs (Webster, 2002). In academic libraries, the most detailed tracking occurs in reserve readings, Web site construction, digital conversions and interlibrary lending (ILL). With so much of the reserve reading and ILL now being transacted and delivered electronically, the suspicions of publishers regarding possible violations are on the increase with new burdens on the technological systems needed to monitor the daily traffic in digital articles. Libraries may need to contract for automatic services with the Copyright Clearance Center, to maintain separate electronic reserves systems with closed servers, to ensure individual identity authentication in campus network access, to post photocopying warnings, implement web controls, collect numerous individual fees and recall the details of the rights specified in scores of specific contracts. The staffing, systems and service costs per year mount steadily. At one mid-size university in 2004, the secure electronic reserves system cost $7,000 annually, not including the costs of hardware, networking, digital conversion, or technical support; fees for copyright compliance for ILL and reserves were almost $10,000 a year, again excluding the labor to monitor this. Moreover, this does not include what may be being paid elsewhere on campus directly by faculty. A constant concern is that the complexity of rules and the labor needed to monitor them may lead many individuals not to bother with compliance.

The extension of copyright terms, the confused status of many works, and the addition of new legislation treating intellectual property in the digital environment combine to create limitations on fair use, or at the very least, to create a perception of such limitations. Within the educational sector, fair use has for years governed such practices as resource sharing, reserve reading, and copying for preservation or personal use. Now, not only are some of those rights being challenged but new educational uses such as the compilation of digital images for teaching, digital archiving for library collections, courseware webs, or re-mix for research products are even more restricted by the difficulties and costs of managing the copyright issues.

FREEDOM OF INFORMATION AND PRIVACY

The interest of large media conglomerates in maintaining their economic base and controlling intellectual property has led not only to difficulties in gaining access to information content, but to serious threats to individual privacy and intellectual freedom. Mechanisms for tracking electronic information use, known as digital rights management (DRM) systems, have been embedded legally in contracts and technologically in hardware and software, and create noticeable risks to the long-held library principle that users' reading preferences are not revealed (Cohen, 2003). Techniques such as digital signatures and digital object identifiers allow publishers to track when an item is being used and by whom; and the anti-circumvention provisions of the DMCA have been used actively both to track uses and to require that content be removed from web sites (Conway, 2003; Webster, 2002). The level of potential invasiveness is very specific, down to every page viewed and every link followed. ISPs, which can be legally interpreted as being the university or the library depending on network architecture, can be required to reveal the names of allegedly infringing users, without warning the user. The information-gathering potential of DRMs may also be exploited by law enforcement utilizing sections of the USA PATRIOT Act. Although that act is unrelated to copyright, the mechanisms publishers have promoted to deal with copyright open up avenues into personal data that did not exist before; libraries trying to protect user borrowing and web use records may run afoul of provisions to comply with copyright tracking.

There is great alarm in several sectors about the assaults on freedom of expression and information wrought by increasingly stringent and punitive approaches to copyright (Brennan Center for Justice, 2005; Electronic Frontier Foundation, 2003; Williams, 2004). Educators, legal advocates, civil libertarians and librarians, are engaged in active debate over threats to traditional expectations of fair use and academic freedom. Combined with the runaway price inflation, the result is an upheaval in the scholarly communications market. There are, however, new models beginning to be tested that challenge both the pricing and copyright restrictions of conventional academic publishers.

ADVOCACY AND ALTERNATIVES

The transition to electronic publishing, and the emergence of Internet-based forms of scholarly communication that go beyond traditional books and journals, have changed some things but not others. There is wide variation by discipline as far as the culture of information production and information sharing, and the amount of credibility attributed to nonstandard venues. In the high-energy physics field, for example, the Xarchiv (http://www.xarchiv.com) has become the widely accepted site for posting papers, well before they appear in standard academic journals; and for researchers in genomics, the contribution of extensive datasets to a disciplinary repository is standard practice. But in fields still strongly wedded to formal monographs and journal articles, or for whom research is less collaborative or multimedia, such newer models may receive not only resistance and lack of citation, but denigration when faculty are considered for awards, promotion and tenure. It is not so much the issue of print versus electronic as it is the authority and perceived status of conventional publications from large well-known publishers, versus formats that may appear to have less rigid peer review or a more informal appearance. New forms of scholarly communication do not imply an absence of peer review, or that copyright is discarded, or that it doesn't take money to produce; electronic journals can be just as high quality and expensive as print ones. The issue is whether more sustainable and responsive economic models can be built, and more flexible intellectual property rights constructed that foster wide academic use.

Over the last 5 or 6 years, various strategies have evolved that are designed to engage university faculty and senior leadership, scholarly societies and funding agencies in taking an active approach to reshaping the scholarly publishing environment. Many of these strategies are linked in advocacy resolutions passed by faculties, which may address pricing, copyright, promotion and tenure, "open access" and the influential roles of faculty editors and authors *vis-à-vis* what might be seen as problematic publishers.[4] Several high-profile scholarly conferences have resulted in internationally publicized sets of principles to be promoted in scholarly communication, such as the Budapest Initiative or the Tempe Principles.[5] Aspects of three broad themes are identified in these resolutions and principles:

1. *Encourage more competition in traditional publishing.* This involves taking a stronger position when negotiating with very large publishers, where the pricing, bundling, and cancellation policies tend to be the most inflexible. Rather than accept boilerplate offers, for example, some libraries and higher education consortia are proactively making proposals with price caps, custom title selection, and guarantees to allow resource sharing and archiving. Libraries and faculty are also working together to support and enhance the academic strength of independent publishers and those who promote reasonable economic models such as the journals allied under the umbrellas of SPARC and BioOne. And there is a re-emphasis of university controlled publications through university presses and by collaborative digital information initiatives on campus.
2. *Show faculty how to manage their intellectual property effectively.* Under the customary model, authors sign forms assigning copyright, so publishers end

up controlling the rights to articles; ironically, universities and public agencies have to pay to buy information originally produced by their own researchers and often at public expense. Encouraging authors to keep more control of copyright is a key component in promoting competition and sustainable economic models. When signing publishing contracts, for example, authors can insert clauses guaranteeing them the right to mount a preprint or postprint of the article on a scholarly or institutional server; granting the publisher copyright only for a limited term of time; or retaining derivative rights such as inclusion in anthologies or coursepacks. Beyond simply educating faculty about the copyright laws, showing them how to take creative control is the focus of Web sites such as that at the University of California's Office of Scholarly Communication (http://osc.universityofcalifornia.edu/manage). There is now a widely known independent nonprofit initiative designed to promote fair use and expand the public domain while still offering protection to authors. Launched in 2002 based on work by groups of scholars and intellectual property attorneys, Creative Commons (http://creativecommons.org) offers a set of model licenses with a range of protections and freedoms. In late 2005, Google released a Creative Commons-enabled search that directs a user specifically to resources licensed under those provisions.

3. *Support open access initiatives.* "Open access" encompasses several publishing philosophies—sustainable business models, generous fair use policies, and scholar-controlled publishing outlets. The term has come to mean the model by which the author (or his or her home institution) pays a publishing fee and access is free for the end-user, but there are actually many flavors of full or partial open access (Willinsky, 2003). In more limited models, access is open after an initial subscriber-only time period, such as 6 months. Open access may be enabled via the original publisher, or by depositing articles in an open repository operated by a university, research organization, or scholarly society. Some of these options can co-exist with conventional publication. Open access is being explicitly mandated in some universities, and, more importantly, by government funding agencies as a condition of grant awards (SPARC, 2005). Major universities are beginning to deploy digital repositories to archive and disseminate courseware, faculty research and administrative documents; some of these have in effect become publishing vehicles, whereas others are more purely for internal records management. Examples include MIT's Dspace, and the University of California's eScholarship; an extensive international registry of open-access repositories and their characteristics is maintained at http://archives.eprints.org.

Although these new initiatives are rapidly changing and very diverse, it is already possible to document the impact. Publishers are introducing their own open-access policies and are responding to pressures from the faculty on whom they rely for editorial labor and reputation. It has been concretely demonstrated that freely available online articles do get cited more widely and thus have greater research impact (Antelman, 2004; Lawrence, 2001). Institutional repositories are growing and access to publicly funded research has become a widely advocated public policy issue.

FEW GUARANTEES AND MUCH VIGILANCE

Public ownership is no guarantee that access will be protected. The lack of copyright on federal documents, for example, does not equate to unfettered dissemination. The very real experience of the last few years is that U.S. government agencies have been stripping the content of open Web sites and recalling previously published government documents, citing national security concerns. It is not clear what the rights of access are for libraries and the general public, and public interest groups are closely watching this trend (PMB Watch, 2005; Susman, 2002). The privatization of government services has changed the scope of information activities and reduced the availability of print and electronic publications in certain agencies such as the Department of Education, which had been the producer of one of the oldest public research databases in the country (Abbott Hoduski, 2004; Corby, 2004).

Even new models cost money. Because most open-access models rely on shifting rather than eliminating the costs, this may simply create problems in other parts of the academy, especially because it is assumed that both traditional and new models will be co-existing. Studies show that the "author-pays" approach, for example, could cost research institutions with large faculties to pay disproportionately high costs, even higher than under traditional subscription models (Davis et al., 2004; Guterman, 2004). The degree of volatility in all of the alternatives outlined here is such that one cannot expect price or access stability for some time to come.

The publishing industry is of great value to the ongoing work of scholarship, and consolidation has ensured persistent growth and effective distribution. But publishers, universities, libraries, and scholars need to be able to work together in a context that does not presume limitless financing and that respects the long traditions of the public domain and the sharing of academic research. Access to information will become more and more locked down unless we work to promote a wide variety of competing models of scholarly communication that genuinely balance the rights of authors, publishers and users (see also chapters by Rice, Lievrouw, and Warner, this volume).

NOTES

1. See, for example, documents listed at the Association of Research Libraries site, http://www.arl.org/osc/marketplace/jnlprices.html.
2. See, for example, the discussion of Gordon and Breach (G&B) Publishers in *Newsletter on Serials Pricing Issues, 43*(August 10), 1992. G&B continued to press suits on this issue for several more years in other countries. For a discussion of media globalization and effects of differing national, cultural and regulatory contexts, see Bielby and Moloney, Chapter 13, this volume.
3. Most universities now maintain extensive Web sites to explain copyright details. Two comprehensive examples are those of Stanford (http://fairuse.stanford.edu/index.html), and the University of Minnesota (http://www.lib.umn.edu/copyright/). The Association of Research Libraries also monitors current and proposed legislation (http://www.arl.org/osc/copyright/index.html).
4. See a number of such resolutions linked at http://www.arl.org/osc/faculty.
5. These are widely cited at various university scholarly communication sites; see for example the University of Maryland Health Sciences Library, http://www.hshsl.umaryland.edu/scholarly/articles/organizations.html.

REFERENCES

Abbott-Hoduski, B. (2004). ERIC has been downsized. *Unabashed Librarian, 130*, 19-21.

American Library Association. (2005). Media concentration. http://www.ala.org/ala/oif/ifissues/mediaconcentration.htm

Antelman, K. (2004). Do open-access articles have a greater research impact? *College & Research Libraries, 65*, 372-382;

Association of Research Libraries. (2000). UCITA: Summary and implications for libraries and higher education. Accessed December 14, 2005, from http://www.arl.org/info/frn/copy/ucitasum.html.

Brennan Center for Justice. (2005). *Will fair use survive? Free expression in the age of copyright control.* New York: New York University School of Law.

Campbell, B. (2002). Media concentration and the role of libraries in promoting a diversity of voices. *Feliciter*, 6.

Cohen, J. E. (2003). DRM and privacy. *Berkeley Technology Law Journal, 18*, 575ff.

Conaway, T. L. (2003). How the Digital Millennium Copyright Act jeopardizes online privacy. *Nebraska Library Association Quarterly, 34*, 12-14.

Corby, K. (2004). Where is ERIC ... and where is ERIC going? *Information Outlook, 8*, 37-38

Davis, P. M., Ehling, T., Habicht, O., How, S., Saylor, J. M., & Walker, K. (2004). Report of the CUL Task Force on Open Access Publishing presented to the Cornell University Management Team, August 9, 2004. *Cornell University Library Technical Reports and Papers.* http://hdl.handle.net/1813/193.

Edlin, A. S. (2004). Exclusion or efficient pricing? The "big deal" bundling of academic journals. *Antitrust Law Journal,* 72, 119-157.

Frazier, K. (2001). The librarian's dilemma: Contemplating the costs of the "big deal." *D-Lib Magazine,* 7(3). Accessed December 11, 2005, from http://www.dlib.org/dlib/march01/frazier/03frazier.html.

Electronic Frontier Foundation. (2003). Unintended consequences: Five years under the DMCA. Accessed December 11, 2005, from http://www.eff.org/IP/DMCA/unintended_consequences.pdf

Greco, A. N. (2001). The market for university press books in the United States 1985-1999. *Learned Publishing, 14,* 97-104.

Guterman, L. (2004). The promise and peril of "open access." *Chronicle of Higher Education, 50,* A10.

Hirtle, P. (2005). Copyright term and the public domain in the United States, January 1, 2005. Accessed December 11, 2005, from http://www.copyright.cornell.edu/training/Hirtle_Public_Domain.htm.

Lawrence, S. (2001). Online or invisible? Preprint; http://ivyspring.com/steveLawrence/SteveLawrence.htm (Edited version appears as "Free online availability substantially increases a paper's impact" in: *Nature, 411,* 521).

McCabe, M. J. (1999). The impact of publisher mergers on journal prices: An update. *ARL Bimonthly Report, 207,* 1-5.

Milliot, J., Mutter, J., & Holt, K. (2004). A vertical leap. *Publishers Weekly, 251,* 25-30.

Munroe, M. (2000). Which way is up? The publishing industry merges its way into the twenty-first century. *Library Administration and Management, 14,* 70-78.

Munroe, M. (2005). The academic publishing industry: A story of merger and acquisition. Prepared for the Association of Research Libraries. Accessed December 15, 2005, from http://www.niulib.niu.edu/publishers.

OMB Watch. (2005). Access to government information post September 11. http://www.ombwatch.org/article/articleview/213/1/229.

Quint, B. (2002). Tasini damage-reporting decisions. *Information Today, 19,* 8-12.

SPARC. (2005). Open access resources: Supportive policies from grant-giving agencies. Accessed December 15, 2005, from http://www.arl.org/sparc/oa/index.html.

Susman, T. (2002). Removal or destruction of Federal depository library documents. Memorandum to the Association of Research Libraries, March 13, 2002. Accessed December 15, 2005, from http://www.arl.org/info/frn/nsletter/Susman.html.

Susman, T. M., Carter, D. J., & Ropes & Gray L.L.P. (2003). *Publisher mergers: A consumer-based approach to antitrust analysis.* Washington, DC: Information Access Alliance.

Webster, D. (2002, December). *The practical realities of the new copyright laws: A librarian's perspective.* Paper presented at the 118th annual convention of the Modern Language Association, New York.

Williams, K. (2004). A call to arms: What in the world is happening to information? *Aplis: Australasian Public Libraries and Information Services, 17,* 56-62.

Willinsky, J. (2003). The nine flavours of open access scholarly publishing. *Journal of Postgraduate Medicine, 49,* 263-267.

18

OPPOSITIONAL NEW MEDIA, OWNERSHIP, AND ACCESS

From Consumption to Reconfiguration and Remediation

Leah A. Lievrouw

THE CHANGING MEDIA ENVIRONMENT

By virtually any measure, information and communication technologies have transformed the media landscape over the last two decades. The "convergence" of traditional and new media technologies and industry structures has become a cliché. Game writers, bloggers, online merchants, and hypertext authors have joined the ranks of movie studios, record labels, book publishers, and newspapers as content providers. Software firms, Internet service providers (ISPs), wireless and cellular operators, and computer hardware manufacturers, as well as broadcasters, publishers, and telephone companies, are media owners. Regional and global regulatory organizations, treaties, and standards bodies govern media policy alongside the U.S. Federal Communications Commission (FCC), patent and copyright offices, state utility commissions, and their counterparts in other nations. "Audiences" have become "users" of media, immersed in a complex media ecology of divides, diversities, and literacies.

Despite this expanding landscape, however, many policy analysts and industry critics continue to view media ownership issues primarily through the lens of traditional mass media production and consumption. The top-down, few-to-many, content-distribution model of mass media dominates the relevant policy debates; new media, when they are considered at all, are framed mainly as novel types of content sources or distribution systems. System owners are assumed to control the content their systems carry, and thus the the ideas, beliefs, opinions, and agendas that they present and that shape public culture and discourse. Of the three main streams of media law and regulation in the United States (First Amendment for press, speech and creative works; common carriage in telecommunications; and broadcast licensing), only broadcasting is subject to any degree of content regulation, insofar as it uses public airwaves. As a practical matter, to influence the diversity of content and assure access for underrepresented groups, we regulate (or attempt to, more or less) media system ownership.

Yet it seems increasingly clear that this model is only relevant for part of the media environment today. Historically, a handful of major firms have dominated American mass media, and they continue to play a major role in mainstream culture and politics. But even as media concentration increases, their dominance has been challenged in the last decade as people have turned to new media technologies to extend their social networks and interpersonal contacts, to produce and share their own content, and to resist, "talk back to," or otherwise engage with and participate in the prevailing media culture.

In this chapter, therefore, media ownership and access are reconsidered in light of a growing tension between a traditional view of the media environment as a site for the production, distribution, and consumption of media products, and an alternative view that sees the environment primarily as a venue for speech, interaction, and creativity. From the first perspective, media ownership is understood in terms of property and gatekeeping, and media access as the reception and consumption of media products. The alternative view considers reputation, credibility, reciprocity, trust, voice, and visibility to be as valuable as property, and sees access as the opportunity to use media technologies to create and communicate as well as to consume.

The tension between these two views is explicit in the growth of *oppositional new media*, people's uses of media technologies to respond to mainstream, homogenizing, mass media culture. An ongoing cycle of capture, co-optation, and subversion of media content and uses between the more concentrated, mainstream "center" and the diffuse but increasingly interactive and participatory "edges," characterizes media culture today.

In the remainder of the chapter, several key genres and features of oppositional new media are described, and illustrative cases are presented. The chapter concludes with a discussion of the implications of oppositional new media for changing ideas about media ownership and access.

1990s MEDIA ECOLOGY: TWO VISIONS

> Once upon a time there were the mass media, and they were wicked, of course, and there was a guilty party. Then there were the virtuous voices that accused the criminals. And Art (ah, what luck!) offered alternatives, for those who were not prisoners of the mass media.
>
> Well, it's all over. We have to start again from the beginning, asking one another what's going on. (Eco, 1986, p. 150)

When the World Wide Web, browser technologies, and relatively inexpensive client-server network architectures were introduced in the early 1990s, technology advocates predicted that Internet use would quickly expand beyond the ranks of elite academic, government, and private-sector corporate users. The Web would deliver on the early "information utility" visions of the 1960s and 1970s (Greenberger, 1964, 1985). The decentralized infrastructure of the Internet, advocates said, was inherently democratic: Communities and groups that had been marginalized in the mass media would finally be able to express their ideas and extend their influence in the wider society. Groups like the WELL in Sausalito, California, Berkeley's Community Memory, and the Electronic Frontier Foundation, with roots in the 1960s counterculture and libertarianism (Barbrook, 2001), espoused ideals of empowerment and participation captured in pundit and *Whole Earth Catalog* founder Howard Rheingold's (2000) phrase, "homesteading on the electronic frontier."

In the mid- to late-1990s, new collaborative environments such as MUDs, MOOs, and chatrooms were flourishing online. E-mail was (re)discovered as the "killer application" for personal and leisure interaction as well as work tasks. Mainstream media content and information services, repackaged for the Web and retrievable with powerful web browsers, as well as new peer-to-peer systems, attracted novices who had never considered using computers before.

The traditional media industries, meanwhile, saw the "frontier" of browsers, servers, networks, and the World Wide Web somewhat differently. Web browsing, search engines, chatrooms, e-mail and games began to compete with television, radio, and theatrical movies for audiences' time and attention. The potentially limitless reproduction and circulation of information online threatened media firms' traditional control over content production and gatekeeping, so, recasting themselves as "content industries," they sought to re-establish the familiar "pipeline" model of mass media production, distribution, and sales in the new media environment. The industry initiated new, aggressive campaigns to enforce and extend intellectual property rights in the new media *milieu.* Media firms merged with one another to enlarge and lock in their market shares, and teamed up with computing firms to bundle their products with other kinds of "software." They built alliances with telecommunications operators and Internet service providers (such as the mergers between America Online [AOL] and Time-Warner and AT&T and Comcast) to gain greater control over the new media infrastructure, particularly the "final mile" of cable or telephone wire into the home. Major increases in bandwidth and even more stringent copy protection regimes were planned, to meet the anticipated surge of demand for entertainment and other industry products online.

Throughout the 1990s in the United States, major media firms lobbied the Congress, Department of Justice, FCC, and other bodies where they wielded influence, to shape a more advantageous regulatory environment. The Congress obliged, in the form of the Telecommunications Act of 1996, the Digital Millennium Copyright Act (DMCA) of 1998, and the Bono Copyright Term Extension Act of 1999, among other legislation. President Clinton's FCC appointees, including Reed Hundt and Michael Powell (later to become FCC chair under President George W. Bush), ushered in a renewed era of media deregulation. In the name of reducing restrictions on corporate "speech," a movement begun under Presidents Reagan and George H.W. Bush, media ownership constraints were substantially weakened under the Telecommunications Act of 1996. Financial interest in syndication (so-called "fin-syn") rules, imposed by the FCC in 1970 to prohibit networks from owning production and syndication companies, and must-carry rules put in place by the FCC in 1966 and requiring cable operators to carry local broadcast channels, were abandoned. The U.S. radio frequency spectrum, formerly considered a scarce natural resource and thus a public good, was redefined as an overabundant commodity, and significant portions of American airwaves were put up for auction, or in some cases, such as for high-definition television (HDTV), given away, to media corporations.

In the rush toward privatization and cross-ownership, the "property" metaphor, borrowed from the invention and patenting culture of high tech, swept the traditional media industries. In what copyright scholar Jessica Litman (1994) called an "intellectual property epidemic" (p. 429), the Data Investment and Intellectual Property Antipiracy Act of 1996 and the Collections of Information Antipiracy Act of 1999 granted property rights to entirely new types of information and expressions, such as databases, page numbers, and telephone listings. The Sonny Bono Copyright Term Extension Act of 1999 extended copyright protection for longer terms than ever before (Lipinski, 1999). Some critics charged that the traditional criteria for awarding patents—novelty and nonobviousness—had been relaxed to the point of absurdity, making it possible for almost anything to be patented in the United States, including previously exempted facts like mathematical algorithms, sequences of genetic code, the click of a computer mouse to order merchandise online, or new animal varieties (Gleick, 2000; "Who owns," 2000; "Patently absurd," 2001).

ISPs such as AOL rejected the service and regulatory models that had governed telephony, postal mail, and publishing, and instead claimed that both the systems they operated and the content they carried (e.g., subscribers' e-mail) were the firms' property and subject to company monitoring and control, and free from government regulation. Contrary to legal precedents in which the U.S. Supreme Court had ruled that information technologies such as VCRs were legal so long as they had significant noninfringing uses, new "anti-circumvention" provisions of the DMCA prohibited any new technology that might conceivably be used to violate intellectual property rights, whether the technology was actually used that way or not (as noted below, the Supreme Court has recently ruled that the peer-to-peer system Grokster violates the DMCA on these grounds). Entertainment firms targeted university campuses—where high-speed networks had been in place for years, and where the Internet was already a routine part of students' studies, entertainment, and interpersonal interaction—as nests of copyright violation, and firms filed punitive lawsuits against student users, their families, and the universities they attended.

The dot-com collapse at the end of the 1990s provoked a precipitous shake out of smaller enterprises and start-ups across the media, telecoms, and computing industries, whose assets—including intellectual property—were sold to larger surviving firms. Although a glut of bandwidth (especially fiber-optic networks) was built in the United States in the 1990s, those networks largely remained dark: major media firms were reluctant to distribute their products online without iron-clad copy protections and revenue guarantees. Broadband services to the home (i.e., cable modem and digital subscriber line [DSL] services) were built "asymmetrically," that is, with much more downstream capacity from the network to subscribers, than upstream from subscribers to the network, reflecting a view of households primarily as consumers, rather than producers, of media. By the turn of the century, the vision of streaming video to the home via high-speed data networks, widely touted by the industry just a few years earlier, had quietly been shelved (Winseck, 2002), and the United States had one of the slowest "high-speed" consumer broadband networks in the world (Belson & Richtel, 2003). The United States still lags most of Europe, Korea, Japan, and other nations in the adoption of broadband services (*"Emerging-market,"* 2005).

In this increasingly concentrated and centralized media climate, worries about personal privacy began to surface in the United States. Major news magazines declared the "end of privacy" ("The end of privacy," 1999). Horror stories about identity theft and telemarketing fraud, the unauthorized "harvesting" of information from personal computers, the spread of viruses online, the lax rules for the circulation of personal financial and medical records, and the high rate and persistence of errors in the major credit databases prompted a popular backlash, new legislation, and the increased use of privacy-enhancing technologies (PETs) by consumers to protect their private data and communications (Phillips, 2004). Widespread public objections and online protests derailed both the sales of the LotusMarketplace consumer database and the implementation of the Clipper Chip decryption technology advocated by the Clinton Administration (Gurak, 1997; Gurak & Logie, 2003). But consumer resistance was countered by private firms and their political allies who claimed that individuals surrendered rights to their personal information once private firms had collected and organized it (Lipinski, 1999).

To a great extent, however, Americans' worries about privacy were forgotten in the wake of the September 11, 2001 terrorist attacks in the United States The rapid passage of the Patriot Act, the establishment of the new cabinet-level Department of Homeland Security, and new restrictions on travel and immigration gave the federal government and its private-sector contractors sweeping new powers to withhold and control public information, and to monitor the movement, activities and communications of citizens and noncitizens alike. Meikle (2002) predicted that the events of September 11 would be seen in retrospect as "the iceberg [that sank] the Internet" (p. 173).

Today, a few years into the new century, the centralized, industrial-style "pipeline" model of mass media content distribution and consumption seems to have survived intact in the new media environment. Media, telecoms, and computing firms have fought any distribution scheme or technology that threatens their gatekeeping, rent-extracting role in the creation and movement of information. As prominent law professor and critic Lawrence Lessig (2003) said, the major media firms have "declared war against rip, mix, burn culture."

MEDIA ECOLOGY TODAY: OPPOSITIONAL NEW MEDIA

> Revolution has to be reinvented, that's all. (Debord, 1961/1981, p. 63)

A number of writers have commented recently on the two visions just outlined, critiquing what they see as the increasing commercialization of the open and supposedly communitarian Internet of the 1980s and 1990s (Atton, 2005; McCaughey & Ayers, 2003; Meikle, 2002). Nonetheless, it is worth asking what has happened to the early "frontier" vision of people seeking to gain greater political and economic influence and participation in the new media environment.

In fact, community groups, cultural and political activists, artists, and ordinary citizens have adapted almost as quickly as technological, economic, and legal barriers to their uses of new media have been erected, creating new genres of *oppositional new media* (ONM), media forms and content that respond to, reflect, critique, parody, rejoin, or subvert situations and events created by and portrayed in mainstream media sources. The term echoes Lovink and Richardson (2001), who described the "media of opposition" used by political activist groups. As used here, "oppositional new media" denotes the use of new media and information technologies by any person or group in response to mainstream media culture. ONM evolve, bottom-up, in response and resistance to popular media culture.

They are typically low-budget, quick-response forms of communication, "art about media with media" (Huhtamo, 2003) that aim to motivate impassioned response and participation. ONM advance the "alternative" philosophy of early Internet proponents and visionaries, and carry forward a longer tradition of "underground," alternative, and radical media (Atton, 2001, 2004; Downing et al., 2001).

Groups like ®™ark and the Surveillance Camera Players (discussed later) may not have the resources of the major media organizations, and some blogs may never reach more than a few dozen loyal readers. But ONM are becoming a more prominent part of the contemporary media environment, contending with and talking back to, and being incorporated into, the mainstream. Online, people have created hospitable spaces to develop and express nonstandard or "fringe" ideas, and to defy homogeneous, mainstream culture. Long before "the Internet" was part of the media landscape, French sociologist Alain Touraine observed:

> a desire to erect communities conceived as a refuge within an increasingly thicker social network. Marginality, considered for so long a failure of integration, becomes thus the hallmark of an opposition, a laboratory in which a new culture and a social counterproject are being elaborated. (Touraine, 1984/1988, p. 106)

GENRES OF OPPOSITIONAL NEW MEDIA

The adaptation of media and information technologies for new and emerging forms of information and interaction poses something of an analytic challenge; the technologies and forms tend to be moving targets, constantly changing in response to a dynamic cultural and technological context. Nonetheless, four main genres of oppositional new media have become common: *culture jamming, alternative computing, mediated sociality/mobilization*, and *indymedia*. This four-genre scheme shares some details with other typologies that have been proposed, such as those by Atton (2005), Meikle (2002), and Vegh (2003), which focus mainly on the uses of the Internet for political activism. However, for the purposes of the present discussion the four types are presented as *genres* of economic, cultural, and social, as well as political, expression online (see Table 18.1 for a summary of the genres, their characteristics, and examples of each). Each has roots in media forms and uses that predate the so-called Internet age, but have moved in new directions in the online context.

TABLE 18.1. *Genres, Characteristics, and Examples of Oppositional New Media*

	Genres (Domains)			
Characteristics	**Culture Jamming (Popular Culture)**	**Alternative Computing (Technology Infrastructure)**	**Mediated Sociality/ Mobilization (Social Networks, Organizing)**	**Indymedia, (News, Information, Opinion)**
Small scale	illegal-art.org sniggle.net	*2600: The Hacker Quarterly* (2600.com) Easter eggs (eeggs.com, mackido.com)	flash mobs, smart mobs (flashmob.com, cheesebikini.com)	Independent Media Centers (indymedia.org) *Slashdot* (slashdot.com)
Interventionist	Naimark laser pointer project (naimark.net/projects/zap/howto.html) Barbie Liberation Organization (rtmark.com/legacy/blo.html)	DeCSS posting cryptome.org denial of service, spam attacks	meetup.com MoveOn.org	Reporters Without Borders (rsf.org) [Tactical Media, watchblogs (mediacritique: watchblog.com)
Subcultural literacy	Deconstructing Beck (rtmark.com/legacy/db.html) Get Your War On Adbusters.com	*A Hacker Manifesto*	P2P: Gnutella, KaZaA, Grokster BitTorrent sites (torrentspy.com)	re-blogs (shey.net, rhizome.org)
Ironic	Peretti–Nike case Yes Men: dow-chemical.com "®™ark Cultural Capital™ Index"	DeCSS Gallery DeCSS Haiku Natural Selection search engine	flashmob.com	"Five *Technically* Legal Signs for Your Library" (librarian.net/technicality.html)

Perishable	billboardliberation.com	war driving (wardriving.com)	dodgeball.com	opinion blogs (boingboing.com)
Collaborative	®™ark (rtmark.org) detritus.net	Free Software Movement (gnu.org), Open source software development, Clipper Chip protests	moblogs (mobile phone images: TextAmerica.com, buzznet.com)	wikipedia.com creativecommons.org ourmedia.org
Heterotopic	Surveillance Camera Players (notbored.org/the-scp.html)	mongrel.org	friendster.com Gay.com MySpace.com	Zapatista movement (ezln.org) digg.com

Note: The categorizations here are intended as a broad descriptive framework, rather than a firm taxonomy. Many sites have several of the characteristics listed, and some have only some of the characteristics and not others. Readers are encouraged to visit the sites listed here and in the text as examples.

Culture Jamming

The term *culture jamming* is often attributed to an influential essay of the same name by media critic Mark Dery (1993), who credits the experimental band Negativland with coining the phrase. Dery defined culture jamming as "media hacking, information warfare, terror-art, and guerrilla semiotics, all in one," that captures and subverts the images and ideas of the mainstream media to make a critical point. It is "a strategy that turns corporate power against itself by co-opting, hacking, mocking and re-contextualizing meanings" (Peretti, 2001b, p. 1), an "explicit attempt to monkeywrench the media machine" (Meikle, 2003, p. 131). Thompson and Sholette (2004, p. 62) call culture jamming a way to "reclaim the streets." New York City activist Andrew Boyd offers "Culture Jamming 101" seminars on university campuses to encourage others to create their own projects (Caldwell, 2003).

Outdoor public spaces are a frequent venue for culture jamming projects. Jill Posener (1982) documented early reworkings of sexist and racist billboards in the United Kingdom. In San Francisco, Craig Baldwin's Billboard Outlaws project and the Billboard Liberation Front created "subvertising" by altering outdoor advertisements for liquor, tobacco, and military recruiting (Thompson & Sholette, 2004). Dery (1993) detailed billboard projects as well as the overnight poster blitzes of Los Angeles poster artist Robbie Conal, and linked them with John Heartfield's anti-Fascist collages during World War II, *samizdat* in the Soviet Union, and the clip-art strategy of *détournement* employed by European Situationist artists and philosophers in the 1960s (discussed later).

Mainstream media have been a longtime culture jamming target (as, e.g., in the advertising and design parodies of the Canadian print magazine *Adbusters*; Lasn, 2000). However, culture jamming entered a new and more vigorous phase with the advent of digital and online media, which are ideally suited to the "cut-and-paste," collage-style approach to designing messages and making art with a point (see Bolter & Grusin, 1999). Activists and artists quickly adopted these tools to reach new audiences, recruit support, and organize artists' collectives. For example, the group Detritus supports artists or "detrivores" who take "existing materials, break them down, and use them as building blocks to form something new" (http://detritus.net). Detritus member Illegal Art creates "the 'degenerate art' of a corporate age: art and ideas on the legal fringes of intellectual property" (http://illegal-art.org).

The nonprofit arts organization ®™ark (pronounced "art-mark"), a prominent player in the culture-jamming world, solicits and distributes philanthropic funding to artists and projects that challenge corporate and government abuses of power and consumerism (http://www.rtmark.org). ®™ark-funded artists have switched voice chips in Barbie and GI Joe dolls and subsequently repackaged and sold them in a major retail toy outlet (the "Barbie Liberation Organization"); have critiqued the popular musician's image and music sources in a CD distributed by

®™ark, *Deconstructing Beck*; and designed and posted GWBush.com, a parody of George W. Bush's 2000 Presidential campaign Web site (for further discussion of ®™ark projects, see Meikle, 2002).

The organization also figured prominently in a dispute in 2002 between the Dow Chemical Company and activist art collective and ®™ark client The Yes Men, who posted a spoof of the Dow Web site on the 18th anniversary of the toxic gas leak at Dow's Union Carbide plant in Bhopal, India, and linked it to the ®™ark home site (Delio, 2002). Alleging copyright violations, attorneys for Dow complained to NTT/Verio, the telecommunications operator for ®™ark's ISP ("The Thing"), which also served numerous other arts organizations in New York City, including P.S. 1, Mabou Mines, and the magazine *Artforum*. Although The Yes Men complied almost immediately with Verio's demand to remove the spoof site, Verio terminated The Thing's service permanently at the end of February 2003, interrupting Internet access for all its clients. Dow thus succeeded in suppressing their critics' speech without a legal challenge. The Thing's owner, Wolfgang Staehle, later told the *Village Voice* that he was considering relocating his service to Europe, beyond the reach of the American DMCA (Carr, 2003).

The growing use of closed-circuit video surveillance cameras is another culture-jamming target. Maps of all the surveillance cameras in certain New York neighborhoods have been posted online (http://www.notbored.org/scp.maps.html). One project uses inexpensive laser pointers to temporarily disable surveillance cameras in public spaces (Naimark, 2002). Experimental dance troupes perform for cameras inside ATM lobbies (Markoff, 2002); the Surveillance Camera Players stage performances of classical and contemporary plays in front of surveillance cameras for the benefit of whoever might be monitoring them (Thompson & Sholette, 2004; see also http://www.notbored.org/the-scp.html). A British television show has sponsored a "competition for the most inventive closed-circuit television performance" (Rimensnyder, 2001).

Ultimately, culture jamming is a technique that "mines" mainstream media culture to criticize it. But the mainstream has become increasingly wise to the strategy and has engaged in some "reverse jamming" of its own. Scholars have noted the counterforces of appropriation of ideas, symbols and identities between "authentic" social and cultural movements and commercial and marketing interests (Gamson, 2003; McCaughey & Ayers, 2003). Some media critics object to commercial interests that co-opt the style and attitude of rebellion, nonconformity, and political resistance to sell their products, arguing that such co-optation closes off the possibility of dissent itself (Frank & Weiland, 1997).

Jonah Peretti's (2001a, 2001b) "Nike Media Adventure" is a case in point. To associate Nike's brand with freedom and creativity, the company invited customers to order customized running shoes inscribed with their choice of words. When Peretti ordered shoes bearing the word "sweatshop," the company refused to fill the order, or to acknowledge the implicit critique of their overseas labor practices. The evasiveness of their e-mail response prompted Peretti to circulate the correspondence to several friends, who passed it along to others; the episode was then reported by online news sites Slashdot.com and Plastic.com. Mainstream print and broadcast news organizations monitoring the online coverage picked up the story. In the end, Nike precipitated exactly the type of negative publicity that they had sought to avoid in the first place. (The Peretti–Nike correspondence is summarized in McCaughey & Ayers, 2003; the transcript is available online at http://www.shey.net/niked.html.)

Alternative Computing

A second genre of oppositional new media, alternative computing, is the province of computer professionals who object to political or commercial restraints on access to information and information technology. It includes critical, but constructive, activities, such as the creation and distribution of "free" or "open source" software (e.g., Free Software Foundation, http://www.gnu.org); planting hidden "Easter egg" codes in software programs that acknowledge the contributions of otherwise uncredited programmers (Temkin, 2003; www.eeggs.com); demonstrating the built-in and usually invisible cultural and political biases of search engines (as in Natural Selection, a search engine hack by the U.K. art and software collective Mongrel; http://www.mongrel.org.uk), or publicly demonstrating the susceptibility of popular software programs to viruses or security breaches, sometimes over the objections of the programs' vendors.

Alternative computing also encompasses more disruptive "Outlaw computer hacking *with the intent of exposing institutional or corporate wrongdoing*" (Dery, 1993, p. 5), such as the development and distribution of data encryption programs or remailing services (e.g., Anonymizer.com) that elude state and commercial surveillance (Ludlow, 2001); disabling or sabotaging digital copy protection schemes; mapping and posting the locations of wireless "hot spots" where users can piggyback on others' broadband access ("war driving"; www.wardriving.com, http://en.wikipedia.org/wiki/wardriving); or launching denial-of-service, spam, or "ping-storm" attacks on the computing systems of organizations that are viewed as engaging in exploitative, unjust or corrupt activities.

Historically, the roots of alternative computing extend to counterculture values that flourished in American society in the 1960s and 1970s, especially among self-styled "hackers," programmers and engineers who saw computing as a force for positive social transformation, and who advocated a

> commitment to total and free access to computers and information, belief in the immense powers of computers to improve people's lives and create art and beauty, mistrust of centralized authority, a disdain for obstacles erected against free access to computing, and an insistence that hackers be evaluated by no other criteria than technical virtuosity and accomplishment (by hacking alone and not "bogus" criteria such as degrees, age, race, or position). (Nissenbaum, 2004, p. 197)

The term *hacker* originated within the computing/software community in the 1960s and 1970s to describe highly skilled programmers who were especially adept at creating elegant, counterintuitive solutions ("hacks") for difficult or intractable programming problems (Levy, 1984; Nissenbaum, 2004; Peterson, 2003). "Hacker culture" combined technological mastery and elitism with libertarian, utopian visions of a more open society (Barbrook, 2001; Thomas, 2002). Student hackers at the Massachusetts Institute of Technology, for example, adhere to a list of "hacker ethics," including:

> Be *safe*. Your safety, the safety of your fellow hackers, and the safety of anyone you hack should never be compromised. . . . Be *subtle*. Leave no evidence that you were ever there. . . . Leave no damage. . . . Do not steal anything. . . . Brute force is the last resort of the incompetent. . . . Do not hack alone (just like swimming). (Peterson, 2003, p. 10)

Most early hacks were typically intended to demonstrate the skill of the programmer rather than to damage or disrupt the system *per se* (Peterson, 2003). Beginning in the 1980s, programmers distinguished between "white-hat" hackers, who might deliberately seek out and expose the vulnerabilities of systems and software programs, for example, and destructive "black-hat" crackers, who might circulate computer viruses or break into computers to destroy or steal data (Leibowitz, 2003).

In recent years, however, the "hacker" label has been taken up by law-enforcement agencies, government and for-profit firms, and widely repeated in the popular media, as a synonym for vandal, thief, or terrorist, in an effort to frame independent technological skill and creativity as dangerous or criminal (Nissenbaum, 2004), often by associating hacking with violations of intellectual property laws. In the widely publicized case of online journalist Eric Corley, Universal Studios and other large entertainment firms sued Corley in federal court, alleging violations of the DMCA, when Corley published the Decrypt Content Scrambling System (DeCSS), a short piece of code written by a young Norwegian programmer, in his online magazine *2600: The Hacker Quarterly* (http://www.2600.com). DeCSS enabled users to decrypt and view legally purchased DVDs on computers running open-source operating systems as well as Microsoft OS-based systems (for further discussions of the case, see Lessig, 2001 and Litman, 2001).

Although Corley invoked his First Amendment speech and press rights as a defense, the court's preliminary injunction in January 2000 directed Corley to remove the program from the *2600* Web site. He complied, but maintained links to other sites where DeCSS was still available. In August 2000, the court ruled that not only could Corley not post the code itself; he was also prohibited from posting links to any other site or source for DeCSS, even in jurisdictions where the program was legal. On appeal in 2002, the U.S. Second Circuit Court of Appeals upheld the original ruling. Corley and other defendants in the case chose not to appeal the case to the Supreme Court.

Despite the 2002 ruling, and the subsequent obsolescence and replacement of DeCSS by other DVD decryption programs, however, DeCSS has remained controversial. Eschenfelder and Desai (2004) found that there was almost no decline in the number of sites, including U.S.-hosted sites, posting or linking to executable forms of DeCSS between January 2001, when the initial *Corley* injunction was issued, and mid-2003. They interpreted the persistent (and illegal) posting of DeCSS as a protest movement against the DMCA within the computing community. Other protesters have converted the code into artworks, on the principle that works of art are unambiguously covered by First Amendment speech protections. An online art gallery at Carnegie-Mellon University exhibits paintings, sculptures, apparel, and other works that incorporate DeCSS (see http://www.cs.cmu.edu/~dst/DeCSS/Gallery). "The History of the DeCSS Haiku" can also be found online (http://www.loyalty.org/~schoen/haiku.html).

Despite setbacks like the Corley case, activist programmers, designers, cultural critics and artists alike have sought to recapture and extend the positive connotation of "hacker" to include creative and knowledge workers of all sorts whose works are subject to commodification as intellectual property. As MacKenzie Wark (2004) put it in *A Hacker Manifesto*,

> Hackers create the possibility of new things entering the world. Not always great things, or even good things, but new things. In art, in science, in philosophy and culture, in any production of knowledge where data can be gathered, where information can be extracted from it, and where in that information new possibilities for the world produced, there are hackers hacking the new out of the old. (§004)

Mediated Sociality and Mobilization

A third genre of oppositional new media involves the use of new media technologies as sites for sociality, participation, and action. Mediated sociality and mobilization supports both mediated interpersonal interaction (e.g., through blogs, chatrooms, or acquaintance-based "social software" sites like Friendster.com and MySpace.com) as well as the coordination of informal "live" meetings (e.g., participation in smart mobs [Rheingold, 2002] or meetups for political or cultural events; see www.flashmob.com, www.meetup.com).

Of course, the use of electronic media in social and political movements is not new. Historian Margaret Graham (2000) showed that radio, sound motion pictures, and television helped promote and mobilize a "national unity culture" of consumption, popular entertainment, wartime solidarity, and economic production in the United States from the 1930s to the 1960s. The uses of radio and motion pictures in the mass political movements of both the right and left before World War II are well documented. Mass media were well suited to the task of presenting consistent, repetitive messages to large, heterogeneous audiences, with the aim of shaping broad-based popular opinion, fostering mass consumption, or mobilizing political movements.

Today, however, as the "mass" view of society has given way to a more complex, dynamic view of social structure as constantly reorganizing, interrelated networks of nodes, links and flows (e.g., Castells, 1996), new media technologies are better suited for helping people seek, find, and assess information and each other than for the wholesale distribution of messages to mass audiences. Mobilization and sociality in today's media environment depends on people's abilities to cultivate relationships, seek and give advice, make recommendations, and amass and trade "reputation capital" (May, 2001) in complex and extensive social, technical, and institutional networks.

The fortunes of peer-to-peer (P2P) computing systems such as Napster richly illustrate the growing interrelationships between social and technological networks and their consequences for sociality and mobilization. There is nothing inherently illicit or "oppositional" about P2P software, which allows people to locate and retrieve materials from one another's computers, rather than from a central repository. As its very name suggests, it is a technological model based on the idea that people seek and give advice and make recommendations in the context of a social network of others with similar interests. Like interpersonal advice networks, P2P systems are social and cooperative environments by definition; they are opportunities for people with similar interests and experiences to share what they know, for "social authorship" (Atton, 2005, p. 103).

However, one of the first P2P systems became a *cause célèbre* in the late 1990s when record companies sued Napster's college-student creator, Shawn Fanning, on the grounds that P2P architecture was designed to facilitate the illegal copying of copyrighted material. Fanning was forced to shut Napster down and sell his software and patents to the media giant Bertelsmann to meet his legal expenses. Napster has since been re-established as a paid-subscription music-download service, with Fanning employed as a spokesman.

Today, other P2P networks, such as KaZaA, Gnutella, and Grokster, with different and internationally distributed architectures, continue to operate, albeit as legal content downloading sites (Woody, 2003). In a recent setback for P2P services, the U.S. Supreme Court ruled that companies that distribute P2P software (in this case, Grokster) are liable for the file-sharing activities of their customers ("A bad week," 2005). As Atton (2005) pointed out, the original Napster was "socially attenuated," that is, not conceived with the purpose of advancing and defending the digital commons idea. Nonetheless, users of the surviving P2P systems have tended to view file-sharing as a form of economic and cultural resistance against an oligopolistic industry that practices an outmoded way of doing business.

Indymedia

A fourth genre of oppositional new media both upholds and critiques the traditions of journalism and the press. *Indymedia* sites, and their close cousins including news and opinion blogs, wikis, and chat sites, provide local news reporting, commentary, alternative information resources, and critique of mainstream news, but are established and run mainly by amateur or volunteer contributors rather than professional reporters and editors. A key characteristic of indymedia is an emphasis on connectivity, interactivity, and community, on breaking down conventional distinctions between information providers, reporters and editors, on the one hand, and readers/citizens on the other (Deuze, 2003). As Kidd (2003, p. 47) put it, Indymedia constitute a "communications commons" whose watchword is "Don't hate the media—become the media" (p. 47)

In 1999, a group of media activists covering the World Trade Organization protests in Seattle established the first independent media center (IMC) to counter mainstream coverage of the protests, and coined the term *Indymedia* (capitalized) (Hyde, 2002). IMCs were soon set up in Boston and Washington, using Web sites as well as print newsletters, radio, and video (http://docs.indymedia.org/view/Global/FrequentlyAskedQuestionEn). By mid-2004 more than 150 such IMCs and Web sites were running in the United States and internationally, constituting "the largest all-volunteer media system in the world" (Meinrath, 2004; see also http://www.indymedia.org).

Strictly speaking, Indymedia (capitalized) applies only to news and opinion sites affiliated with IMCs. However, in the last couple of years the term indymedia (lowercase) has been applied more broadly to a variety of web-based alternative, radical, or critical news sites that have adopted the practices and/or philosophies of public, civic, participatory, or "open-source" journalism, using the "indymedia news model" (Platon & Deuze, 2003, p. 336). Critics charge that by employing journalistic conventions and practices and Internet technology, indymedia in fact reproduce a Eurocentric, technology-driven cultural and political ideology of progress (Hoofd, 2004). However, indymedia proponents claim that they provide an alternative to the agendas, politics, economics, reporting practices, and ethics of mainstream news organizations, to advance minority or marginalized points of view in a global arena; therefore, they are the direct descendants of "alternative" or "radical" news media (Atton, 2002, 2003, 2005; Downing et al., 2001).

FEATURES OF OPPOSITIONAL NEW MEDIA

Culture jamming, alternative computing, mediated sociality/mobilization, and indymedia, then, are four distinct genres or "faces" of ONM in the online environment today. Although each relates to a particular social/technical domain (popular culture, technology infrastructure, social networks and organizing, and news and opinion, respectively), all four genres share a number of important characteristics or features: they are small-scale, interventionist, subculturally literate, ironic, perishable, collaborative, and heterotopic.

The first characteristic is the relatively *small scale* of ONM projects. They are "micromedia" (Peretti, 2001a), or put differently, they are "what happens when the cheap 'do it yourself' media. . . . are exploited by groups and individuals who feel aggrieved by or excluded from the wider culture" (Garcia & Lovink, 1997). They are run with minimum investment. For example, ®™ark distributes no more than a few thousand dollars of donor money per project that it supports; indymedia sites are sustained by volunteer staff, and donated equipment and operating funds. Blogs are frequently produced and hosted by private individuals or small teams. The small scale of most oppositional sites therefore tends to give visitors and participants a sense of intimacy, of being an insider or initiate.

The second important characteristic of ONM is that they are *interventionist.* They "introduce noise into the signal" (Dery, 1993), parodize, invert or distort media images (Hebdige, 1979), and "create situations in the world at large" (Thompson & Sholette, 2004, p. 10). As Chris Atton (2005) said of alternative Internet sites, they are ". . . a range of media projects, interventions and networks that work against, or seek to develop different forms of, the dominant, expected (and broadly accepted) ways of 'doing' media" (p. ix).

Their creators aim to subvert commonsense or taken-for-granted meanings and situations: "Culture jammers do not exist without corporate billboards" (Lovink & Richardson, 2001).

ONM can either constitute intervention and action in themselves (e.g., performances by the Surveillance Camera Players; the persistent posting of DeCSS code on U.S. Web sites), or invite and motivate intervention by others (e.g., ®™ark's "Mutual Funds," which solicit "investors" for prospective artworks, http://www.rtmark.com/legacy.funds.html, or Michael Naimark's posted instructions for disabling surveillance cameras; see also Braman, 2002). Dery (1993) argued that those who make counterintuitive interpretations of the dominant culture are practicing the "guerrilla semiotics" advocated by Umberto Eco in 1967:

> What must be occupied, in every part of the world, is the first chair in front of every TV set (and naturally, the chair of the group leader in front of every movie screen, every transistor, every page of every newspaper) . . . the battle for the survival of man as a responsible being in the Communications Era is not to be won where the communication originates, but where it arrives. (Eco, 1986, p. 142)

The third key feature of ONM is that they both manifest, and expect from their audiences, a relatively high level of *subcultural literacy* (Kahn & Kellner, 2003), a "hyper self-reflexivity about the nature of pop culture" (Collins, 1995, p. 2), an "aesthetic of poaching, tricking, reading, speaking, strolling, shopping, desiring" (Garcia & Lovink, 1997, n.p.). ONM sites reinterpret the familiar with an acute sense of awareness of and referentiality to everyday events, styles, and ideas, expecting that audiences will "get it," too. Popular issues, images, buzzwords, and attitudes are selected, captured, subverted, co-opted, fragmented, recombined and re-presented in unexpected (and ideally felicitous) ways. For example, Deuze (2003) emphasizes the role of *bricolage* in indymedia sites, of bringing diverse materials together with a pastiche effect. Atton (2004) emphasized that alternative sites on the Internet are essentially subcultural.

Subcultural literacy thus links oppositional new media to a longer legacy of radical, activist and subversive art, like that made by Soviet avant-garde artists in the 1920s and 1930s, Fluxus in the 1950s and 1960s, and the Situationists in the 1960s (Sholette, 2004). Guy Debord's Situationist concept of *détournement*, the "strategy of diverting elements of affirmative bourgeois culture to revolutionary ends, of distorting received meanings" (McDonough, 2002, p. xiv), has been particularly influential in ONM. Debord and his colleagues used images from popular comics, pornography, film stills, and advertising in the posters, statements, films, and manifestoes of the French student uprising in May 1968 (Ford, 2005; Gray, 1974/1998; Viénet, 1967/1981; for overviews of situationism and *détournement* relative to culture jamming, see Meikle, 2002, and Kahn & Kellner, 2005). Virtually the same technique is employed today in a wide array of contemporary online sites such as *Get Your War On* (www.mnftiu.cc/mnftiu.cc/war.html).

ONM also tend to be *ironic*, playful, humorous, campy, or parodic (Braman, 2002). "[Culture jamming] is immature . . . [it] celebrates the possibility of ironic, humorous and contradictory political actions" (Peretti, 2001a). ONM "do not take themselves that seriously . . . they know how to laugh" (Lovink & Richardson, 2001, n.p.). Although "Nothing can suck the air around it like political art: so many words, so much ideology worn on so transparently on the sleeve, so much certainty, and so little of interest to look at" (Thompson, 2004, p. 10), many ONM sites employ humor as a rhetorical tool. The absurd quality of much of mainstream culture, economics and politics is exploited by the Surveillance Camera Players, The Yes Men, ®™mark, and instigators of smart mobs, for example, whose aim is to inspire outrage by pointing out and mocking the absurdity. Even situations that are not humorous or ironic to begin with can be reframed ironically, as the artworks in the Carnegie-Mellon DeCSS Gallery do with the *Universal Studios et al. v. Reimerdes (Corley)* case.

A fifth characteristic of oppositional new media is that they are *perishable*. They are ephemeral responses to rapidly changing cultural contexts and meanings, and as short-lived as electronic media and information technologies are generally. Like the social and technological networks that support them, they organize, disorganize, and reorganize more or less continuously. Neither the infrastructure nor the content of networked new media have stable form or fixed structure. They are notable for their "mobility, [their] flexible response to events and changing contexts" (Meickle, 2000, n.p.), and "capable of taking risks, even if this means they might self-destruct in the process" (Lovink & Richardson, 2001, n.p.). Smart mobs or flash mobs, for example, are organized online on the spur of the moment, and disperse almost as soon as they form (see www.flash-mob.com; www.cheesebikini.com; also see Rheingold, 2002). In this respect, ONM resembles earlier movements, like conceptual and performance art, in which works were deliberately made not to leave material (and thus collectible or commodifiable) traces or remnants.

Two final characteristics are closely related. First, ONM are *collaborative*; they are often group efforts. Community and interactivity are assumed in the design and organization of oppositional sites and projects: the power of new media "lies less in the information that it [sic] carries than in the communities it creates" (Schrage, n.d.). For example, wikis are collaboratively built, independent, online collections of reference information and advice (see, e.g., the Wikipedia project, www.wikipedia.org). In their analysis of indymedia, Platon and Deuze (2003) noted, "The potential for immediate interaction between users on the internet surpasses all other media . . . [Indymedia] also has an increased potential for (re)connecting media formats and forms of journalism with different types of community" (p. 337) The point was made previously regarding the "social authorship" of P2P systems (Atton, 2005). Even projects created by individuals, such as personal or opinion blogs, typically aim to elicit response and participation from their readers.

Second, this capacity of ONM to reach and organize groups of people with specific interests makes them *heterotopic*. That is, oppositional sites often constitute diverse "other spaces" for personal and group expression, affiliation, and creativity, apart from the mainstream or dominant culture (Lievrouw, 1998). Critics have noted the potential for social fragmentation, or the loss of civic or public culture, in the diversity of online sites (e.g., Mitchell, 2003; Sunstein, 2001). However, such "cyber-separatism" or "nanoaudiences" (Kahn & Kellner, 2005) may be balanced by the growing tendency of underrepresented groups to seek greater visibility or legitimacy for their views in the global arena by going online, as in the widely cited and intensely studied use of Web sites by the Zapatista movement in Mexico in the 1990s (see, e.g., Atton, 2005; Garrido & Halavais, 2003; Meickle, 2002; Wright, 2004).

ONM, OWNERSHIP, AND ACCESS

As the preceding discussion suggests, the contemporary media environment is no longer dominated exclusively, or even primarily, by traditional media industries, content, or forms. Increasingly, mass media must contend with small, localized, innovative, and counter cultural "micromedia" that sample, capture, co-opt, criticize, hack, and parody the mainstream culture the industries produce. For every Universal Studios, there is a *Hacker Quarterly*; for every Bertelsmann, Dow

Chemical, or *New York Times*, a Napster, Yes Men, or *Slashdot*. What are the implications of this contested landscape for media ownership and access?

On one hand, as illustrated in the "two visions" discussion, the governing principle underlying the traditional media environment has been the notion of *property* and the extension of property rights to ever more types of expressions, ideas, practices, and relationships. Mass media firms have tended toward concentrated, centralized ownership in order to exploit industrial-style economies of scale and to maximize return on the mass production and distribution of tangible products such as movies, software, music CDs, and books. Their success and profitability have depended on the efficient control and use of fixed, capital-intensive, hierarchically organized production and distribution facilities and other assets (e.g., radio and television stations, motion picture studios and theaters, printing plants), and on strict gatekeeping control over the distribution and sale of their products. As new media have challenged and eroded their markets, major media firms have waged a forceful campaign to establish progressively more restrictive intellectual property regimes that expand the definition of property and property rights, and commodification and control, beyond tangible assets and products and into the domain of intangible cultural practices, relationships, and resources (Cohen, 2004). To some observers, the extent of commodification now defies common sense:

> the rules that copyright lawyers take for granted are so very counterintuitive that people commonly refuse to believe that they could possibly be the rules ... Members of the general public commonly find copyright rules implausible, and simply disbelieve them. (Litman, 2001, p. 29)

"Media access," under these conditions, remains more or less synonymous with reception or consumption, whether of tangible (books, television programs) or intangible (rights, licenses) media products.

In contrast, the governing principle underpinning the new media environment is that it is as much an arena for speech, interaction, and affiliation as for the production and consumption of goods. Reputation, credibility, voice, visibility, trust and affiliation have become the essential "currencies" online (e.g., Kollock, 1999; May, 2001). Instead of economies of scale, network externalities determine the value of new media: The larger and more heterogeneous media systems are, the more valuable they are to each individual user. Value is not determined by how much of the system is owned or by whom, but by the system's scope (the variety of users and sources available), connectivity (how well and how easily it links users to each other and to informational resources), and utility (how well it helps people do what matters to them). Computers, servers, wireless networks, and Internet access are much less capital- and skills-intensive than broadcasting facilities, movie studios, publishing houses, or cable systems, giving users the means to connect to the network, to create, move and receive content from any point in the system, and to locate and contact any other user in the system.

Moreover, this diversity of uses and users, and the dynamic, point-to-point architecture and global scale of new media systems, are recombinant: That is, they are the product of an "ongoing cycle of human action and available technical and cultural resources" (Lievrouw & Livingstone, 2006, p.5). Thus, the new media environment tends to resist fixation and stabilization, and new media ownership tends to be relatively decentralized and distributed, at least at the still-developing "edges." For example, no single entity, group, or industry controls the greatest part of the Internet, from the backbone out to users' desktops, with anything like the degree of integration seen in the traditional motion picture, recorded music, or broadcasting industries. Access in the current media environment, then, requires the ability to express, create, tinker, play, search, interact, and organize—as well as to consume.

These characteristics suggest that policies based on the property model, and on the economics of production and consumption, by themselves cannot assure media access. Certainly, it is essential to safeguard access in the traditional sense of consumption, for example by preserving universal service obligations and extending them beyond the traditional arenas of voice telephony and public/state broadcasting, and to expand the public domain and related rights such as first sale and fair use, as persuasively argued by Lessig (2001) and others (see also Lessig's Creative Commons project). However, people engage with media differently than they did even a decade ago: Although mass media products are still produced and consumed, people also use media to create, navigate, search, interrogate, retrieve, and play with those products and with each other. To foster effective media access under these conditions, policymakers must not only insure the broad availability of centrally controlled media systems, products, and services; they must also ensure that people have the opportunities, affordances, and capabilities to use those products and services effectively to achieve their own ends.

Media access in this sense entails two main processes. First, effective access requires that people should be able to *reconfigure* media technologies as needed to suit their particular purposes. The continuing process of innovation, adaptation, and reinvention of media and information technologies distinguishes new media from older mass media systems. As recent history demonstrates, many of the most interesting and useful media technology innovations have been created by independent inventors like Shawn Fanning, who combined existing techniques and materials to solve a particular problem in a new way. The scope, connectivity, and utility of his solutions prompted others to adopt them. Heavily capitalized, established industries, in contrast, are more likely to have a stake in the existing technological base, which tends to discourage such small-scale, quick-response types of innovation.

Second, access in the new media environment requires that people should be able to *remediate* content and relationships to construct new meanings and forms of interaction. Bolter and Grusin (1999) defined remediation as "the representation of one medium in another" (p. 45), including straightforward repackaging (e.g., storing still photographs or taperecorded music on CD-ROM), augmentation or enhancement (e.g., online newspapers or encyclopedias that combine hyperlinking, audio, and motion images with conventional text), refashioning (e.g., "tiled" windows on a desktop or television screen, perceived by the user as a composite whole), and absorption, where one medium fully captures another (e.g., as video games use many of the cinematic and plot devices of motion pictures). These techniques can be used alone or together, but either way, existing content and forms are borrowed, adapted, sampled, or remixed to create new expressions, new relationships, and new content. Of course, there is nothing new about this process from a creative point of view, but Bolter and Grusin argued that digital media tools have helped make remediation the hallmark of contemporary creative work and media culture.

The lesson of oppositional new media with regard to media ownership and access, then, is that they are "laboratories" where users resist the fixity of traditional media, and negotiate and shift the boundaries of old and new media in an ongoing process of co-optation and reinvention. The property metaphor that has framed the understanding of media production, distribution, and consumption processes in the past is no longer sufficient for understanding the multiple, networked, and recombinant character of the contemporary media environment, where speech, interaction, and affiliation are as sought after as intellectual property. To foster effective access today, media policy must insure that people can continue to reconfigure and remediate culture as well as consume it.

ACKNOWLEDGMENTS

Some material in this chapter has appeared in Lievrouw (2003, 2006), and in presentations at the 2004 annual meetings of the International Communication Association and the Society for Social Studies of Science, and the 2006 Participatory Design Conference. The author acknowledges the research assistance of Pam Jackson, graduate student in the Department of Information Studies at UCLA, in the background research for this chapter, and the support of the UCLA Academic Senate Faculty Research Grant Program.

REFERENCES

Atton, C. (2002). *Alternative media.* London: Sage.

Atton, C. (2003). What is "alternative" journalism? *Journalism, 4*(3), 267-272.

Atton, C. (2005). *An alternative Internet.* New York: Columbia University Press.

A bad week for pirates. (2005a, July 2). *Economist,* pp. 57-58.

Barbrook, R. (2001). HyperMedia freedom. In P. Ludlow (Ed.), *Crypto anarchy, cyberstates, and pirate utopias* (pp. 47-58). Cambridge, MA: MIT Press.

Belson, K., & Richtel, M. (2003, May 5). America's broadband dream is alive in Korea. *New York Times*, pp. C1, C4.

Bolter, J. D., & Grusin, R. (1999). *Remediation: Understanding new media.* Cambridge, MA: MIT Press.

Braman, S. (2002). *Defining tactical media: An historical overview.* Retrieved August 2, 2005, from http://www.nyu.edu/fas/projects/vcb/case_911/reverberations/braman2.html

Caldwell, R. (2003, January 18). The guru of conscientious consumption. *Toronto Globe & Mail*, p. R11. Retrieved August 10, 2005, from http://www.globeandmail.com/.

Carr, C. (2003, January 17). *Dow v. Thing.* A free-speech infringement that's worse than censorship. *Village Voice.* Retrieved August 2, 2005, from http://www.villagevoice.com/issues/0304/carr.php.

Castells, M. (1996). *The rise of the network society — The information age: Economy, society and culture* (Vol. 1.) London: Blackwell.

Cohen, J. (2004). *Normal discipline in the age of crisis.* Unpublished manuscript, Georgetown University Law Center, Washington, DC.

Collins, J. (1995). *Architectures of excess: Cultural life in the information age.* London: Routledge.

Debord, G. (1981). Instructions for taking up arms. In K. Knabb (Ed.), *Situationist International anthology* (pp. 63-65). Berkeley, CA: Bureau of Public Secrets. (Original work published 1961)

Delio, M. (2002, December 31). DMCA: Dow what it wants to do. *Wired.* Retrieved August 2, 2005, from http://www.wired.com/news/print/0,1294,57011,00.html.

Dery, M. (1993). *Culture jamming: Hacking, slashing and sniping in the empire of signs.* Westfield, NJ: Open Magazine Pamphlet Series. Retrieved August 2, 2005, from http:///www.levity.com/markdery/jam.html.

Deuze, M. (2003). The web and its journalisms: Considering the consequences of different types of newsmedia online. *New Media & Society*, *5*(2), 203-230.

Downing, J. D. H., with T. V. Ford, G. Gil, & L. Stein (2001). *Radical media: Rebellious communication and social movements.* London: Sage.

Eco, U. (1986). *Travels in hyperreality.* New York: Harcourt Brace Jovanovich.

Emerging-market indicators: Broadband penetration. (2005b, July 2). *Economist*, p. 90.

The end of privacy. (1999, May 1). *Economist*, pp. 15-16, 21-23.

Eschenfelder, K. R. & Desai, A.C. (2004). Software as protest: The unexpected resiliency of U.S.-based DeCSS posting and linking. *The Information Society*, *20*(2), 101-116.

Ford, S. (2005). *The Situationist International: A user's guide.* London: Black Dog Publishing.

Frank, T., & Weiland, M. (Eds.). (1997). *Commodify your dissent.* New York: W.W. Norton.

Gamson, J. (2003). Gay Media, Inc.: Media structures, the new gay conglomerates, and collective sexual identities. In M. McCaughey & M.D. Ayers (Eds.), *Cyberactivism: Online activism in theory and practice* (pp. 255-278). New York and London: Routledge.

Garcia, D., & Lovink, G. (1997). *The ABC of tactical media.* Retrieved August 2, 2005, from http://subsol.c3.hu/subsol_2/contributors2/garcia-lovinktext.html.

Garrido, M., & Halavais, A. (2003). Mapping networks of support for the Zapatista movement: Applying social-networks analysis to study contemporary social movements. In M. McCaughey & M.D. Ayers (Eds.), *Cyberactivism: Online activism in theory and practice* (pp. 165-184). New York & London: Routledge.

Gleick, J. (2000, March 12). Patently absurd. *New York Times Magazine*, pp. 44-49.

Graham, M. (2000). The threshold of the information age: Radio, television and motion pictures mobilize the nation. In A. D. Chandler, Jr. & J. W. Cortada (Eds.), *A nation transformed by information* (pp. 137-176). Oxford & New York: Oxford University Press.

Gray, C. (Ed.). (1974). *Leaving the 20th Century: The incomplete work of the Situationist International.* London: Rebel Press. (Original work published in 1974 by Free Fall Press)

Greenberger, M. (1964). The computers of tomorrow. *Atlantic Monthly, 213*(5), 63-66.

Greenberger, M. (Ed.). (1985). *Electronic publishing plus.* White Plains, NY: Knowledge Industry Publications.

Gurak, L. J. (1997). *Persuasion and privacy in cyberspace: The online protests over LotusMarketplace and the Clipper Chip.* New Haven, CT & London: Yale University Press.

Gurak, L. J., & Logie, J. (2003). Internet protests, from text to web. In M. McCaughey & M.D. Ayers (Eds.), *Cyberactivism: Online activism in theory and practice* (pp. 25-46). New York & London: Routledge.

Hebdige, D. (1979). *Subculture: The meaning of style.* London & New York: Routledge.

Hoofd, I. (2004, April 2). Problems within and lessons to learn from the Indymedia project. Paper presented at *The Other Internet: Exploring Digital Resistance.* Conference sponsored by the A. W. Mellon Foundation. American University of Paris, Department of International Communications. Retrieved February 28, 2007, from http://www.aup.fr/news/pastconf/programs/2004_otherinternet.htm

Huhtamo, E. (2003, February 20). Perspectives on media, art and interactivity. Colloquium, Department of Information Studies, University of California, Los Angeles.

Hyde, G. (2002). Independent media centers: Cyber-subversion and the alternative press. *First Monday, 7*(4). Retrieved August 1, 2005, from http://firstmonday.org/issues/issue7_4/hyde/index.html.

Kahn, R., & Kellner, D. (2003). *Internet subcultures and oppositional politics.* Unpublished manuscript, Graduate School of Education and Information Studies, University of California, Los Angeles.

Kahn, R., & Kellner, D. (2005). Oppositional politics and the Internet: A critical/reconstructive approach. *Cultural Politics, 1*(1), 75-100.

Kidd, D. (2003). Indymedia.org: A new communications commons. In M. McCaughey & M.D. Ayers (Eds.), *Cyberactivism: Online activism in theory and practice* (pp. 47-70). New York & London: Routledge.

Kollock, P. (1999). The economies of online cooperation: Gifts and public goods in cyberspace. In M.A. Smith & P. Kollock (Eds.), *Communities in cyberspace* (pp. 220-242). London & New York: Routledge.

Lasn, K. (2000). *Culture jam: How to reverse America's suicidal consumer binge—and why we must.* New York: HarperCollins.

Leibowitz, B. (2003). Hack, hacker, hacking. In T. F. Peterson, *Nightwork: A history of hacks and pranks at MIT* (p. 4). Cambridge, MA: MIT Press. (Original work published 1990)

Lessig, L. (2001). *The future of ideas.* New York: Random House.

Lessig, L. (2003, May 8). *The Eugene Garfield Lecture.* Department of Information Studies, University of California, Los Angeles.

Levy, S. (1984). *Hackers: Heroes of the computer revolution.* New York: Doubleday.

Lievrouw, L.A. (1998). Our own devices: Heterotopic communication, discourse and culture in the information society. *The Information Society, 14*(2), 83-96.

Lievrouw, L.A. (2003). When users push back: Oppositional new media and community. In M. Huysman, E. Wenger, & V. Wulf (Eds.), *Communities and technologies* (pp. 391-405). Dordrecht: Kluwer Academic.

Lievrouw, L.A. (2006). Oppositional and activist new media: Remediation, reconfiguration, participation. In G. Jacucci, F. Kensing, I. Wagner & J. Blomberg (Eds.), *Proceedings of the Participatory Design Conference, PDC 2006: Expanding Boundaries in Design* (pp. 115-124). Trento, Italy, July 31-August 5. Palo Alto, CA: Computer Professionals for Social Responsibility and Association for Computing Machinery.

Lievrouw, L.A., & Livingstone, S. (2006). Introduction. In L. A. Lievrouw & S. Livingstone, (Eds.). *Handbook of new media* (updated student edition, pp. 1-14). London: Sage.

Lipinski, T. A. (1999). The commodification of information and the extension of proprietary rights into the public domain: Recent legal (case and other) developments in the United States. *Journal of Business Ethics, 22*, 63-80.

Litman, J. (1994). Mickey Mouse emeritus: Character protection and the public domain. *University of Miami Entertainment and Sports Law Review, 11*(2), 429-435.

Litman, J. (2001). *Digital copyright.* Amherst, NY: Prometheus Books.

Lovink, G., & Richardson, J. (2001). *Notes on sovereign media.* Retrieved August 2, 2005, from http://subsol.c3.hu/subsol_2/contributors0/lovink-richardsontext.html.

Ludlow, P. (Ed.). (2001). *Crypto anarchy, cyberstates, and pirate utopias.* Cambridge, MA: MIT Press.

Markoff, J. (2002, October 7). Protesting the Big Brother lens, Little Brother turns an eye blind. *New York Times*, C1.

May, T. C. (2001). Crypto anarchy and virtual communities. In P. Ludlow (Ed.), *Crypto anarchy, cyberstates, and pirate utopias* (pp. 65-79). Cambridge, MA: MIT Press.

McCaughey, M., & Ayers, M. D. (Eds.). (2003). *Cyberactivism: Online activism in theory and practice.* New York & London: Routledge.

McDonough, T. (2002). Introduction: Ideology and the Situationist utopia. In T. McDonough (Ed.), *Guy Debord and the Situationist International* (pp. ix-xx). Cambridge, MA: MIT Press.

Meikle, G. (2000). gwbush.com: Tactical media strike. *M/C Reviews*, April 12. retrieved August 10, 2005, from http://reviews.media-culture.org.au/sections.php?op=viewarticle&artid=50.

Meikle, G. (2002). *Future active: Media activism and the Internet.* London & New York: Routledge, in association with Pluto Press Australia.

Meinrath, S. (2004, May 31). *Studying the rise of Indymedia: Confusions and conclusions.* Paper presented at the annual conference of the International Communication Association, New Orleans, LA.

Mitchell, W. J. T. (2003). *Me++: The cyborg self and the networked city.* Cambridge, MA: MIT Press.

Naimark, M. (2002, October). *How to ZAP a camera: Using lasers to temporarily neutralize camera sensors.* Retrieved August 2, 2005, from http://www.naimark.net/projects/zap/howto.html.

Nissenbaum, H. (2004). Hackers and the contested ontology of cyberspace. *New Media & Society, 6*(2), 195-217.

Patently absurd? (2001, June 23). *Economist*, pp. 40-42.

Peretti, J. (2001a). *Culture jamming, memes, social networks, and the emerging media ecology: The "Nike Sweatshop Email" as object-to-think-with.* Unpublished manuscript, MIT Media Lab. Retrieved January 25, 2003, from http://xenia.media.mit.edu/~peretti/nike. [NB: As of August 2005 this URL is no longer available. Original Peretti/Nike correspondence is available at http://www.shey.net/niked.html.]

Peretti, J. (2001b, April 9). My Nike media adventure. *The Nation.* Retrieved August 2, 2005, from http://www.thenation.com/doc.mhtml?i=20010409&s=peretti.

Peterson, T. F. (2003). *Nightwork: A history of hacks and pranks at MIT.* Cambridge, MA: MIT Press.

Phillips, D. (2004). Privacy policy and PETs: The influence of policy regimes on the development and social implications of privacy enhancing technologies. *New Media & Society, 6*(3), 425-440.

Platon, S., & Deuze, M. (2003). Indymedia journalism: A radical way of making, selecting and sharing news? *Journalism, 4*(3), 336-355.

Posener, J. (1982). *Spray it loud.* London: Pandora.

Rheingold, H. (2000). *The virtual community: Homesteading on the electronic frontier* (rev. ed.). Reading, MA: Addison-Wesley.

Rheingold, H. (2002). *Smart mobs: The next social revolution.* Cambridge, MA: Perseus.

Rimensnyder, S. (2001, September 6). Mugging for the cameras: Tune in to International Surveillance Camera Awareness Day. *Reason.* Retrieved August 2, 2005, from http://reason.com/hod/sr090601.shtml.

Schrage, M. (n.d.) The relationship revolution. *Merrill Lynch Forum.* URL (consulted August 2, 2005): http://www.ml.com/woml/forum/index.htm.

Sholette, G. (2004). Interventionism and the historical uncanny. In N. Thompson & G. Sholette (Eds.), *The Interventionists: Users' manual for the creative disruption of everyday life* (pp. 133-142). North Adams, MA: MASS MoCA Publications.

Sunstein, C. (2001). *Republic.com.* Princeton, NJ: Princeton University Press.

Temkin, E. (2003, May). *Defiant programming: The culture of Easter eggs and its fandom.* Paper presented at the annual conference of the International Communication Association, San Diego, CA.

Thomas, D. (2002). *Hacker culture.* Minneapolis & London: University of Minnesota Press.

Thompson, N., & Sholette, G. (Eds.). (2004). *The Interventionists: Users' manual for the creative disruption of everyday life.* North Adams, MA: MASS MoCA Publications.

Touraine, A. (1988). *The return of the actor.* Minneapolis: University of Minnesota Press. (Translation of *Le retour de l'acteur: Essai de sociologie.* Paris: Librairie Arthème Fayard, 1984.)

van de Donk, W., Loader, B. D., Nixon, P. G., & Rucht, D. (Eds.). (2004). *Cyberprotest: New media, citizens, and social movements.* London: Routledge.

Vegh, S. (2003). Classifying forms of online activism: The case of cyberprotests against the World Bank. In M. McCaughey & M.D. Ayers (Eds.), *Cyberactivism: Online activism in theory and practice* (pp. 71-96). New York & London: Routledge.

Viénet, R. (1981). The Situationists and the new forms of action against politics and art. In K. Knabb (Ed.), *Situationist International anthology* (pp. 213-216). Berkeley, CA: Bureau of Public Secrets. (Original work published 1967)

Wark, M. (2004). *A hacker manifesto.* Cambridge, MA & London: Harvard University Press.

Who owns the knowledge economy? (2000, April 8). *Economist* p. 17.

Winseck, D. (2002). Wired cities and transnational communications: New forms of governance for telecommunications and the new media. In L.A. Lievrouw & S. Livingstone (Eds.), *Handbook of new media* (1st ed.; pp. 393-409). London: Sage.

Woody, T. (2003, February). The race to kill Kazaa. *Wired* 11.02. Retrieved August 2, 2005, from http://www.wired.com/wired/archive/11.02/kazaa_pr.html.

Wright, S. (2004). Informing, communicating, and ICTs in contemporary anti-capitalist movements. In W. van de Donk, B. D. Loader, P. G. Nixon, & D. Rucht (Eds.), *Cyberprotest: New media, citizens, and social movements* (pp. 77-93). London: Routledge.

ABOUT THE CONTRIBUTORS

Ann Louise Bardach is the author of *Cuba Confidential: Love and Vengeance in Miami and Havana,* a finalist for the New York Public Library Helen Bernstein Award for Excellence in Journalism and the PEN USA Award for Best Nonfiction, and named one of Ten Best Books of 2002 by *The Los Angeles Times.* She is the editor of *Cuba: A Traveler's Literary Companion.* She won the PEN USA Award for Journalism in 1994 for her reporting on Mexican politics, and was a finalist in 1993 for her coverage of women in Islamic countries. Ms. Bardach is a former reporter for SLATE.

Denise D. Bielby is professor of sociology and affiliated faculty, Center for Film, Television, and New Media at the University of California, Santa Barbara. Her current research focuses on the global market for television. Her scholarly publications on the television and film industries have appeared in journals that include *American Journal of Sociology, American Sociological Review, Journal of Broadcasting & Electronic Media, Gender & Society, Journal of Poplar Culture, American Behavioral Scientist,* and *Work and Organizations.* She is co-author, with C. Lee Harrington, of *Soap Fans: Pursuing Pleasure and Making Meaning in Everyday Life* (1995), and *Popular Culture: Production and Consumption* (2001). Professor Bielby is co-recipient of two national awards for her research, the Ruben Hill Research and Theory Award from the National Council on Family Relations, and the Kathleen Gregory Klein Award for Excellence in Feminist Studies from the Popular and American Culture Associations. Her research has been funded by the National Science Foundation and the National Institute of Mental Health.

Shawn Chang joined the office of Congresswoman Tammy Baldwin (D-WI) as a legislative assistant at the start of 2006. In that capacity Shawn advises Ms. Baldwin on commerce, judiciary, international relations, trade, education, homeland security, telecommunications, and veteran issues. Shawn graduated from Georgetown University's School of Foreign Services, where he majored in international human rights. After graduating from college Shawn worked for NOW Legal Defense and Education Fund as a legislative assistant on human services and welfare reform issues before accepting his first Capitol Hill job as a legislative assistant for Congresswoman Patsy T. Mink (D-HI). After Mrs. Mink passed away in the fall of 2002, Shawn joined the office of Congresswoman Diane E. Watson (D-CA) as a legislative assistant on judiciary and telecom issues before accepting an offer to work for Ms. Baldwin. He is currently completing his J.D. at George Washington University Law School.

Benjamin M. Compaine (MBA, Harvard; PhD, Temple University), the former president and chief executive of Nova Systems Inc., a software firm, is now a researcher and writer on issues related to the economics and technologies of the information and mass media industries. He has been a research consultant to the Program on Internet and Telecoms Convergence at MIT. Previously, he had been executive director of the Program on Information Resources Policy at Harvard University. He also runs an e-commerce business. From 1994 to 1997, he was the Bell Atlantic Professor of Telecommunications at Temple University. He is the author, coauthor or editor of 10 books. His best known book is the award-winning *Who Owns The Media?* Other books include *The Digital Divide: Facing a Crisis or Creating a Myth?* (2001) and *The Information Resources Policy Handbook* (1999). His articles have appeared in trade, popular, and scholarly journals, including *Telecommunications Policy, Science Digest, Foreign Policy, Reason, Daedalus* and the *Journal of Communication.* His research and teaching interests include Internet and telecommunications policy, mass media economics, and the social and cultural implications of changing information technologies. He has been a consultant and invited speaker for clients and at venues in Europe, South America, Asia, Australia as well as in the United States and Canada.

John Dunbar was project manager for the Center for Public Integrity. Before that, he was chief investigative reporter for the *Florida Times-Union* in Jacksonville. While at the Center, Dunbar worked on the award-winning 50 States Project, the Center's Enron and Harken Oil investigations and a report on the California energy crisis. He managed the Center's 3-year investigation of the telecommunications industry. Dunbar's work has received numerous awards recognizing his work at the Center. His 2002 co-authored book, *Capitol Offenders: How Private Interests Govern Our States*, won the Investigative Reporters and Editors (IRE) national book award. He is a graduate of the University of South Florida in Tampa, where he earned a bachelor's degree in mass communications. Dunbar is currently a reporter for Associated Press.

Mark Epstein (JD, Columbia University; PhD, University of Michigan) is a professor of law at Southwestern University School of Law. While in law school, he served as book review editor of Columbia Human Rights Law Review. Later, he received a Public Interest Law Foundation Fellowship and served an internship at the Media Access Project in Washington, DC. He worked in one law firm as an associate focusing on media mergers and acquisitions, bank refinancing, leveraged leasing and alternative energy projects, and a second firm in the areas of bankruptcy, corporate and real estate law and lobbying efforts before Congress and federal agencies on behalf of clients. Professor Epstein returned to academia to earn his MA and PhD degrees, and taught courses such as Communication, American Culture, media law and theory, communication and society, at Michigan and at Syracuse University's S.I. Newhouse School of Public Communications. He serves as a faculty advisor to the Media Law Forum and the Entertainment & Sports Law Society and assists students in arranging entertainment law externships. He is the author of numerous articles, most recently on copyright law and regulating converging industries.

Nancy Gillis (MA, Georgetown) is a graduate student in Fordham University's International Political Economy and Development program and a research associate at Fordham's Donald McGannon Communications Research Center. Her recent focus has been on the use of information technology for economic development. Ms. Gillis co-edited the book *A Community Dialogue for Meeting the Millennium Development Goals* in which she wrote a chapter "Adaptive vs. Prescriptive Policymaking: Community Empowerment through the Framework of Complexity Science." Professionally, Ms. Gillis leads the research and consulting work on international development for Paradigm Research International.

Kenneth Harwood (PhD, USC) writes on media ethics. He studies media economics, including economics of the web, motion pictures, video, audio, newspapers, and books, and the history of media awards such as the Emmy in television. He suggests economic theory to support further funding of public television in the United States. Dr. Harwood was professor of Communication at University of Alabama, University of Southern California, Temple University, and UC-Santa Barbara. He was president of the International Communication Association, president of the Broadcast Education Association, and director of the National Association of Broadcasters. His awards include Distinguished Education Service Award, Broadcast Education Association, 1986, and Frank Stanton Fellow, International Television and Radio Society, 1979.

Jennifer Holt (PhD, UCLA) is an assistant professor at the University of California, Santa Barbara, in the Film and Media Studies department and has also taught at UCLA. She teaches critical media studies and specializes in the political economy and industrial history of American film and television. Her current research looks at the effects of deregulation and current policy on the industrial structure and entertainment products of today's global media conglomerates. She has published articles in various journals and anthologies including ***Film Quarterly***, *Film & History*, *Quality Popular Television* and the forthcoming series Screen Decades: American Culture/American Cinema. Presently, she is finishing a manuscript entitled *In Deregulation We Trust: The Business of Entertainment in the New Hollywood.*

Leah A. Lievrouw (PhD, USC), is a professor in the Department of Information Studies, Graduate School of Education and Information Studies, at the University of California, Los Angeles. Her research and writing focus on the social and cultural changes associated with technologies and knowledge. Dr. Lievrouw is co-editor (with Sonia Livingstone of the London School of Economics) of *The Handbook of New Media: Social Shaping and Consequences of ICTs* (2002). Her other books include *Mediation, Information and Communication: Information and Behavior,* vol. 3 (co-edited with Brent Ruben, Transaction, 1990) and information and communication technologies and the relationship between new *Competing Visions, Complex Realities: Social Aspects of the Information Society* (co-edited with Jorge Reina Schement, 1987). She is also an editor of the journal *New Media & Society* published by Sage London.

Amy S. Mitchell is associate director for the Project for Excellence in Journalism where she has been since its inception in 1997. Her primary focus is directing all research, including the Project's Annual Report on the State of the News Media as well as other more specific studies of the news media. She also works on the Project's teaching instruments including editing a case study curriculum for journalism teaching, titled Case Studies in Journalistic Decision Making. Prior to this occupation, Ms. Mitchell was a congressional research associate at the American Enterprise Institute where she researched public policy and the election process, co-authored several articles and contributed to books including *Debt & Taxes* and *Vital Statistics on Congress.* Ms. Mitchell also spent 2 years working in the publishing industry. Originally from the midwest, she now lives in Silver Spring, MD with her husband and three children.

Molly Moloney is a PhD candidate in the Department of Sociology at UCSB. She studies cultural sociology, media culture, gender, and technology. Her dissertation research analyzes struggles over the future of television vis-à-vis digital technologies, including debates about and transformations in conceptualizations of property, ownership, and audiences. Her previous work includes the chapters "Aesthetics of Television Criticism: Mapping Critics' Reviews in an Era of Industry Transformation" (2005), and "Performance and Accomplishment: Reconciling Feminist Conceptions of Gender" (2002).

Philip M. Napoli (PhD, Northwestern University) is the director of the Donald McGannon Communication Research Center, and an associate professor of Communications and Media Management in the Graduate School of Business, Fordham University. He is the author of *Audience Economics: Media Institutions and the Audience Marketplace* (2003), the recipient of the 2004 Robert Picard Award for the best book in media management and economics, awarded annually by the

Association for Education in Journalism & Mass Communication. He also wrote *Foundations of Communications Policy: Principles and Process in the Regulation of Electronic Media* (2001), and edited the forthcoming *Media Diversity and Localism: Meaning and Metrics*. Professor Napoli is also the author of more than 30 articles and book chapters that focus on media policy and media institutions. He has testified before Congress on the issue of media ownership and has spoken before the Federal Communications Commission on media policy issues on multiple occasions. He has been interviewed on media policy issues in publications such as *The Los Angeles Times*, the *Baltimore Sun*, and the *Christian Science Monitor*. His research has been funded by organizations such as the Ford Foundation, the Benton Foundation, the National Association of Broadcasters, and the National Association of Television Programming Executives.

Eli Noam (PhD, Harvard) has been professor of economics and finance at the Columbia University Business School since 1976. In 1990, after having served for 3 years as commissioner with the New York State Public Service Commission (CITI), he returned to Columbia. He is the director of the Columbia Institute for Tele-Information, a university-based research center focusing on strategy, management, and policy issues in telecommunications, computing, and electronic mass media. In addition to leading CITI's research activities, Noam initiated the MBA concentration in the Management of Media, Communications, and Information at the Business School and the *Virtual Institute of Information*, an independent, web-based research facility. Besides the more than 400 articles in economics, legal, communications, and other journals that Noam has written on subjects such as communications, information, public choice, public finance, and general regulation, he has also authored, edited, and co-edited about 25 books. The most recent include *Interconnecting the Network of Networks* (2001); *Internet Television* (2004); *Competition for the Mobile Internet* (2004); *Mobile Media Content and Services for Wireless Communications* (2005); and *Media Ownership and Concentration in America* (2005).

Sarah M. Pritchard (MA and MLS, University of Wisconsin-Madison) was the university librarian at the University of California, Santa Barbara, 1999–2006, and is now University Librarian at Northwestern University. She had previously worked at the Library of Congress, and the Association of Research Libraries. She is active in major initiatives of the Association of Research Libraries, the Coalition for Networked Information, and the University of California library system. In the American Library Association she has held numerous positions; she currently chairs the Committee on Professional Ethics, and has just completed three terms on the ALA Council. In California, she was a gubernatorial appointee to the state board of the Library of California multitype cooperative network. In 2001, she was selected by the Association of College and Research Libraries to receive the Career Achievement Award in Women's Studies Librarianship. She has served on several grant and editorial review panels and is the author of more than

70 articles and reviews on library management, information technology, women's studies librarianship, and consortial library projects. At UCSB, she oversaw a staff of 175 and acquisitions of print and electronic resources costing more than $6 million annually from publishers and other providers all over the world.

Ronald E. Rice (PhD, Stanford University) is the Arthur N. Rupe Chair in the Department of Communication, and Co-Director of the Carsey-Wolf Center for Film, Television and New Media at UCSB. He has co-authored or co-edited *The Internet and Health Care* (2006); *Social Consequences of Internet Use* (2002); *The Internet and Health Communication* (2001); *Accessing and Browsing Information and Communication* (2001); *Public Communication Campaigns* (three editions); *Research Methods and the New Media* (1988); *Managing Organizational Innovation* (1987); and *The New Media: Communication, Research and Technology* (1984).

Mark Rose (PhD, Harvard University) is a professor in the English Department at the University of California, Santa Barbara, where he has taught since 1977, and, through summer 2005, was associate vice chancellor for academic personnel. He has taught at Yale University and the University of Illinois as well as at UCSB. From 1989 to 1994 he was director of the system-wide University of California Humanities Research Institute, located on the Irvine campus. He is the author of *Heroic Love* (1968), *Golding's Tale: A Novel* (1972), *Shakespearean Design* (1972), *Spenser's Art* (1975), *Alien Encounters* (1981), *Authors and Owners: The Invention of Copyright* (1993), *The Norton Shakespeare Workshop* CD-ROM (1998), and editor of *20th Century Views of Science Fiction* (1979), *20th Century Interpretations of Antony and Cleopatra* (1977), *Bridges to Science Fiction* (1980), and *Shakespeare's Early Tragedies* (1994). *Authors and Owners: The Invention of Copyright* was a finalist for a National Book Critics Circle Award. He frequently serves as a consultant and expert in litigation involving allegations of copyright infringement. In 2002 he was the first layman chosen to deliver the important Melville B. Nimmer Memorial Lecture on Copyright at the UCLA Law School. His current interests include both Shakespeare and the history of intellectual property.

Christopher Sterling (PhD, University of Wisconsin-Madison) is professor of media and public affairs, and of public policy and public administration, George Washington (GW) University. Before coming to GW, he served as a special assistant to one of the members of the Federal Communications Commission from 1980 to 1982. Through the 1970s he was on the communications faculty at Temple University in Philadelphia. His primary research interests focus on the history of and policy for both electronic media and telecommunications. He has authored or edited nearly 20 books. He is general editor of a three-volume and multiauthor *Encyclopedia of Radio* (2004), edits *Communication Booknotes Quarterly*, and serves on the editorial boards of six scholarly journals. His most recent books are *Shaping American Telecommunications: A History of Technology, Policy and Economics* (co-author, 2006); *Stay Tuned: A History of American broadcasting* (co-author, 3rd ed., 2002); and

History of Telecommunications Technology: An Annotated Bibliography (2000). Among his earlier monographs are *The Focal Guide to Electronic Media* (editor of this CD-ROM, 1998), and *Broadcasting in America: A Survey of Electronic Media* (co-author with others, several editions).

William B. Warner (PhD, Johns Hopkins University) is professor of English, and Director of the University of California Digital Cultures Project, at UCSB. Dr. Warner conducts research on the Enlightenment, the novel, the history of media culture from the 18th century to the present, and free speech and censorship. His books include *Licensing Entertainment: The Elevation of Novel Reading, 1684-1750* (1998), *Chance and the Text of Experience: Freud, Nietzsche, and Shakespeare's Hamlet* (1986), *Cultural Institutions of the Novel* (edited, 1996), and *Reading Clarissa: The Struggles of Interpretation* (1979). His current research project is focused on Early American networks. It is part of a book project entitled, *American Networks: From the Continental Congress to the Internet.*

Diane E. Watson (PhD, Claremont Graduate School) is Congresswoman from the 33rd California Congressional District. Her lifetime commitment to education stems from her involvement in the Los Angeles public schools where she worked as an elementary school teacher and school psychologist. She has lectured at both California State Universities at Los Angeles and Long Beach. During her tenure in the California State Senate, Congresswoman Watson became a statewide and national advocate for health care, consumer protection, women, and children. In 1993, she authored the California Birth Defects Monitoring Program Act, which led to pioneering research into the causes of birth defects, and the Residential Care Facilities Act, to ensure that senior citizens receive quality care in nursing and assisted living homes. In 1997, she introduced legislation to toughen food health safety requirements for restaurants. In 1998, Congresswoman Watson served as the United States Ambassador to the Federated States of Micronesia until 2001, when she was sworn in as a Member of Congress after the death of Congressman Julian Dixon, who held the seat for 22 years. In January 2003, Congresswoman Watson was sworn in as a member of the 108th Congress.

AUTHOR INDEX

SUBJECT INDEX

G

J

V

W

X

Y

Z

Printed in the United States
106154LV00002B/320/A